THE ORGANIC DAIRY HANDBOOK

A COMPREHENSIVE GUIDE FOR THE TRANSITION AND BEYOND

THE ORGANIC DAIRY HANDBOOK

A COMPREHENSIVE GUIDE FOR THE TRANSITION AND BEYOND

Edited by Katherine Mendenhall

Project coordinated by NOFA-NY
Northeast Organic Farming Association of New York, Inc.

NOFA-NY, PO Box 880, Cobleskill, NY 12043
(607) 652–6632
office@nofany.org, www.nofany.org

Project funded by the United States Department of Agriculture Risk Management Agency (RMA)

Cover & book design by Stacy Wakefield, Evil Twin Publications.
Cover photos by Bethany Wallis.
Printed in Canada
First printing, August 2009
Printed on 100% post-consumer, recycled paper.

Library of Congress Control Number: 2009931009
ISBN 978-0-9820850-1-1
304 pages

TABLE OF CONTENTS

Acknowledgments........10

Introduction........11

CHAPTER ONE

Planning for the Transition, Certification, and Beyond........14

Determine farm goals
Assess current practices, resources, and facilities
Identify needed changes
Implement changes
The certification process
The organic system plan
Resources

Farmer Case Study: Chuck and Mary Blood........24

CHAPTER TWO

The Organic Milk Market........26

Growth of the organic milk market and market drivers
How the organic market works
Considerations when choosing an organic milk market
Alternatives to marketing your milk as a commodity

Farmer Case Study: Brian and Liz Bawden........40

CHAPTER THREE

Organic Soil Management........44

Key principles of organic soil management
Assessing soil health
Organic soil management practices
Pasture soil management
Soil management in tilled cropland
Evaluating organic soil management
Resources

Farmer Case Study: Paul and Maureen Knapp........62
Farmer Case Study: Dave and Susan Hardy........64

CHAPTER FOUR

Pasture Management........66

The benefits of good pasture management
Pasture pitfalls in poorly managed grazing systems
Overview of types of grazing systems and methods
What you need to know to set up a successful grazing system
Calculating paddock sizes, DMI, and acreage needed
Designing your grazing system
Pasture plants
Pasture soil health
Setting up a new grazing system
Ration balancing and pastures
Converting to low or zero grain feeding
Pasture and internal parasite management
Pasture requirements of milk buyers and organic standards
Improving your pasture
Resources

Farmer Case Study: Brent and Regina Beidler........95
Farmer Case Study: Earl Fournier........99

CHAPTER FIVE

Organic Crop Management........102

Pros and cons of growing crops other than permanent pasture

Meeting feed needs through an organic crop plan
Developing a sound crop rotation
Crop options and their management
Organic pest management
Evaluating and improving your crop production plan
Resources
Farmer Case Study: Kevin and Lisa Engelbert.......131

CHAPTER SIX
Managing Dairy Nutrition....... 134
Assessing your current herd for organic management
Overview of nutritional concepts
Identify nutritional goals
Different feeding systems and methods
Requirements and rations
Managing seasonal diet shifts
Evaluating your feed components, ration balance, milk production, and herd health
Resources
Farmer Case Study: Rob and Pam Moore.......179
Farmer Case Study: Jim Gardiner.......183

CHAPTER SEVEN
Organic Dairy Herd Health....... 186
Concepts of organic dairy health
Risk assessments and best management practices
Alternative and complementary treatment and medicines
Youngstock
Cow comfort
Reproduction
Care and management of the fresh cow
Udder health and milk quality
Hoof health and lameness
External parasites and diseases of the skin
Internal parasites of cattle
Adult digestive disorders
Adult respiratory disease
Basic cattle immunology
Strategic vaccination programs
Biosecurity
Conclusion
Resources
Farmer Case Study: Bob, Rick, and Kathie Arnold.......234
Farmer Case Study: Tom and Sally Brown.......237
Farmer Case Study: Vaughn and Susan Sherman.......240

CHAPTER EIGHT
Organic Recordkeeping.......244
Keeping organic records
Evaluate your current recordkeeping system
Choosing a recordkeeping system
Records support the organic system plan
Evaluating the organic recordkeeping system
Resources
Farmer Case Study: Les Miller.......255

CHAPTER NINE
Organic Support Personnel.......258
Work as a team with your family and farm staff
Seek help and information from many sources
Tips on forming a good mentor-mentee relationship
Evaluating the costs and benefits of paid consultants

The buck stops with you
Conclusion
Farmer Case Study: Jamie and Sarah Huftalen....265

CHAPTER TEN
Full Farm Case Study: Siobhan Griffin, Raindance Farm....268

APPENDIX A
Organic Certification Agencies....274

APPENDIX B
Resource Organizations....275

APPENDIX C
Organic Seed and Soil Amendment Sources....277

APPENDIX D
Organic Grain and Mineral Sources....279

APPENDIX E
Soil, Forage, and Manure Testing Services....282

APPENDIX F
Northeast Organic Milk Markets....283

APPENDIX G
Dairy Milk Production per Cow....284

APPENDIX H
Body Condition Scoring....285

APPENDIX I
Locomotion Scoring Guide....287

APPENDIX J
Cow Assessment....289

APPENDIX K
Individual Animal Health Record....291

APPENDIX L
Dimensions for Tiestall and Freestall Facilities....292

APPENDIX M
Hock Scoring Guide....293

APPENDIX N
Female Cow Anatomy and Estrous Cycle....294

APPENDIX O
Raw Milk Bacteria Counts....296

APPENDIX P
Your Animal Health Toolbox....297

APPENDIX Q
Pasture Management Troubleshooting Guide....298

APPENDIX R
Nutrition Troubleshooting Guide....299

APPENDIX S
Organic Dairy Herd Health Troubleshooting Guide....301

Index.. 302

LIST OF TABLES

1.1 Common conventional farm practices that must change during the transition to organic..16
2.1 Overview of pay price in New England, 2006–2009 (in USD)........34
2.2 Component payments by company..35
2.3 Quality premiums offered by company.......................................35
3.1 Rules of thumb on how sand, silt, and clay influence soil properties and behavior...48
3.2 Testing and monitoring biological and physical soil health on the farm.. 50
3.3 Taking soil samples in pasture and tilled cropland...................... 51
3.4 Potential nitrogen contributions from legumes........................... 54
3.5 Major nutrient composition of dairy manure.............................. 55
3.6 Average secondary macronutrient and micronutrient content of dairy manure..55
4.1 Comparing rotational, continuous, and MIG grazing systems....... 85
4.2 Measuring 30% average daily DMI per animal from pasture........ 91
4.3 Pasture quality and condition score chart.................................. 93
5.1 Principles for building a sound rotation.................................... 109
5.2 Characteristics and management of specific crops......................112
5.3 Lifecycles of common weeds.. 122
6.1 Management considerations for assessing milk production goals under organic management..144
6.2 Feeding systems and considerations...148
6.3 Minimum recommended forage quality for organic milk production...150
6.4 Lactating cow nutritional requirements..................................... 152
6.5 Protein and energy values for common organic feed grains........155
6.6 Nutrient guidelines for early dry, close-up dry, and fresh cows..... 159
6.7 Post-weaning requirements for various ages and breeds of heifers... 161
7.1 Commonly used botanicals on organic farms.............................192
7.2 Checklist for evaluation of new treatments................................ 194
7.3 National list of allowed synthetics for livestock health................ 196
7.4 Advantages and disadvantages of calf housing options...............198
7.5 Routine preventive healthcare practices for calf management.... 199
7.6 Scours management...201
7.7 Respiratory disease management... 202
7.8 Advantages and disadvantages of commonly used bedding materials...205
7.9 Summary of reproductive challenges, potential causes and management considerations.. 208
7.10 Recommended fresh cow procedures..209
7.11 Common fresh cow problems and management considerations.... 210

7.12 Estimating production lost and infected quarters by bulk milk SCC....211
7.13 Options for individual animal somatic cell count testing.... 212
7.14 Common diseases of the teats and external udder....213
7.15 Microorganisms most commonly responsible for bovine mastitis... 214
7.16 Common causes of lameness and management considerations.... 218
7.17 Common flies in the Northeast....219
7.18 Common causes of common skin diseases and management considerations.... 220
7.19 Vaccine types, needed action, benefits, and cautions.... 227
7.20 Diseases to be considered in designing a strategic vaccination program....228

8.1 Required organic dairy certification records.... 253

LIST OF DISPLAYED FEATURES

FIGURES

1.1 Sample product label.... 21
1.2 Sample seed tag.... 22
3.1 Soil after one year of a sweetclover sod crop.... 45
3.2 A cover crop of rye and hairy vetch, planted in early September, photographed in mid-October....59
3.3 Aerating the windrow.... 63
4.1 A Holstein reaching for a bite of good quality pasture with her tongue....70
4.2 Example of a correctly sized paddock with a high stocking density on Morvin Allen's farm in Massachusetts....73
4.3 High tensile fence and interior subdivisions with portable fence posts and polywire on Stonypond Farm in Vermont.... 77
4.4 Portable pasture water tub with float valve....78
4.5 The main lane from the barn.... 78
4.6 Earl Fournier's high-quality pasture.... 80
5.1 Corn yield during transition.... 107
5.2 Soil compaction benefits weeds....121
5.3 Some type of early-season cultivation tool is needed for organic row crops....124
6.1 Generalized milk energy to feed energy relationship.... 152
6.2 Happy Lines.... 176
7.1 Mastitis treatment decision tree.... 216
7.2 Nematode lifecycle.... 221
7.3 Unvaccinated herd.... 226
7.4 Vaccinated herd.... 226
8.1 Farm map examples for organic records.... 247
8.2 Illustration of a complete audit trail.... 248
8.3 Example of feed storage record from ATTRA.... 250
8.4 Animal identification records....250
8.5 Breeding wheel used for reproduction records.... 251

8.6 Example of feed ration record form from ATTRA....252
8.7 Example of individual animal health record form from ATTRA... 252

EXAMPLES

1.1 Organic dairy transition timeline.... 18
4.1 Using worksheet 4.1.... 74
4.2 Grazing acreage needed....74
6.1 Tiestall barn with component feeding.... 145
6.2 Tiestall barn with molasses component feeding.... 146
6.3 Tiestall barn with TMR.... 147
6.4 Freestall barn with TMR.... 151
6.5 Forage only, no grain ration.... 151
6.6 Balancing a ration for peak production using Table 6.4....153
6.7 Developing a dry cow ration for 45–60 days prefresh....160
6.8 Ration comparison of 6-month Holstein heifer to breeding-age Holstein heifer.... 161
6.9 Ration comparison of 16-month Holstein heifer to heifers in the last third of gestation.... 163
6.10 Ration calculation for a 250-lb., newly weaned 4-month-old calf.... 165

WORKSHEETS

4.1 Calculating paddock sizes and number of acres needed....75
5.1 Assessing the cost of purchased organic feed....104
5.2 Assessing farm fields....106
5.3 Developing a sustainable organic crop rotation.... 111

BOXES

3.1 Benefits of the addition of organic matter.... 46
3.2 How to build up and maintain soil organic matter.... 47
3.3 Reducing nutrient loss and environmental damage in manure storage and handling.... 56
3.4 Benefits and drawbacks of compost compared to manure.... 57
4.1 Common situations that cause overgrazing....69
4.2 Grazing guidelines summary....79
4.3 Foraging behavior and diet selection.... 86
4.4 Common grazing nutrition pitfalls....88
4.5 Parasite management and prevention guidelines....90
4.6 Grazing recordkeeping....92
5.1 NOP crop rotation rule.... 108
6.1 Adjust production expectations....136
6.2 Basic dairy nutrition.... 140
6.3 Forage testing.... 142
6.4 Minerals....156
6.5 Nursing.... 163
7.1 Homemade electrolyte solution.... 200
8.1 From the inspector....249

ACKNOWLEDGMENTS

The Northeast Organic Farming Association of New York, Inc (NOFA-NY) would like to thank the Risk Management Agency of the USDA. Funding for this project was provided by a USDA Research and Development Risk Management Research Partnerships Grant. The project has created two books, *Transitioning to Organic Dairy: Self-Assessment Workbook* and *The Organic Dairy Handbook: A Comprehensive Guide for the Transition and Beyond,* to help dairy farmers manage the risks during the transition to organic dairy-farm management.

NOFA-NY could not have completed the project without the collaboration of the authors who contributed to the components of this *Organic Dairy Handbook* and who helped shape these books. We also acknowledge the many farmers who shared their farm stories with us for the farmer case studies throughout the book. NOFA-NY would also like to acknowledge the many experienced and successful organic dairy farmers throughout New York state who donated their time to identify the principles crucial to a successful transition to organic dairy. These farmer-identified principles served as the foundation to both books. The farmers include Charles Blood, Joan Burns, Kevin Engelbert, Jim Gardiner, David Hardy, Maureen Knapp, Les Miller, and Susan Sherman.

NOFA-NY would also like to acknowledge the conventional dairy farmers, transitioning dairy farmers, and certified organic dairy farmers who helped field-test this handbook. They include Regina and Brent Beidler, Lisa and Kevin Engelbert, Jonathan Falby, Peter Miller, Dave Hardy, Rob and Pam Moore, Bill Slate, Kimberly Ward, and Sam Weber. Their time and contribution to the project helped ensure that the book is user-friendly and meets the needs of transitioning farmers.

Finally, we acknowledge our project advisory panel, which reviewed the book and offered its expert advice and recommendations for this project. They include Kathie Arnold, Twin Oaks Dairy, LLC; Lisa Engelbert, NOFA-NY Certified Organic, LLC; Dorothea Fitzsimmons, Alfred State College; James Gallons, USDA Risk Management Agency; Roberta M. Harrison, Cornell Cooperative Extension; Peter Miller, Organic Valley/CROPP Cooperative; Sumner Watson, Cold Springs Farm; Greg Swartz, NOFA-NY; and Robert Wiederhold, First Pioneer Farm Credit.

Dairy cows grazing on Hidden Opportunities Farm.

INTRODUCTION

Over the last decade, organic dairy has been the fastest-growing organic agriculture sector. This thriving market has attracted many conventional dairy farmers to transition to organic production. The lush green pastures of the Northeast are a good fit for farming organically, and the region's ever-increasing population of consumers interested in healthy, nutritious foods keeps expanding the regional market demand. The Northeast Organic Farming Association of New York (NOFA-NY) began this educational project at a time when there was a flurry of conventional dairy farms transitioning to organic production because of this strong market and weak conventional prospects. At the core of organic dairy farming must be a focus on sustainability—both for the environment and for the farm business.

This book outlines the practical knowledge and systematic understanding necessary for conventional farmers to make a successful transition to organic production. Additionally, it provides the necessary tools for new farmers to develop a successful organic dairy business. With in-depth discussion of the following areas, it serves as a comprehensive guide to all aspects of organic dairy farming.

PLANNING Chapter 1 outlines the planning required to begin a transition to organic dairy, including specific information for creating a realistic transition timeline and completing the National Organic Program required organic system plan.

MARKETING Chapter 2 addresses the history of the Northeast organic milk market, introduces the market players, and explains how to research a fair organic milk contract.

SOIL Quality organic soil management is the foundation for all successful and healthy organic farms. These principles and practices, outlined in Chapter 3, are helpful for all organic farms, regardless of how many years they have been in organic production.

PASTURE Chapter 4 highlights another pillar of successful organic dairying: effective pasture management. Consumers want organic milk from cows that graze on lush green pastures. Effective pasture management not only fulfills this consumer demand and expectation, but is also crucial to raising healthy cows and maintaining a positive bottom line.

CROPS Chapter 5 outlines organic crop management—an important skill as input costs continue to rise. Many farmers plant different crops under organic management than they had as conventional growers; this chapter outlines the potential for diversifying the crop rotation.

NUTRITION Many farmers find that as they transition to organic, their feed rations change. Many adjust to a more pasture-based feeding program, and most philosophically begin to emphasize a healthy organic system rather than milk production levels alone. Chapter 6 outlines the nutritional concepts needed to effectively manage an organic-dairy feeding program.

HERD HEALTH Healthy soils lead to healthy crops and pastures, which result in healthy animals. Chapter 7 focuses on organic dairy herd health and the basic management principles that result in healthy and happy cows. This chapter also suggests alternative treatments and how to create an effective organic herd-health toolbox.

RECORD KEEPING Often, a complaint about organic certification is the multitude of records that farmers are required to maintain, but the organic farmers interviewed in this book find the recordkeeping requirements critical for their businesses' success and planning. Chapter 8 outlines these organic recordkeeping requirements.

SUPPORT Chapter 9 describes the ideal support team that transitioning and new organic dairy farmers should organize to help them as they learn a new method of farming. The organic community is famous for its collaborative spirit and willingness to share information, and mentoring has long been a strong point.

The final chapter of this book, Chapter 10, provides the reader with a comprehensive look at one successful organic dairy farm.

Many of the chapters include lists of additional resources for gaining additional insight into specific areas of organic dairy production. These resources make up the reference libraries on many organic dairy farms throughout the country and have proven to be worthy investments. The appendixes at the back of the book are helpful tools that farmers will refer to again and again. The troubleshooting guides for pasture management, nutrition, and herd health will be especially helpful to transitioning farmers as they become familiar with farming organically.

In addition to the comprehensive practical and technical information outlined above, we have interspersed case studies of successful Northeast organic dairy farms throughout the book. This collection of farm stories demonstrates that there is no cookie-cutter approach to successful organic dairy farming. Rather, this diversity of farm models and techniques helps make the organic dairy community unique and exciting. The many farmers' voices throughout this book testify to the large number of possibilities and models for a successful transition. We are grateful for their expertise and the time they shared with us to pass along their organic farming knowledge to you.

This book's hands-on companion, *Transitioning to Organic Dairy: Self-Assessment Workbook,* introduces the basic requirements for managing a certified organic dairy. It helps farmers evaluate whether or not organic production is a good match for their farm and business partners. The activities in the workbook help identify the needed farm infrastructure and management changes and where to source information along the way. This *Handbook* goes into further depth than the *Self-Assessment Workbook*, serving as a holistic reference for farmers to answer common questions encountered throughout the transition process, during the first years as a certified organic dairy, and beyond.

CHAPTER ONE

PLANNING FOR THE TRANSITION, CERTIFICATION, AND BEYOND

ROBERT PERRY

Planning is an important skill that is essential for organic management. Taking the time to plan before, during, and after you transition will help create a road map to guide the farm through its transition and develop it into a successful organic dairy farm. The transition to organic dairy farming is exciting and creates opportunities to learn new methods of farming. Having a clear plan for how to approach these new learning opportunities will help ensure a rewarding transition.

There are multiple levels of planning to consider before transitioning to organic dairy. A formal business plan or a holistic farm plan will help identify your farm goals, prioritize the action steps you will take to implement the transition, and evaluate the financial impact the transition period will have on your farm. At the very minimum, creating an organic farm plan is a requirement for organic certification called the organic system plan (OSP); however, a more comprehensive whole farm plan is worth considering. This chapter will outline the planning steps needed to write an effective organic system plan as well as how to prepare for a successful transition to organic dairy.

KEY ORGANIC TRANSITION POINTS

- The organic system plan (OSP) is a required part of your application to organic certification. Everything you plan to do or use on your organic farm must be included in the OSP.
- Assess your farm and abilities realistically due to the cost and time involved in transition.
- Your start date for the transition period is determined by the certifier, not when you *think* you began. Contact your certifier early to open a line of communication during transition.

DETERMINE FARM GOALS

The first step in planning your transition to organic dairy is to identify both your personal and your farm goals. These goals will help you create a business plan or holistic farm plan that will guide the decisions you will make throughout the transition and as a certified organic producer. By writing down your goals, you are more likely to achieve them. They will also provide you with a framework within which to monitor your farm transition. Sit down with your family members and other farm partners. Brainstorm together about what the goals of the farm are and what you all would like to see the farm achieve through the organic transition process. Answer these basic questions to get started.

- Why do you want to transition to organic dairy management?
- What do you hope a transition to organic dairy management will provide for your family?
- What are your financial goals for the transition to organic?
- What are your milk production goals for the transition?
- What are your crop production goals for transition?
- What are your short-term goals for the transition?
- What are your long-term goals for the transition?

The end of this chapter lists resources that focus on determining farm goals and creating extensive business or holistic farm plans.

ASSESS CURRENT PRACTICES, RESOURCES, AND FACILITIES

It is important to assess the resources your farm currently has and how they will work in an organic system as well as your own management strengths and those of your farm employees. For some individuals, the land takes several years to acclimate to an organic system. Another farmer might have land that is immediately certifiable, so he or she needs only one year for the dairy herd transition. Others may choose to purchase an existing organic herd to shorten the transition process. Develop a long-range plan that meets your farm goals, actual needs, and financial situation. It is difficult to recover from mistakes partway through the transition period without financial loss, so planning ahead is essential. Use many resources to educate yourself, your family, and your employees about organic agriculture. Everyone working on the farm should read and understand the National Organic Standards which outline the rules and regulations for organic certification.

IDENTIFY NEEDED CHANGES

Identifying the conventional farming practices you will have to change to qualify for organic certification is an important stage in the planning process. The NOFA-NY *Transitioning to Organic Dairy: Self-Assessment Workbook* is an excellent reference for evaluating the needed changes for an individual farm. For many organic dairy farmers, changing their farming practices and philosophy was the biggest hurdle during the transition. Many practical changes also require thinking about dairying differently. Organic dairy farming focuses on creating the best management practices to build healthy soil, healthy plants, and healthy cows. Table 1.1 outlines common conventional farming practices that require mandatory changes for organic management. What farming practices will you have to change? How will you have to shift your management practices?
In order to meet the criteria for organic farm management, a tran-

sitioning farm may have to make additional farm changes that are not specifically mandated by the National Organic Program (NOP). Such changes may include the following:

- rethinking farming philosophy
- including more pasture acres to accommodate dry matter intake requirements
- switching to management intensive grazing (MIG)
- changing herd size
- increasing the farm's land base to grow additional crops
- growing new alternative crops that better suit organic management
- locating organic consultants and product dealers
- using preventative herd health practices and implementing best management practices
- reducing grain feeding and improving forage quality

TABLE 1.1 COMMON CONVENTIONAL FARM PRACTICES THAT MUST CHANGE DURING THE TRANSITION TO ORGANIC

COMMON CONVENTIONAL PRACTICES	ORGANIC PRACTICES
Confinement housing	Year-round access to outdoors and pasture management
Antibiotics	Alternative health treatments (i.e., homeopathy, herbal therapy, and preventative maintenance)
Routine worming	Most conventional dewormers are not allowed. Best management practices, good soil health, and good grazing management
Treated seeds and chemical fertilizers	Organic seeds, compost, manure management, cover crops, and approved soil amendments
Dry cow treatment	Most conventional medications are not allowed; good dry cow nutrition and best management practices
Roundup Ready® crops and conventional spraying	Organic seeds, cultivation, natural weed suppression through crop rotation, and cover cropping

IMPLEMENT CHANGES

Due to the vast differences in Northeast dairy operations, researching many information sources to find the right information to suit your farm and your resources will be most helpful. This handbook is a great place to start gathering that information, but the learning process should not end here. Once you feel comfortable with the information you have gathered on organic management and the changes you need to make to your farm, it is time to take the leap. The section Developing a Transition Timeline will help plan the appropriate timing for each needed change in farm management. It may be beneficial to implement some farm changes before actually starting the organic transition.

THE CERTIFICATION PROCESS

CHOOSING A CERTIFIER

Not all certifiers are the same. Find a certifier that provides the services you are looking for within your price range. Some certifiers operate regionally only and certify only specific types of operations. For a current list of all accredited organic certifiers, visit the USDA NOP website: (http://www.ams.usda.gov/nop/). Many certifiers are affiliated with a nonprofit educational organization that can provide information and technical support to transitioning and organic farmers. This may be an attractive feature. When researching certification agencies, ask other organic farmers in your region for feedback on their certifier, why they chose it, and the quality of the services it provides. See Appendix A for a list of certifiers in the Northeast and Appendix B for a list of organic educational organizations.

CERTIFICATION COSTS

Individual certification agencies have their own pricing structures which typically include an initial application fee, annual renewal fees, and inspection fees. Once certified, farmers are required to update their information annually, pay an annual recertification fee, and have at least one inspection per year.

Often certifiers base their pricing on a percentage of annual gross organic sales, and costs vary depending on the size of the operation. Funding may be available to help alleviate the cost of organic certification. Certification fee reimbursement has been included in past U.S. Farm Bills and is awarded through states' departments of agriculture. In 2009, farms were eligible for reimbursement of 75% of their organic certification fees up to a maximum of $750 per year. The Natural Resources Conservation Service (NRCS) Environmental Quality Incentive Program (EQIP) may also help cover certain costs of a transition to organic production. For more information about these government programs, ask your certification agency, contact your state department of agriculture, or contact your county NRCS office.

DEVELOPING A TRANSITION TIMELINE

The timeline for transitioning to organic can be different for every producer. Taking time to plan your transition will save time and money in the end. Often the response to the application and planning process is, "Too much paperwork!" However, developing a thorough organic system plan creates a benchmark for your operation that will improve with time and learned experiences. Keeping detailed plans and records will also help you determine the areas that need improvement over the transition period and in the future. Use Example 1.1 as a template for creating a transition plan for your own farm.

THE INSPECTION

Prior to granting certification, the certifier requires an on-site inspection of every transitioning farm. A trained inspector walks the farm, examines the farm records, and asks the farmer questions to verify the information submitted to the certifying agency. The inspector writes a report based on his or her observations and the information gathered during the inspection, which is then submitted for the final review process of the application. The inspector does not determine if the farm is certified, but rather reports on whether the on-farm practices match what was entered in the organic system plan and application, and that they follow the NOP standards. Maintaining early contact with the certifier will help make this final step in the certification process successful and on time prior to the end of your one-year dairy transition.

THE ORGANIC SYSTEM PLAN

The organic system plan (OSP) is a required component of every organic application. The National Organic Program (NOP) defines

EXAMPLE 1.1 ORGANIC DAIRY TRANSITION TIMELINE

This example outlines the needed transition steps for a farm that needs three years of land transition, prior to the one-year transition for the dairy herd. Your individual situation may allow you to start the process in the first, second, or third year depending on your past management practices.

TRANSITION YEAR 1

- Determine your farm goals. Think about how this transition will benefit you, your family, and farm.
- Assess your current practices, resources, and facilities.
- Identify the needed changes you will have to make on your farm. The companion to this handbook, *Transitioning to Organic Dairy: A Self-Assessment Workbook*, helps determine a farm's ability to transition and assists in the transition process.
- Decide on a certifying agency and begin talking with its staff.
- Implement the needed organic management changes.
  - Begin the land transition and follow all the NOP regulations for organic crop production (organic seed, approved soil amendments, crop rotation, etc.). Cropland transition takes three years (36 months) from the last use of any prohibited materials immediately preceding a harvest. For an example, if treated corn was planted in early May, but then the crop was sprayed on June 1, the transition date would be three (3) years from the June application of the prohibited material.
  - Document all farm production practices (planting dates, harvesting, yields, etc.).
  - Keep detailed records and begin your organic audit trail. Document all farm production practices even if you are not seeking certification for all areas of production (see Chapter 8, "Organic Recordkeeping").
- Actively source a market for your milk for when you will end your transition, and review the contracts. Keep in contact with potential milk marketing agents.
- Gather information: attend field days and other workshops about organic production.
- Seek out an organic mentor as well as other organic farmers to share your questions and concerns.
- Begin to experiment managing your herd organically and develop an alternative herd health plan.

the organic system plan in section §205.201 as, "A plan of management of an organic production or handling operation that has been agreed to by the producer or handler and the certifying agent and that includes written plans concerning all aspects of agricultural production or handling described in the Act and the regulations."[1]

Most certifiers consider the full certification application to be the OSP because the application addresses all areas of farm management. This document serves as a legal contract between the certifier and the farm operator, creating a set of guidelines for the management of the farm under the NOP. The certifier will evaluate the OSP to determine the eligibility of the farm operation for or-

TRANSITION YEAR 2

- If you have not determined which certification program you will contract with, finalize that decision.
- Submit your application. It is not mandatory that the entire farm be eligible for certification prior to submitting an application. However, you cannot market your farm products as organic until you receive your organic certificate from your certifier! This process may take several months. Submit the application prior to your third year of transition.
- Actively source a market for your milk and, if you have not already done so, sign a contract that coincides with your transition end date.
- Continue to experiment with managing your herd organically and develop an alternative herd health plan if you have not already done so.
- Develop a grazing plan that will meet the NOP and your certifier's pasture requirements.
- If you are not currently grazing, transition your herd to pasture and grass. This will reduce the stress of the transition on the animals and give you additional time to learn more about grazing.
- Determine the amount of buffer zones needed and identify adjoining land use concerns.
- Plan how you will store organic crops to keep the certifiable crops separate from any remaining conventional crops. Noncertifiable crops cannot be fed in the final year.

TRANSITION YEAR 3

- Finalize and submit all pieces of the full application for organic dairy certification and talk to your certifier to make sure you are on track.
- Transition your cows according to your approved herd transition date. Plan with your certifier. You must manage your cows 100% organically for this full transition year.
- Feed 100% certified organic feed *or* feed from your third-year transition cropland.
- Compile three-year histories for each farm field including all amendments, sprays, and crops grown.
- Maintain all the required documentation for your organic audit trail and make sure it matches your organic system plan.
- Prepare for your inspection. Is everything on the farm ready to have an inspector assess it?

ganic certification and identify areas that may need to be changed before the application process continues. The OSP is verified by an on-farm inspection; therefore, descriptions of current farming practices and organic management plans for the transition period should be accurate. You, the producer, are responsible for providing a realistic and accurate profile of your farm practices.

DESCRIPTION OF PRACTICES

The first component of the OSP starts with the basics including general information like the farm's address, number of years farming at the current location, type of operation (dairy), and the crops to certify. Most certifiers have a basic application for crop production and additional application sections for dairy livestock producers.

It is important that applicants document all farm production practices, even if they are not seeking certification for their entire farm. The certifier needs a clear picture of what activities take place on the farm. If some parts of the farm will remain under conventional management, the producer must show how he or she will separate the nonorganic livestock, grains, and crops from the organic. If some current practices do not meet the requirements for organic certification, the producer must describe what changes will be made on the farm to bring these areas into organic compliance before the transition begins.

Practices described in an application

- Intended tillage practices
- Soil management
- Crop management
- Seed sources
- Fertility and manure management
- Herd health practices
- Dairy nutrition program
- Pasture management (including year-round access to outdoors)

LIST OF SUBSTANCES USED IN PRODUCTION

Each producer must list all of the products planned for use in dairy and crop production. The certifier reviews all products as part of the application process and determines if they are allowed according to the NOP's National List of approved substances. Always obtain your certifier's approval of a product *before* using it. The certifier will consider the use of a prohibited product on a transitioning or certified farm as a noncompliance, which could jeopardize the farm's certification process.

Which products are approved?

The NOP's National List sets the benchmark for allowed substances in organic production; however, it is not a brand-name reference list and offers limited guidance for products offered in the marketplace. The Organic Materials Review Institute (OMRI) publishes an annual approved product list on their website (http://www.omri.org)

and in print (the *OMRI Generic Materials List* and the *OMRI Products List*). Ask whether your certifier maintains its own list of approved materials, provides its own brand-name material reviews, and honors other lists.

Labels: The product label is a helpful tool to determine whether a product complies with the NOP. However, cross-referencing all of the product's ingredients with the National List is not a simple task. Ask your certifier—it is much quicker and more reliable! Some certifiers require that farmers file copies of the labels from all farm inputs with the certifier. Contact your certifier for their product recordkeeping requirements. Your inspector may ask to see product labels, so file all product labels and seed tags. See Figures 1.1 and 1.2.

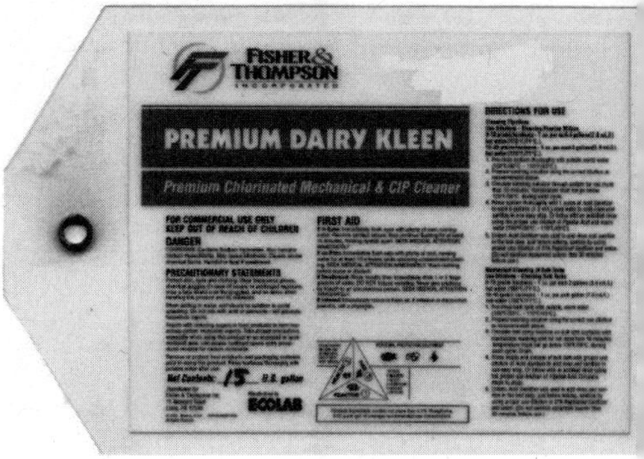

FIGURE 1.1
Sample product label that can be filed with farm input records to document a product intended for use

Seeds: The NOP requires certified organic seeds for organic production. There are many alternative crop choices available today that were not an option a decade ago, including many outstanding certified-organic hybrid seed varieties. The organic system plan must provide documentation of the origin of all seeds the farmer plans to use (including grass and cover crop seed). The farm records must document the purchase date and where the seed was purchased. Farmers must save both seed tags and organic certification certificates for seeds used during transition and after certification to verify the seeds' organic and untreated status.

If certified organic seed is not commercially available, the NOP standards allow organic growers to purchase untreated nonGMO seed varieties. (See NOP Section §205.204 on seeds and planting stock.) The definition of commercial availability may vary some by certifier; check with your certifier to make sure you meet the commercially available specifications for organic seed before sourcing conventional untreated, documented-nonGMO seed. If a quality organic seed variety is not available, the farm records must document that you contacted at least three organic seed distributors in your search to find organic seed. See Appendix C for a list of organic seed companies.

Always confirm with your certifier before purchasing or using a new product. Trust your certifier that it is allowed, not your salesperson!

Livestock Feed and Forages: The OSP must document that the farm can produce or purchase sufficient certified organic feedstock to sustain the entire herd, including youngstock, for one year. A list of the livestock feed rations, sources of ingredients, and descriptions of the pasture system's forage production, grain production, or source of purchased organic grain is required to support this feed plan. Third-year transitional crops from *your* farm are allowed for the one-year transition of the dairy herd, but any purchased feedstock, including any potentially edible bedding materials, must be 100% certified organic. Salt and mineral

LAKEVIEW ORGANIC GRAIN
BOX 361, 119 HAMILTON PL.
PENN YAN, NEW YORK 14527

Organic Keuka Oat Seed
LOT # O-Keuka-AB-08

PURE SEED:	96%
WEED SEED:	1.0%
INERT MATTER	2.0%
CROP SEEDS	1.0%
GERMINATION:	85%

DATE TEST COMPLETED: 4/09

ORIGIN: NY. NET WT. 50 LBS

FIGURE 1.2
Sample seed tag that can be filed with the farm input records to document organic seeds planted

formulations and any other supplements must not contain NOP-prohibited ingredients.

Emergency Feed Plans: All forages and grains fed to transitioning and organic animals must be certified organic regardless of the situation. Within the organic farm community, grain is often supplied on a regional basis. The OSP requires an emergency feed plan that outlines how the farm will feed the herd if faced with unexpected drought, flooding, hail damage, or other unplanned challenges.

Whole-Herd Transition for Dairy: The whole dairy herd transition is a one time opportunity for a conventional dairy operation to transition the entire herd to organic production. Whole herd is defined by the animals on the farm at the beginning of the official transition. From this date, the farm must follow all organic regulations and NOP standards completely for one year, including but not limited to the following.

- All of the animals to transition to organic must be on the farm at the beginning of the transition. Conventional animals cannot be added during the one-year transition.
- Animals should be identified with permanent identification and records must be kept for each animal.
- If an animal is treated with a prohibited substance, it must be removed from the herd before the end of transition. This animal cannot be transitioned again. (The NOP prohibits the use of antibiotics and hormones, except for an emergency. See Chapter 7, "Organic Dairy Herd Health.")

MONITORING PRODUCTION PRACTICES

Your organic system plan should describe how the farm will evaluate its soil and animal health, product quality, and contamination control. Soil tests, milk testing, forage analysis, visual observation, crop yields, and the use of an agronomist are examples of monitoring practices (see Appendix E for a list of testing resources).

RECORDKEEPING SYSTEM FOR THE AUDIT TRAIL

The organic system plan is the foundation for the recordkeeping and audit trail program. Everything mentioned in the OSP should have documentation to support its use on the farm. This is why recordkeeping is such an important component of organic farm management. See Chapter 8, "Organic Recordkeeping" for details on how to build an effective organic recordkeeping system and audit trail.

ADDITIONAL INFORMATION, ADDENDA, AND FARM PHILOSOPHY

Your certifier may require additional forms or addenda for specialty crops or on-farm processing in the basic organic system plan. Every product you plan to sell as certified organic must be included in your organic system plan and will be listed on the organic certificate for your farm. Certification is not a blanket certificate for farm products not included in the application. Your certifier may also ask you to include other contracts or agency documentation with your organic farm plan (e.g., nutrient management plans, NRCS program contracts, wetlands protection plans, etc.). Organic certification may not be compatible with all government programs, so check with the government agencies and inform them of your organic status.

NOTE

1. National Organic Standards, Rule §205.2 Terms defined. http://www.ams.usda.gov/nop, 3/24/2009.

CHUCK AND MARY BLOOD, ROCKY TOP ACRES

ROBERT PERRY AND BETHANY WALLIS

In the heart of Madison County, New York, near Moscow Hill lays the town of Hubbardsville. There, perched against the Brookfield State Forest, is Rocky Top Acres, the home and farm of Chuck and Mary Blood.

In 1982 Chuck and Mary moved to Rocky Top Acres to work as hired farmhands. They raised their children in a small rented house on the farm. When the owner of the farm passed away, Chuck and Mary were able to obtain the farm. In June of 1997, after a 90-day transition for the dairy, they certified the farm organic. The shallow rocky soils on the farm naturally favored having a permanent grass-based dairy. They had been grazing the cows for many years prior to going organic, feeding plenty of hay, and already challenging the feed salesperson, so the dairy transition was not a real issue for them.

Chuck Blood

Chuck and Mary milk 40–60 cows on a year-round basis and raise all their own dairy replacements. They also raise organic beef and sell organic hay and baleage. Chuck manages his extensive pasture system to complement his nutrition plan, and believes grazing is the foundation of herd health. They currently operate 700 acres of owned and rented crop- and pasturelands, with 250 acres of permanent pasture with the balance in hay. He grows BMR sorghum and BOP (barley, oats, and peas) for baleage and reseeding hay land. They purchase cornmeal to supplement the winter rations for the high producers. They also use molasses in the feed ration. Prevention is the basis of their health-care philosophy, which they supplement with excellent forage, free-choice kelp, and Redmond salt.

Following transition, Chuck connected with other farmers from his local organic community, who were always ready for discussions about organic dairy production and to serve as his farmer mentors. Chuck has never been a quiet bystander. He has always asked, “Why? Prove to me this will make my farm better.” This quest for answers has kept him seeking information every day and not taking the first explanation someone has offered. Having this network of successful

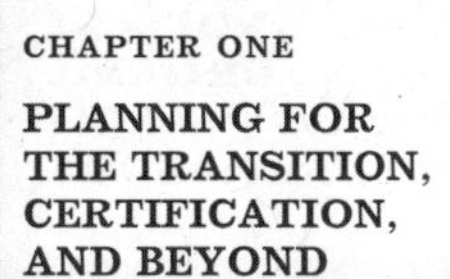

Rocky Top Acres's dairy herd grazing

organic dairy farmers has challenged him to continue learning and improving. However, the most influential person for Chuck remains his partner, Mary, who continually asks the question, "Are you sure you want to do this?" They have continued this successful journey together. Chuck has become an influential spokesperson for organic milk marketing while developing and mentoring other organic family farms.

Chuck's certifiers were helpful in creating Chuck's original organic systems plan and he has continued to modify his original organic systems plan over the years. His philosophy for success is being open to change. "If you are not flexible and not trying new ideas the organic system isn't working." Originally, the Bloods grew row crops, but now manage a completely forage-based system. Including sorghum and small grains in their crop rotation to reseed grasslands, frost seeding, and experimenting with a rotovator have all been modifications of the original organic systems plan. Other modifications include adding organic beef to the system. Chuck has also branded his own chocolate milk, which he processes off the farm. Most of the major changes in farm production evolved from organic farmer-to-farmer networking and sharing information learned from their own personal gains and disappointments. The Bloods have found the organic farm community to be outstanding in its openness, interaction, and ability to determine solutions without mainstream advice.

The Bloods have found that their detailed audit trail records have allowed them to determine the real costs of the operations. The records have been crucial to their business success, providing better documentation for personal financial decisions as well as documenting monthly production profiles for banking needs. With strong family support and mastering the art of organic dairy farming, Rocky Top Acres continues to reflect the self-sustaining family farm philosophy Chuck and Mary Blood strive for.

CHAPTER TWO

THE ORGANIC MILK MARKET

ED MALTBY

The organic milk market has grown quickly over the past decade and has experienced growing pains as the market develops. A decision to transition to organic production must be made with a very clear knowledge about the intricacies of the organic dairy industry plus careful budgeting. This chapter will provide you with a snapshot of the organic milk market, insights into how it functions, and things to consider when choosing a milk company.

KEY ORGANIC TRANSITION POINTS

- Research your organic milk marketing and certification options. Talk with other farmers in your area and meet with multiple processors.
- Run the numbers! Any budget needs to assume that it will take at least three years to reach average organic production and that your pay price may drop, so allow for a $2/cwt safety net.
- Make sure you read and understand your milk contract and that its provisions match your production system before signing.
- Secure an organic milk contract before you transition your herd.

GROWTH OF THE ORGANIC MILK MARKET AND MARKET DRIVERS

The commercial organic milk market began in the mid-1980s, but did not gain national market share until after 2000 when Horizon Organic and Organic Valley/CROPP Cooperative established national distribution and procurement networks. From 2002 to 2007, the organic milk market saw incredible growth, which brought ample opportunity for producers to maximize their return from organic production and for processors to establish an infrastructure for future profitability while meeting their costs and profitability goals. During 2007 and 2008, supply increased at a faster rate than demand. At times there was a surplus of supply, which spurred an increase in manufactured dairy product. In 2009, the economic depression dramatically affected sales of organic dairy. Sales of manufactured organic dairy product dropped by as much as 20% and total fluid sales increased by only 3-4%. With limited plans in place for supply management, the companies lowered their pay price by up to $2/cwt. HP Hood wrote new contracts with pay price varying monthly and Organic Valley instituted a new quota system mandating 7% less production. These measures caused many organic dairy producers to experience financial difficulties.

The future of producer profitability will require maintaining a sustainable pay price while maximizing home-produced organic feed either by production practices (intensive grazing, small grain production, annual forage crops) or by management decisions (seasonal dairying, calving patterns to meet seasonal payments). It will also be crucial that the market continue to ensure the integrity of organic standards in order to maintain a value-added price in the marketplace, as well as to establish a level playing field for the interpretation and implementation of these standards.

Consumers first enter the organic market by purchasing organic milk and produce. These markets saw record annual growth averaging 25% during 2004–2008. From 2005 to 2006, demand exceeded supply, creating a competitive market for new producer entrants. From 2005–2007, organic dairying was profitable for producers and attractive for new entrants as the pay price responded to increased consumer and manufacturing demands plus competition for supply. In 2006–2007, the market saw a 50–60% increase in the supply of organic milk from hundreds of farmers transitioning in response to low conventional milk prices and a June 2006 USDA rule change that increased the cost of feeding the transitioning herd. During 2006, the major organic dairy brands paid large signing bonuses and transition payments because demand was high. This aggressive drive to increase supply led to a slight surplus in the spring and summer of 2007.

In 2007 the drive for corn-based ethanol and an increased demand for organic livestock feed (principally from poultry operations) increased the price of organic grain by more than 30%. While these organic feed prices increased, the cost of fuel, labor, property taxes, and many other farm inputs also increased dramatically. Unfortunately, as farm input prices increased, there was no significant increase in pay price for organic milk, and organic dairy farm profitability plummeted in 2008.

In early 2009 the effects of the recession caused sales of organic fluid milk to drop by 4% compared to 2008, with an unprecedented drop of 12% in February 2009. This was the first time the organic dairy industry experienced serious oversupply. Organic dairy was no longer immune to downward fluctuations of price, and processors reacted to the loss of sales. After a decade of exceptional growth, Organic Valley instituted a quota system in 2009 that paid producers an organic price for 93% of milk produced based on monthly production records from the last three years, and $15/cwt for anything above this base. Horizon Organic cut their Market Adjustment Program (MAP) payment by $1/cwt and asked for a 7% voluntary reduction in production. HP Hood canceled contracts in New England and the Midwest, causing over 80 dairies to leave organic production. In July 2009, HP Hood instituted new contracts that ended seasonal and market payments and mandated that pay price would change every month based on Hood's internal assessment of how much milk was sold as organic. In the latter part of 2009, the market remained unstable and unpredictable.

The most significant sector for growth of organic milk has been in store-brand and private-label milk, which grew to nearly 20% of total consumption of organic dairy products from 2006 to 2009. Store-brand organic milk is packaged for a retailer's store brand (Trader Joe's, Whole Foods 365 brand, Safeway's O Organic) and private label is milk packaged for a private distributor (Woodstock brand for natural foods distributor United Natural Foods, Inc.). All the major brands process and package some of their milk supply into the private-label/store-brand market to increase their ability to negotiate with retail outlets for shelf space for their *own* brands.

All consumer research points to the consumers' willingness to pay extra for dairy products produced without the use of herbicides, pesticides, growth hormones, or antibiotics from cows that are pasture based.

Store-brand milk is usually sold at 10–15% less than branded product and does not carry the high overhead necessary to gain market share for the brands. Aurora Dairy,[1] a vertically integrated dairy and processing operation, has been able to capture a significant share of this market with its unique combination of a milking herd and a state-of-the-art processing and packaging facility and has become a model for other large dairies. Aurora Dairy also qualifies as a producer handler and is exempt from paying into the Federal Milk Marketing Order Pool, which gives them a competitive advantage of $2–3 per cwt over other processors. National brands that also package store brands provide consumers with less expensive alternatives to their own brand and therefore threaten any increase in sales for their national brand. For example, OV/CROPP packages the Whole Foods 365 brand, which sells for up to $1.20 less per half gallon than CROPP's own brand, despite the fact that it is the same milk packaged at the same plant.

The growth of store-brand organic fluid milk has increased consumption beyond the core organic consumer into a consumer category influenced by price as well as the benefits of an organically certified product. While store brands have been a vehicle for the growth of consumer demand, it has had a mixed effect on long-term profitability for organic producers and processors. This has become evident in the following ways.

- *Decreased profitability for processors*: The competition for supplying and packaging store brands has undercut prices in what some have described as a "race to the bottom" among the major companies. For some processors, store brands are now up to 30% of their total milk sales. The decreased profitability for processors has provided them an excuse for not increasing the pay price to producers, even though the decision to undercut the retail price aims to gain market share and increase the profitability of the processors.

- *The growth of store-brand milk:* and the demand for lower-priced wholesale organic milk has encouraged the growth of large dairies that have exploited the inability of the National

Organic Program (NOP) to enforce universally applied standards for grazing.

Ultimately, the long-term future of the organic milk market is based on providing an authentic, third-party-certified product that can easily be differentiated by the consumer and that holds the qualities consumers value. All consumer research points to the consumers' willingness to pay extra for dairy products produced without the use of herbicides, pesticides, growth hormones, or antibiotics from cows that are pasture based. Organic production has always met all those qualities with the exception of a consistent and universal interpretation of how much pasture is required. Ongoing rulemaking will provide the legal language for certifiers to enforce a pasture-based system and give the NOP the ability to identify dairies in noncompliance. While consistent application of the rules will provide a level playing field for producers and guarantee consumers a product that meets their expectations, it does nothing to ensure a farmgate price that will provide economic sustainability for all sizes and types of organic dairies. The conventional milk market has shown the importance of producers leveraging their power with processors and consumers to maintain an equitable return for their time, labor, and capital investments.

HOW THE ORGANIC MARKET WORKS

Many organic farmers have found operating in the organic milk market beneficial to their farm business. The organic milk market has been less volatile than the conventional market because farmers lock in a higher base pay price under a one- to two-year contract, or an annually determined price set by a cooperative board. Whereas the conventional base pay price fluctuates with the world market, the organic pay price has historically remained the same over the course of the year. This has helped producers plan financially and focus on managing the factors they can control to help their bottom line. The organic market has added benefits such as low or no hauling fees, a field staff to help farmers through the transition, and access to a supportive community of successful organic farmers.

Unlike conventional milk prices, organic milk prices are not determined by any central authority. Despite the fact many independent companies compete for the same milk, they all offer approximately the same regional base price for each cwt of milk and tend to raise their prices by the same level at the same time. Organic milk is not pooled when it is picked up from the farm (although it is pooled if sold on the spot market), and processors keep their milk separated during transportation and initial processing. Producers contract directly with companies that have brands in the marketplace, such as WhiteWave Foods, HP Hood, and Organic Valley/CROPP, or to handlers like Dairy Marketing Services (DMS) that sell to manufacturers. These contracts are usually for two years and vary depending on the purchaser of the raw milk. Farmers are paid through one of their existing cooperatives (for example: DFA, Dairylea, LOFCO,

Mount Joy Farmers Coop, NFO, ODFC, St. Albans Cooperative Creamery) or LLCs (Agri-Mark, DMS[2]). If farmers are not already cooperative members, they may need to join a cooperative and pay equity to that cooperative.

In the Northeast, many companies purchase raw organic milk in significant quantities. These companies either purchase milk on behalf of processors or process it themselves. Initial negotiations with transitioning farmers occur between the individual farmer and either the end user having a brand in the market (WhiteWave, HP Hood, and CROPP) or representatives from DMS on behalf of many customers. Some of the larger companies and ownership structures are as follows.

Farmer-controlled cooperatives with a board of farmers
- Upstate Niagara Cooperative, Inc.
- Dairy Farmers of America (DFA)
- Organic Valley Family of Farms/(CROPP Cooperative)
- Lancaster Organic Farmers Cooperative (LOFCO)
- PastureLand Cooperative
- Organic Dairy Farmers Cooperative (ODFC)
- Dairylea Cooperative, Inc.

Limited liability companies (LLC)
- HP Hood LLC (Stonyfield Farm brand)
- Organic Choice
- Dairy Marketing Services, LLC (DMS)

Regional private companies without shareholders
- Natural Dairy Products Corporation

National companies with shareholders
- WhiteWave Foods (Horizon Organic and Organic Cow brands), a subsidiary of Dean Foods

ORGANIC FARMGATE PRICE

Unfortunately, there is no rational explanation for how the different companies set the organic farmgate price. Some claim it is set by supply and demand, others by individual negotiation, but there are no consistent criteria applied by all companies to set prices. In the past, competition for supply has increased the price as has advocacy by producers. More recently, however, those tactics have not been as effective. What we can highlight is what criteria are ***not*** consistently used to set the organic farmgate price.

- *Not by government*: Organic milk is neither directly subsidized nor supported, although organic producers are eligible for MILC payments. The farmgate price is not set by the federal government, but all processors, with the exception of Aurora Organic Dairy, pay into the federal pool.
- *Not by the retail price*: The farmgate price bears no relationship to the retail price, which is set by supermarkets based on competitive wholesale pricing among processors, in-store

promotions, and an average margin of 31% for their operating expense and profit.

- *Not as a percentage of retail price*: In most years, organic dairy farmers receive a smaller percentage of the consumer dollar than the conventional farmer, 34% for organic compared to 41% for conventional (based on a farmgate price of $15 and a retail gallon price of $3).
- *Not by parity pricing*: Parity price[3] for November 2008 was $41.20 per cwt for Midwestern milk and $45/cwt in the Northeast with regional premiums.
- *Not by comparison to conventional dairy*: Conventional farmgate prices rose by as much as $10 per cwt in 2007 and dropped by the same amount in 2008 without any reaction in the organic farmgate price.
- *Not by supply and demand*: In 2008, consumer demand for organic dairy grew at 15–20%, slightly down from the previous 25% per year from 2005–2008, but the pay price increased an average of only 3.8% per calendar year during 2005–2008.
- *Not by national negotiation*: Most contracts are with individual farms, except DMS contracts, and many producers have confidentiality clauses that prohibit sharing information among producers.
- *Not by costs of production*: Farmers received little increase in their pay price despite a rapid increase in production expenses such as organic feed, diesel fuel, petroleum-based agricultural products, insurance, taxes, and labor.[4]

MAJOR PURCHASING COMPANIES

Each company that purchases raw milk has a different process for determining its contracted price with producers, depending on their corporate structure. Smaller cooperatives such as LOFCO (PA), Upstate Niagara (NY), and ODFC (NY) purchase milk from a small number of producers and then either sell directly to a local bottler or market the product themselves, plus balance their milk by selling excess on the open market. The following large companies purchase the majority of raw organic milk in the Northeast.

Horizon Organic and Organic Cow

WhiteWave Foods, a subsidiary of Dean Foods, owns both Horizon Organic and Organic Cow brands and is the largest processor and distributor of milk and other dairy products in the United States. WhiteWave Foods is also the nation's leading manufacturer of organic milk, with approximately a 40% share of the organic dairy market. WhiteWave reports directly to its parent company Dean Foods, which in turn reports to its shareholders. WhiteWave has several processing plants but mostly contracts for processing with other Dean Foods plants. Currently the WhiteWave organic brands are subsidized by soymilk products (Silk) and International Delight artificial coffee creamer. WhiteWave works closely with Dairy Farmers of America (DFA) and Dairy Marketing Services (DMS), both of which handle the logistics of moving raw milk from farm to processing plant. WhiteWave

Foods has individual contracts with each farmer, negotiated sometimes with a confidentiality clause. WhiteWave is usually the price leader, but it does not provide additional services beyond its dedicated field staff, producer meetings, producer newsletters, and occasional funding for attending events. Individual contracts vary depending on the size and location of the farm, and the price is set in individual negotiations based on a nationwide base price, as dictated by management in Boulder, Colorado. WhiteWave has its own field staff that works directly with its producers.

Organic Valley/Cooperative Regions of Organic Producer Pools (CROPP)

Organic Valley/CROPP is a marketing cooperative of organic farmers created in Wisconsin in 1988 to market produce. It has grown into a national cooperative that markets mostly organic dairy products with its own brand in the retail market. CROPP grew rapidly from 2005 to 2009 and currently has about 800 organic dairies and approximately 30% of the organic dairy market. Producers who sell their milk to CROPP must purchase a base, or the ability to sell CROPP a certain volume of milk, as their equity in the cooperative. This equity can either be paid in one lump sum or deducted from the producer's milk check. The price that CROPP pays farmers for their organic milk is set by the CROPP Board with advice from regional farmer committees and management. CROPP contracts with National Farmers Organization, DMS, and other cooperatives to move their milk from farm to plant. It owns only one cheesemaking plant, choosing to contract with plants owned by DFA, Dean Foods, and some independent plants for processing and packaging its product. CROPP has a defined committee structure to solicit input from its member-owner producers. Leadership from its top-level management and the board has remained stable. CROPP has profit sharing in the following ratio: 45% to farmers, 45% to employees, and 10% reinvestment, in addition to seeking outside capital, with nonvoting preferred shares carrying a 6% dividend. CROPP provides dedicated field staff, veterinary support staff, newsletters, producer meetings, and assistance purchasing organic feed, plus a growing organic cull cow and dairy steer program.

HP Hood LLC

HP Hood entered the market in 2006 with the Stonyfield brand, which it has grown to about 15% of the market. It leases the brand name from Gary Hirshberg, Stonyfield Farm, and is aggressively establishing market share based on the Stonyfield name and low wholesale pricing to retail outlets. The LLC consists of HP Hood and National Dairy Holdings, a subsidiary of DFA. DMS procures all of HP Hood's organic milk and provides all field-staff support. HP Hood has not invested in an organic market infrastructure and uses its own processing plant and distribution systems for its Stonyfield brand and private label milk. Its entry into the market three years ago stimulated increased competition and a subsequent increase in pay price, but it has since set its contract pay price to match its competition, CROPP and WhiteWave Foods.

WHAT TO LOOK FOR WHEN NEGOTIATING CONTRACTS

Strong partnerships between processors and producers built the organic dairy industry with the knowledge that an economically sustainable pool of producers provides the security processors need to expand their markets and product lines. With the growth of the organic dairy market, many of those personal relationships have weakened. Contracts offered by the different companies are similar, and choosing a company most often comes down to personal philosophical preferences for company ownership and relationships with field staff and organic dairy neighbors.

Navigating the contract process is complex and can be confusing. It may be helpful to discuss how to negotiate a potential contract with a farmer mentor who is familiar with the process. You may consider discussing potential contracts with your farm lawyer. Some things to consider as you evaluate a contract.

- Are there seasonal bonuses and benefits? How will this affect your farm?
- Your projected production costs both during and following transition—will the base price cover your costs of production?
- Will coop agreements change the terms of a contract after it is signed?
- Are animal welfare standards defined in the contract; are they well defined?
- Are there confidentiality clauses, and what are the limitations? Has a lawyer evaluated this with you?
- What supporting services does the company offer (technical assistance, organic vet, etc.)?
- How does the farmgate price compare to other companies'?
- Are there any hauling fees and what are they? Are they cheaper at a closer processing facility?
- Can you still be profitable if there is a $2 per cwt drop in pay-price?

BREAKDOWN OF CONTRACT PRICING

In the first few months of 2008 the following New England farmgate pay prices consisted of a base price plus market adjustment premiums (MAP)[5]: WhiteWave ($26.25), HP Hood ($26.90), and CROPP, which does not have a MAP but has a base price plus a regional premium ($27). WhiteWave and Hood have a regional pricing system that is not shared publicly, and CROPP posts its base price plus regional adjustments on its website. All three companies have discretionary programs that can be negotiated, with seasonal bonuses that vary from $3 for four months to as little as $0.50 for two months. The effect on net income of the seasonal bonuses is difficult to assess on a regionwide basis because the impact is subject to many variables including unpredictable weather, crop conditions, and the price of purchased feed. In the Midwest, farmgate prices for producers drop by as much as $4.50 per cwt even though USDA studies show that operating costs are around the same as in the Northeast. Producers in the Midwest and the West have an average base price of approximately $22.50 per cwt. The determination of pay price in other parts of the country follows the same pattern as in the Northeast, with little variance among processors. Processors also "exchange" milk with each other from areas in surplus (the West for example) to satisfy markets in the East.

CROPP pay price is easier to evaluate with only a base price plus regional payments, but the WhiteWave and Hood prices can give a higher return based on the farm's production system. Since 2006, WhiteWave has led the market slightly, but only by less than a dollar (see Table 2.1).

TABLE 2.1 OVERVIEW OF PAY PRICE IN THE NORTHEAST, 2006–JUNE 2009 (IN USD)

	HP HOOD				HORIZON ORGANIC				ORGANIC VALLEY[a]			
	2006	2007	2008	2009[b]	2006	2007	2008	2009[b]	2006	2007	2008	2009[b]
Base price	26.00	26.00	24.90	25.30	24.00	24.00	25.00	25.00	26.00	26.00	28.25	27.25[c]
MAP	—	—	2.00	2.00	2.00	2.00	2.50	2.00				
Short (3–4 months)	2.00	2.00	2.00	2.00	1.50	1.50	3.00	4.00	—[c]	—[c]	—[c]	—[c]
Short for 2009												(1.00)[d]
Long (8 months)					0.75	0.75	1.50	1.50				
Trucking charge/yr	—	—	—	—	—	—	—	—	900	900	900	2,160
AVERAGE YEAR-ROUND PRICE[e]												
Long					26.50	26.50	28.50	28.00				
Short	26.50	26.50	27.40	26.80	26.50	26.50	28.50	28.33	26.00	26.00	28.25	27.17

a Organic Valley requires producers to purchase preferred stock equivalent to 5.5% of their annual base gross income. Historically this investment has an 8% return on the required amount in Class B Stock. Profit sharing is activated if Organic Valley's 2.2% profit goals are met or exceeded.

b In July 2009, HP Hood eliminated their MAP payments. Organic Valley introduced a quota program that gave producers an "active base" (different from the base price they purchased as equity in the coop) of 93% of their last three years of milk production on a monthly basis. Over-quota milk takes a $15 pay reduction.

c $1.00/cwt for milk produced in Oct., Nov., Dec., provided the average is greater than the average for May, June, July.

d Seasonal deduction for May, June, July.

e Seasonal bonus paid is multiplied by the number of months and divided by a complete calendar year. The trucking charges are not included in the OV calculations as the allocation to a per-cwt charge varies with the size of the herd.

TABLE 2.2 COMPONENT PAYMENTS BY COMPANY ($/LB.) IN 2008

	HP HOOD	HORIZON ORGANIC*	ORGANIC VALLEY	FMMO
BUTTERFAT	FMMO	1.82	2.00	1.77
PROTEIN	FMMO	1.56	1.86	3.13
OTHER SOLIDS	FMMO	0.25	1.65	0.60

*Horizon offers a choice of 3 programs, one just for butterfat ($0.13 BFD/.1 point ±3.5%BF), one as above, and one using the FMMO.

TABLE 2.3 QUALITY PREMIUMS OFFERED BY COMPANY ($/CWT)

VOLUME PREMIUM	HP HOOD	HORIZON ORGANIC	ORGANIC VALLEY[a]
Less than 750 cwt/month	$0.15	$0.15	none
Less than 1,500 cwt/month	$0.30	$0.30	none
Less than 3,000 cwt/month	$0.50	$0.50	none
LOW STANDARD PLATE COUNTS			
9,000–16,000	$0.05	$0.05	11–20,000 @ $0.10
5,000–8,000	$0.15	$0.15	0–10,000 @ $0.25
6,000 or less	—	$0.28	0–10,000 @ $0.25
4,000 or less	$0.25	$0.37	0–10,000 @ $0.25
LOW PRELIMINARY INCUBATION COUNTS (P.I.)			
17–32,000	$0.20	$0.28	$0.25
9–16,000	$0.35	$0.56	0–15,000 @ $0.50
Less than 12,000	—	$0.84	0–15,000 @ $0.50
Less than 8,000	$0.75	$1.25	0–15,000 @ $0.50
LOW SOMATIC CELL COUNTS			
251,000–300,000	$0.25	<300 @ $0.38	—[b]
201,000–250,000	$0.50	<225 @ $0.75	—[b]
151,000–200,000	$0.75	<175 @ $1.13	—[b]
Less than 150,000	$1.00	$1.50	—[b]

a OV has a Lab Pasteurization Count (LPC) payment of 0–51 @ $0.50/cwt; 51–100 @ $0.25/cwt. LPC is a laboratory test involving the heat treatment of bulk milk samples at 145°F for 30 minutes.

b FMMA-type formula yields $0.48 cents per 100,000 on a linear premium basis. The hinge point is 350,000. The deduction formula is $0.48 per 100,000 over 350,000 on a linear basis. Maximum deduct level is at a cell count of 1,000,000. As an example, a somatic cell count of 250,000 would result in a $0.48/cwt premium, while a somatic cell count of 450,000 would result in a $0.48/cwt deduction.

COMPONENTS AND QUALITY PRICING

Components and quality pricing varies with each company. With components, pricing varies from using the Federal Milk Marketing Order (FMMO) to a fixed price within a contract. Depending on farm operating practices and the breed of cows, components, seasonal, and quality payments can increase gross income by as much as $3 per cwt (see Tables 2.2 and 2.3).

TRANSITION INCENTIVES

Each company has its own transition incentives and programs that provide benefits other than cash. Space limits preclude a full description of these programs, but they include Horizon Organic's HOPE program and Organic Valley's centralized grain purchasing and veterinary helpline, plus support staff from all companies. Specific information on these programs is available online.

- http://www.horizonorganic.com/site/forfarmers/community.html
- http://www.farmers.coop/farmers-wanted/how-to-become-an-organic-farmer

CONSIDERATIONS WHEN CHOOSING AN ORGANIC MILK MARKET

As you transition to organic dairy, deciding who to sell your milk to involves many issues. During 2005–2007, processors actively sought new farms, but since 2008 they have become more selective, signing farms that meet their criteria based on location and size. While demand for organic milk is projected to increase, it is no longer certain that processors will take all the organic milk that is available. The size of the organic market now justifies its own infrastructure. Therefore there is little risk of being unable to sell your milk if your processor goes out of business (except in areas where trucking charges are prohibitively expensive). Dairy Marketing Services (DMS) controls most of the transportation for organic milk across the country and is able to move milk to different manufacturers. (Appendix F lists many milk markets in the Northeast.)

When choosing a processor there are three areas you should investigate before starting detailed discussions with any one specifically.

- Talk with your organic dairy neighbors and attend meetings with other organic dairy producers to find out which processors provide the best service in your area, have the most qualified support staff, arrange the most productive producer meetings, and have enough farms to justify a load of milk even if a couple of producers change processors or go out of business.
- Arrange meetings with all the major companies and independent organic processors, and ensure that they leave written details of what they are willing to offer. While the pay price is fairly even among all processors, there are many differences in incentives and deductions among processors. You should share this information with other organic producers, lawyers, or family members to ensure that you understand the implications of

what you are being offered, as the economics of organic production are different from those of conventional.

- Run the numbers and check your assumptions with other producers. Although long-term production levels can match or exceed those you achieved under conventional production, it may take three to five years to reach them and you may need to change your production practices to achieve the best return. You can then take these numbers back to the processors and explain what you need by way of transition or signing bonus to sign with them. Your budget should always take into account a lowering of the pay price by $2 per cwt.

The following sections explore two areas to consider when choosing with which processor to market your milk.

PRACTICAL AND ECONOMIC CONSIDERATIONS

1. BEFORE STARTING YOUR TRANSITION, ensure you have a market for your milk on the date you *finish* your transition,
2. Work with your certifier to calculate the date you will finish your transition.
3. All companies pay some money to assist with transition expenses or as an enticement to change processors. All processors advertise the assistance they can give to help with transition costs, but many will give much more in the form of a signing bonus if you are already close to an existing organic trucking route, are close to where the milk is processed, have sufficient volume to fill a truck, or have a record of producing clean milk. You can negotiate this and should get it in writing.
4. If you are transitioning to organic, you will need the support of well-qualified field staff. Some companies do not provide any field staff, or rely on staff that is only knowledgeable about conventional milk production. Talk to other organic farmers about the quality of the field staff.
5. Some companies require that you be a member of an existing cooperative to process payments to you, while others do not. If you are not a member of a cooperative, this might increase your costs and deductions as it might require equity payments.
6. Make sure you understand the details of the pricing structure.
   - What constitutes the base price?
   - What regional premiums you will receive?
   - Will you pay for trucking and how is that prorated?
   - What is the MAP?
   - What criteria will be used to pay components? This can sometimes amount to an extra $3–4 per cwt.
   - Who will test your milk for somatic cell counts and how independent will the tester be?
   - When will deductions for equity start?
   - How long will the contract be for, and what are the renewal conditions?

This is a legal contract and the details are important. You may find it helpful to have your lawyer review the contract with you.

PHILOSOPHICAL OR PERSONAL PREFERENCE

1. You have every type of corporate structure to choose from in the major processors. The three major processors are the most secure but provide limited involvement in decision making. CROPP is a marketing cooperative so it does allow some involvement in the decision making, but do not expect to make a great impact. WhiteWave is a large corporation that must look hard at the bottom line and the return to its shareholders. HP Hood has been a major player in conventional milk on the East Coast for many years and now operates nationally as an LLC designed to make a profit for its owners. In many areas, there are smaller cooperatives to join as well as the ability to sell directly to individual manufacturers. This will give you more control, but it also has increased risk that they will be manipulated by the major companies.
2. Some companies may want you to sign a confidentiality clause.[6] While many producers refuse to do so and still sign up with the company, this can be used to intimidate the producer or as a bargaining chip ("Sign this clause and we will increase your signing bonus") in reaching a deal. Cooperatives do not ask for one. This will be a legal contract enforced by law and companies retain the right to enforce a confidentiality clause in any breach of contract action, so it is not wise to ignore it.
3. Organic farming has a mission to improve the earth as well as an economic advantage to add value to a wholesale product. You might want to choose a processor that shares your mission or one with whom your neighbor has had a good experience over a number of years.

ALTERNATIVES TO MARKETING YOUR MILK AS A COMMODITY

VALUE-ADDED PROCESSING

Because consumers wanting a healthy, environmentally friendly dairy product drive the organic market, and local foods are growing in popularity, value-added processing may be a successful business model. There are many stories of successful organic dairy processing businesses in the Northeast such as the following.

- Butterworks Farm in Westfield, VT (http://www.butterworksfarm.com)
- Evans Farmhouse Creamery in Norwich, NY
- Hawthorne Valley Farm in Ghent, NY (http://hawthornevalleyfarm.org)
- New England Organic Creamery in Dracut, MA (http://www.neorganiccreamery.com)
- Strafford Organic Creamery in Strafford, VT (http://www.straffordcreamery.com)

Beginning a value-added dairy venture requires a diverse set of business skills that are different from running an organic dairy farm. While there are many examples of successful organic dairy processors, there are also many examples of failed businesses. If value-added processing is something that appeals to you, take the time to create an effective business plan that will help you be successful. If you have never run a value-added processing business before, consider first transitioning your farm to organic and then begin phase two by implementing your own processing facility. Many great organizations can assist you in creating a solid business plan if you are interested in pursuing this opportunity. Check with your county extension office for guidance on finding the best organization to help you create a business plan.

RAW MILK

Consumer demand for environmentally sustainable organic raw milk has increased in popularity over the past decade. It requires almost no extra infrastructure, but does require an added attention and commitment to milk quality. Each state has its own raw milk regulations. In the Northeast, Connecticut and Maine allow in-store retail selling of raw milk. Other states allow registered raw milk sales on the farm or through cow-share programs. To find out what your state requires, contact your state department of agriculture. Many online resources contain information about raw milk, including the following.

- http://www.ftcldf.org
- http://www.rawusa.org
- http://www.realmilk.com/happening.html
- http://www.westonaprice.org/

NOTES

1 Aurora Dairy has five farms with a total of more than 12,000 cows. All their milk is processed and packaged in their own processing plant on their farm in Boulder, CO.

2 DMS is an LLC comprised of many different cooperatives set up by DFA. It coordinates the movement of conventional and organic milk across the country.

3 The Agricultural Adjustment Act of 1938 states that the parity price formula is the "average prices received by farmers for agricultural commodities during the last ten years and is designed to gradually adjust relative parity prices of specific commodities."

4 Despite requests from individual farmers and farmer organizations like NODPA and FOOD Farmers, organic dairy farmers received no increase on their base price from January 2006 to December 2007, with the exception of one company that gave a small cost of living increase. From January 2008 to December 2008, organic dairy farmers received an average increase in their farmgate price of 9% based on their December 2007 price. That is an average of 3.8% per calendar year, approximately 8¢ per gallon/per year for the period January 2006 to December 2008.

5 Market Adjustment Premium and seasonal payments are paid at the discretion of the processor.

6 By signing a contract that contains a confidentiality clause, the producer agrees not to disclose the specific terms of the agreement to any other party other than the processor without approval from the processor. This means that the producer agrees not to share the specific contract terms with parties such as family members, lawyers, accountants, lenders, other producers, another processor, or farmer organizations without first seeking approval from the processor. The restriction on the producer is not limited to sharing the written version of the contract with another party. The producer agrees not to "disclose the specific terms." This means that a producer who orally discusses the terms of the contract, or part of the contract such as the pay price or premiums offered, is likely violating the contract by disclosing its specific terms.

Liz, Nathan, and Brian Bawden.

BRIAN AND LIZ BAWDEN

BETHANY WALLIS

Brian and Elizabeth Bawden's farm is a stone's throw from the Saint Lawrence River in northern New York, and on a clear day you can watch the masts of the ships hauling cargo as they float down the river. The Bawdens began farming in Canada, but due to the constraints of the Canadian quota system, high land prices, and low milk prices they researched the U.S. market as a possibility for their future in dairy. When faced with this option, the U.S. was truly "the land of opportunity" for them to get started in.

Brian first learned about organic management before there was an organic market available in Canada. He sees organics as a way to make a dent in reversing the trend of declining small family farms and to provide a future for his family's next generation. Brian emphasizes the need to get a return on his family's investments in order to build a successful business. They continue to maintain a short-term debt structure by remaining small and avoiding additional investments in land and buildings.

Brian and Liz both work full-time on the farm and employ a part-time milker. They produce hay, oats, and corn as well as manage a custom-harvest operation. Their future vision for the farm is to become feed independent. The Bawdens grow mostly oats for their grain source due to their farm's climate and soil conditions. They need a minimum of 10 acres of forage and grain per cow to maintain her for the year, since they typically take only one hay cutting. Their land has shallow clay soils that rest on flat rock and they utilize aged manure to increase soil fertility. They have conducted soil samples in the past and now balance their soil fertility with crop rotations and green manure.

The Bawdens feed grain solely to maintain body condition. Their milking herd's feed ration includes pasture, dry hay, a 16% protein grain (5–10 pounds/milking), a high-calcium mineral, free-choice Redmond salt, and kelp. Liz and Brian monitor the cows for production and watch manure consistency to determine if they need to make adjustments in their feeding program. Approximately 100 acres surrounding the farm are pasture with 7 acres in permanent pasture. They rotate the cows among large paddocks, moving them once a week from May to October. They maintain high-quality forage by clipping pastures and using their permanent pasture to allow recovery time for the system. The herd has access to water in each pasture and in the barnyard.

Liz and Brian recall that during the transition they experienced herd health problems they had never before encountered. They relate this to the cows going through detox. Since transitioning, their herd health problems have declined and mastitis is rarely an issue, which they believe is due to reduced production and less stress on their system. To maintain good udder health they change inflations monthly and use the best cleaners available.

Liz has learned how to make her own tinctures and utilizes alternative health treatments. Some resources she has found helpful over the years include Dr. Paul Dettloff's publications, Dr. Hubert Karreman's writings, attending Northeast Organic Dairy Producers Alliance (NODPA) field days and other conferences, as well as joining NODPA's O-Dairy online chat group.

Brian and Liz are very active in their organic milk market. When they first began shipping organic milk, the market was small and encouraged camaraderie among producers. Liz states, "We all really knew each other, which made it possible to have more of a united front. Processors wanted farmers to learn from one another about organic management. They set up farmer meetings as well as held an annual update meeting. At these meetings the processors were more available for producers to ask the tough questions about the state of the organic market and milk price." Liz Bawden stays informed about her milk market because she feels there is power in knowledge. "If you don't stay informed of your own business, you are doomed to fall into someone else's blunders." Liz feels that the organic market is unique due to the tone set by the original folks in the industry. "Organics have always had a basis of sharing information, and the farmers bend over backward to help one another. There is a great deal of communication from farm to farm."

Organic farmers can make a difference in their milk market. Liz notes that farmers drawn to organic management in the first place are more likely to get involved. "They tend to be risk takers who

Nathan Bawden with cows on fall pasture.

do not follow the beaten path or care about what their neighbors think, and they seem to be more prone to ask questions." Liz suggests that transitioning farmers get involved in the industry by joining their local organic dairy producers association, attending organic conferences, field days, and workshops; and talking with other organic farmers. She emphasizes the importance of farming organically for a while to gain knowledge and the respect of your peers in the industry. She suggests that farmers should then join other groups where organic issues may overlap, such as local boards and other organic organizations. Finally, she stresses getting involved with your processor. Depending on the processor, the role you can play will vary, so research your involvement options when looking for a market.

One benefit of organic marketing is a relatively stable price, which allows organic farmers to better plan for the future. Contracted milk sets a base price that farms can rely on, unlike the conventional industry, which can fluctuate throughout the year. This stable price has allowed Liz to continue working on the farm as opposed to most 60–cow conventional dairies that have to send a spouse to work off the farm. Organic processors like Horizon Organic and Hood, LLC have the flexibility to increase or lower the market adjustment premiums (MAP), and Organic Valley can change its regional payment.

In 2009, however, the stable base price had not increased at the same rate as the costs of production. Liz warned, "We are running on the same treadmill as conventional, paying more for grain without an increase in pay price. The organic market has enough milk

so we are going to take less for our milk and pay more for inputs. Suddenly, we are on a very undesirable path, and those of us who are going to remain successful are going to have to devise a method to deal with the situation. The guy who is going to be successful is the one who isn't afraid to experiment."

For many organic farmers, focusing on growing all their own feed was a solution to this problem, and many turned to seasonal dairies. Liz cautioned, however, in 2009 "the organic market saw the largest spring flush it had ever experienced. Low cost of production was the push in organic management, leading to low inputs and feeding less grain with the natural tendency toward seasonal herds. You can make milk cheap seasonally with no grain and good pasture." However, when conventional price is low, there is no profitable place for organic processors to dump excess spring flush milk, and the farmers pay the price. "[In 2009], it was a huge problem since there is a large disparity between conventional and organic price, approximately $14. With the big spring flush and low conventional price, processors were in a pinch." As a result, processors lowered their MAP and regional payments.

Liz's hope for the future is that the organic market will continue to grow and continue to have a large impact on conventional farm management practices. Someday, it could become the norm and the organic label would eventually disappear, allowing organic practices to benefit the environment and consumers. In the meantime Liz says, "We are just going to have to put our heads down and bull through it. The market will change, the real question is will it get worse before it gets better?"

CHAPTER THREE

ORGANIC SOIL MANAGEMENT

ELIZABETH DYCK

Healthy soil is the not-so-secret weapon of organic farming. A healthy soil produces high-quality pasture and crops that in turn create and sustain a healthy and productive herd. Healthy soil also buffers the farm against risk—drought, flood, and pest and disease attack. On a larger scale, healthy soil stores carbon, working to reduce CO_2 levels in the atmosphere. Ask successful organic growers where to start when transitioning the farm, and they will all give the same answer: start with sound soil management. This chapter will outline the basic principles of organic soil management. It will detail methods to assess and monitor soil health and then discuss management practices that build up and sustain soil productivity on transitioning and organic farms.

KEY POINTS FOR TRANSITIONING FARMERS

- Become familiar with NOP rules §205.202–§205.206, §205.601, and §205.602.
- Understand the concept of soil health and how to assess it on the farm.
- Shift focus from fertilizers to building soil organic matter.
- Effectively use the farm's own resources, e.g., N-fixing legumes, manure, and cover crops, to build and maintain soil productivity.

KEY PRINCIPLES OF ORGANIC SOIL MANAGEMENT

Transitioning to organic soil management often requires some major shifts in perspective, including attention to overall soil health, a focus on soil organic matter, and learning from natural ecosystems.

MANAGING OVERALL SOIL HEALTH, NOT JUST SOIL FERTILITY

Soil management in conventional farming focuses largely on soil fertility, which is maintained mostly through inputs of synthetic fertilizers. In transitioning to organic soil management, farmers need to broaden their perspective to think about overall soil health or soil quality, not just soil fertility. What is meant by soil health? A healthy soil is one that has these qualities.

- Optimum levels of essential plant nutrients and pH.
- Good soil structure: a mixture of varying sizes of soil crumbs or aggregates (stuck-together soil particles) that resist erosion

and help to create channels for good air circulation and water infiltration and retention in soil (see Figure 3.1). This creates an environment in which soil organisms, including plant roots, can thrive.

- An active community of diverse soil organisms, both microscopic (e.g., bacteria, fungi, algae, and nematodes) and visible (e.g., earthworms, sowbugs, centipedes, and beetles). These organisms perform numerous vital functions, including sticking together soil particles, cycling nutrients into forms available to plants, and suppressing disease pathogens and insect pests. Plant roots enhance biological activity in the soil by exuding materials that support the growth of many microorganisms.

FIGURE 3.1
Soil after one year of a sweetclover sod crop. Good soil structure is shown by the mix of sizes of soil aggregates, or crumbs.

Soil health or soil quality, then, actually has three components. In addition to a chemical component (soil fertility), there are physical and biological components. To optimize soil health, all three components need to be monitored and managed.

FOCUSING MORE ON SOM (LESS ON NPK)
The key factor in organic soil management is building up and maintaining soil organic matter (SOM). The major benefits of adding organic matter to the soil are listed in Box 3.1. Note that SOM plays critical roles in all three components of soil health.
SOM consists of a number of different fractions, all of which are essential for maintaining a healthy soil. A useful way of understanding these components is to think of them in the words of soil scientists Magdoff and van Es, as the "living," "the dead," and the "very dead."[1]

- The "living" consist of soil organisms.
- The "dead" consist of fresh and relatively fresh residues (e.g., dead organisms, crop residues, manure). *The decomposition of these residues provides the major source of nutrients in organic farming.* Compounds released during residue decomposition also help to stick soil particles together, improving structure.
- The "very dead" consist of humus, very well decomposed and thus stable material that has a critical role in the soil's ability to hold onto to certain essential nutrients and to create good soil structure.

There are no substitutes for soil organic matter in organic farming. No purchased inputs can supply the benefits listed in Box 3.1. Sufficient levels of soil organic matter can only be built up and maintained by sound management, including the practices listed in Box 3.2.

LEARNING FROM NATURAL SYSTEMS

Natural ecosystems generally exhibit high levels of stability and are self-sustaining. Fueled by the sun and acted on by natural physical processes, they typically contain a complex web of organisms. Nutrients are conserved within the ecosystem by the process of nutrient cycling: plants and animals die and decay in the soil, the nutrients in their residues are eventually converted into forms that can be taken up by plants (and from them by animals), and the cycle repeats.

Farms can never be natural systems because farm operations disrupt the natural landscape and its community of organisms and because farm products (containing nutrients) leave the farm. Nevertheless, by observing natural ecosystems and trying to mimic the characteristics that help to sustain them, organic farms can become more sustainable themselves. Key organic management principles derived from properties of natural systems include these practices.

- Protect the soil resource by
  - keeping the ground covered—not only in pastures but also in cropland (with sod crops, intercrops, and cover crops), and
  - leaving under permanent cover soils that are not suited for tillage, i.e., steeply sloping, rocky, droughty, or poorly drained soils.

BOX 3.1 BENEFITS OF THE ADDITION OF ORGANIC MATTER

Adding organic matter to the soil (which increases the diversity and activity of soil organisms) provides these benefits.

- Increases the supply of available plant nutrients (through a variety of processes) including, N, P, S, Ca, Mg, K, and micronutrients
- Helps suppress such crop pests as soilborne diseases and parasitic nematodes
- Forms humus and other growth-promoting substances
- Detoxifies harmful substances
- Increases water-holding capacity
- Improves aeration
- Improves the soil's resistance to erosion and compaction
- Darkens the soil, causing the soil to warm earlier in spring
- If added as a surface mulch, provides direct protection of the soil from erosion and moderates soil temperature extremes.

Sources: N. C. Brady and R. R. Weil. 1999. The Nature and Property of Soils, 12th ed. Prentice Hall, Upper Saddle River, NJ: p. 469; and F. Magdoff and H. van Es. 2000. Building Soils for Better Crops, 2nd ed. Sustainable Agriculture Network, Beltsville, MD: p. 22.

- Rely as much as possible on natural processes occurring on the farm for the energy and nutrients needed to grow crops and livestock: e.g., photosynthesis, biological N fixation, and nutrient cycling.
- Include on the farm a diversity of plants—e.g., a variety of crop species, hedge rows, tree groves, riparian zones—that provide habitat and food for a multitude of other organisms that contribute to nutrient cycling and protect against crop and livestock pests.

BOX 3.2
HOW TO BUILD UP AND MAINTAIN SOIL ORGANIC MATTER

- Leave crop residues in the field.
- Include sod crops (e.g., forage legumes, grasses, or their mixture) and high-residue crops (e.g., small grains, sorghum-sudangrass, Japanese millet) in crop rotations.
- Add manure or compost.
- Grow cover crops.
- Avoid intensive tillage and cultivation, which burn up SOM.
- Practice good pasture management.

ASSESSING SOIL HEALTH

Early in the planning stages for transition, it is important to do a thorough assessment of soil health in each of the farm's pastures and in crop fields. This information is critical in developing a soil management plan that will sustain farm productivity through the transition period and beyond. It also establishes a baseline to which farmers can compare the results of subsequent testing to see whether their soil management has improved soil health.

IDENTIFYING AND MAPPING SOILS

If not already done, a good place to start the soil assessment is to identify and map the soil types that occur on the farm. An excellent, free resource is the *County Soil Survey*. Soil surveys for most counties are available to farmers at their local Natural Resources Conservation Service (NRCS) office (formerly known as the Soil Conservation Service) or online (see the Resources section at end of this chapter). The survey contains maps that identify the dominant soils on the farm and their location. This information can also be transferred to the farm map required for certification. The survey gives a detailed description of each soil, including slope, topsoil and subsoil textural classification, depth and color, and drainage. Also included are estimates of how the soil will yield when planted with a variety of crops and recommendations for its use.

Soil texture

The soil survey's identification of a soil's textural classification (determined from the percentage of sand, silt, and clay in the soil) is a big help in managing the soil to increase its health. For example, as transitioning farmers learn about the importance of increasing soil organic matter, they often ask, "What is the level of soil organic matter I should aim for?" The answer can vary, depending on the soil's texture and drainage. Capacity for soil organic matter buildup is lowest in coarse-textured sandy soils and highest in fine-textured clay soils. SOM is more easily conserved in a poorly drained soil than in well-drained soil. A SOM content of around 2% may be good for a sandy loam but somewhat low for a silt loam. A SOM content of 4% would be very good for a silt loam but somewhat low for a poorly drained clay soil. Table 3.1 shows the many soil properties in addition to SOM level that soil texture influences, along with management implications.

TESTING AND MONITORING SOIL BIOLOGICAL AND PHYSICAL HEALTH

Soil physical properties can be assessed by sending soil samples to land-grant university soil testing labs. Some private labs will test for biological activity and type of soil microorganisms present. (See Appendix E for a list of soil testing resources.) Apart from the expense, the problem with using soil labs to monitor soil biological and physical health is that it deprives farmers of an opportunity to develop expert knowledge of their soils and to get actual, hands-on feedback on how their soil management is working. Table 3.2 describes simple, quick, but highly informative tests that farmers can do themselves to monitor their soils' biological and physical health.

TESTING AND MONITORING SOIL FERTILITY

Regular monitoring of soil fertility is critical to ensure that pasture and crop plants have the nutrients they need and to avoid nutrient buildup. Soil in both crop fields and pastures should be tested at least every three to five years. Consider testing a set of fields and pastures one year, another set the next year, and so on to distribute the cost of soil testing over time. Soil sampling methods for both pastures and tilled cropland are described in Table 3.3.

TABLE 3.1 RULES OF THUMB ON HOW SAND, SILT, AND CLAY INFLUENCE SOIL PROPERTIES AND BEHAVIOR

PROPERTY/BEHAVIOR	SAND	SILT	CLAY
Water-holding capacity	low	medium-high	high
Aeration	good	medium	poor
Drainage rate	high	slow-medium	very slow
Soil organic matter level	low	medium-high	medium-high
Decomposition of organic matter	rapid	medium	slow
Warm-up in spring	rapid	moderate	slow
Susceptibility to compaction	low	medium	high
Susceptibility to water erosion	low[a]	high	low[b]
Suitability for tillage after rain	good	medium	poor
Leaching potential	high	medium	low
Ability to store plant nutrients	poor	medium-high	high
Resistance to pH change	low	medium	high

a Unless fine sand, which is more erodible.

b If there is good soil structure. Clay soils with poor structure are highly erodible.

Adapted from N. C. Brady and R. R. Weil. 1999. The Nature and Property of Soils, 12th ed. Prentice Hall, Upper Saddle River, NJ: p. 125.

Standard soil analyses usually include pH, organic matter content, phosphorus (P), potassium (K), and often magnesium (Mg) and calcium (Ca). Analyses are also often done for sulfur (S) and some micronutrients (often zinc, manganese, iron, copper, and boron). Cation exchange capacity (CEC), a measure of the soil's ability to hold nutrient cations (positively charged ions like K+, Mg(2+), and Ca(2+) that are readily available for plant uptake), is also often tested. Although nitrogen (N) is usually the most limiting nutrient to crop production, soil tests for N are not very useful because plant-available N is in constant flux in the soil. Instead, N requirements should be determined by considering the field's cropping history, what soil amendments have been applied over the last several years, and the N required for the crop to be grown.

Choosing a soil testing lab

There is a bewildering variety of soil testing labs. Soil laboratories differ in what they include in standard tests, the procedures they use, and the recommendations they make. Two rules of thumb can help sort out which soil testing laboratory to choose.

- Use a soil lab in your state or region since it will use tests and procedures developed for the soils, crops, and climate of your farm.
- Stick with the same soil lab. Because different labs use different extraction procedures, the same soil sample sent to two different labs will likely show different results. By sticking with the same lab over the years, confusion will be avoided and trends in soil fertility will be more clearly evident.

It is harder to suggest a rule of thumb for choosing between labs that use different interpretations of soil analyses to make recommendations for nutrient needs. This is because there is an unresolved disagreement between some organic growers, many of whom are long-time successful organic farmers, and most soil scientists.

Some organic farmers are proponents of a theory of soil management that is known by a variety of different names—the Albrecht, Walters, or Kinsey approach or the balanced cation approach. According to this soil management theory, basic cations must make up certain percentages of a soil's CEC. These percentages vary somewhat from source to source but typically are given as follows: Ca, 60–70%; Mg, 10–20%; K, 2–5%. Proponents of this theory maintain that when the soil cation content is balanced in this way, soil tilth is improved and crop yield and quality are optimized. Soil labs following this approach make recommendations for nutrient additions to bring the soil in line with the desired nutrient ratios. They also closely monitor and make recommendations for trace minerals (e.g., boron, manganese, zinc, iron, and copper).

Many soil scientists and land-grant university soil testing labs follow a different approach, which maintains that there is a critical or sufficiency level of a nutrient needed for optimum » *continued on page 51*

TABLE 3.2 TESTING AND MONITORING BIOLOGICAL AND PHYSICAL SOIL HEALTH ON THE FARM

Not all of the following tests have to be done, but it is important to repeat the tests you choose every three to four years. Do the tests at two or more sites in each field. Use a landmark or GPS to locate each sampling site so that you can return to the same location in future years. Assessments for most of the indicators will improve over time with sound soil management, especially those practices listed in Box 3.2. For severe erosion problems, consult with your local NRCS office.

INDICATOR	SAMPLING SITE/ CONDITIONS	HOW	ASSESSMENT
BIOLOGICAL			
BIODIVERSITY OF SURFACE "CRITTERS"	Pasture and hayfields Do in the spring or fall. Soil should be moist but not wet or cold.	1. Construct a 12" diameter ring from flexible wire. 2. Lay ring on ground at sampling site. 3. Clip vegetation to soil surface and remove from ring. 4. Work from the outer area of the ring to its center, pulling aside surface litter and counting the number of different organisms present.	Record the number of different species of organisms found in the ring. The greater the diversity of organisms present, the better initial decomposition of organic matter will proceed.
EARTHWORMS	Pasture, hay, and tilled fields Do in the spring or fall. Soil should be moist but not wet or cold.	Take a full shovelful of soil. Break apart the soil and count the worms, looking for casts and holes.	POOR: 0–1 worms in shovelful. No worm casts or holes. MEDIUM: 2–10 worms in shovelful. A few casts and holes present. GOOD: 10 or more worms in shovelful. Lots of casts and holes present.
SOIL COLOR (ORGANIC MATTER CONTENT)	Pasture, hay, and tilled fields Do in the spring or fall. Soil should be moist but not wet or cold.	Take a slice of soil that extends into the subsoil layer.	POOR: Topsoil color similar to subsoil. GOOD: Topsoil clearly defined and darker than subsoil.
SOIL SMELL	Pasture, hay, and tilled fields Do in the spring or fall. Soil should be moist but not wet or cold.	Grab a handful of topsoil and smell.	POOR: putrid/chemical/sour MEDIUM: no smell GOOD: fresh/earthy/sweet
ROOT HEALTH: SHAPE AND LENGTH/ DISEASE	Pasture (spring or fall) hay and tilled fields (late spring)	Dig up plants to a depth of 12"–18". Brush or wash off soil from the roots and examine.	POOR: Roots few in number, thickened, stunted; no subsoil penetration; discolored or blackened. GOOD: Roots branched and extensive, reaching into subsoil. Exterior and interior of roots are white.
PHYSICAL			
EROSION	Pasture, hay, and tilled fields	Observe after rainfall.	POOR: Large gullies over 2" deep, thin or no topsoil, runoff is color of soil. MEDIUM: A few rills or gullies less than 2" deep, runoff is lightly colored. GOOD: No gullies or rills; no runoff or runoff is clear.
DRAINAGE/ INFILTRATION	Pasture, hay, and tilled fields	Observe a day after rain.	POOR: Soil is still puddled (always very wet ground). MEDIUM: Soil is puddled for a short time, but eventually drains. GOOD: No puddling and no runoff; soil not too wet or too dry.
WATER-HOLDING CAPACITY	Tilled fields	Observe two days after a rain, then one week after the rain.	POOR: Plant stress two days after a good rain. MEDIUM: Plant stress a week after a good rain. GOOD: No plant stress after a week (ground does not puddle).
COMPACTION	Pasture, hay, and tilled fields	Construct a homemade penetrometer out of a 3' length of 1/4" rod sharpened to a point at one end. Use a file to make inch marks up from the pointed end. (This method will not work in a very stony soil.)	Record the depth and ease of penetration. For comparison, probe an undisturbed area (e.g., under a fence line) nearby.
AGGREGATE STABILITY (RESISTANCE TO EROSION AND COMPACTION)	Pasture, hay, and tilled fields Do in the spring or fall. Soil should be moist but not wet or cold.	From a handful of soil, select a soil crumb this big. ● (Make sure it is not a pebble.) Drop the crumb into a small clear container of water. Allow it to stand one minute. If the crumb is intact after one minute, gently swirl the container several times. If the aggregate is still intact, swirl the bottle vigorously.	POOR: Soil crumb broke apart in standing water. MEDIUM: Soil crumb broke apart after gentle swirling. GOOD: Crumb broke apart after vigorous swirling. VERY GOOD: Crumb remained intact even after vigorous swirling.

continued from page 49 » yield of a particular crop. Additions of a nutrient above this level will not increase yield. Critical or sufficiency levels of nutrients are determined through experiments that determine the effect of increasing soil test levels of each nutrient on crop yield.

TABLE 3.3 TAKING SOIL SAMPLES IN PASTURE AND TILLED CROPLAND

	PASTURE	TILLED CROPLAND
TOOL	Soil corer, spade, or shovel	Soil corer, spade, or shovel
DEPTH OF SAMPLE	3 inches	6 inches (or to depth of plow layer)
NUMBER OF SAMPLES	15–20 cores or subsamples per 5 acres	15–20 cores or subsamples per 20 acres. If the field is larger and very uniform, 15–20 subsamples per 40 acres.
SAMPLING AREA	Take separate soil samples of hilltops, hillsides, and valleys (foot slopes). These soils often have different characteristics and may have received varying amounts of dung and urine due to grazing patterns. Note: Combine sampling of all hilltops of same soil type, all side slopes of same soil type, etc., instead of sampling each hilltop, side slope, and foot slope separately.	Take within a uniform area of a field—a section of the field with the same soil type and that has been under the same management.
AREAS TO AVOID OR TO SAMPLE SEPARATELY	Areas around feeding areas, shade trees, watering sites, loafing areas, and other areas where dung and urine are apt to be concentrated. Also avoid wet spots.	Headlands, low or wet spots, old fencerows or any area that is unusual in terms of topography or management.
TIME OF SAMPLING	Take at a consistent time over the years (e.g., in the fall or in the spring). Avoid taking shortly after lime, manure/compost, or any other fertilizer application.	Take at a consistent time over the years (e.g., in the fall or in the spring). Avoid taking shortly after lime, manure/compost, or any other fertilizer application.
PROCEDURE	1. Mark sampling areas on a map and label them with a unique name or number. Be sure to store the map where you can easily find it once soil testing results come back! 2. Collect the 15–20 cores to make sure the sampling area is well represented. Collect samples in a zigzag pattern throughout the sampling area, sampling at the end of each zig and zag. 3. Before taking each subsample, scrape off any plant or manure residue from soil surface. 4. Mix the subsamples thoroughly in a clean plastic pail. 5. If you have a sample box or bag from a soil testing lab, fill it to the line indicated. Otherwise, collect approx. ½ lb. from the mixing bucket for a composite sample. 6. Label the sample as the area was marked on the map. 7. Keep sample cool until delivered to the soil testing lab.	

Sources: P. G. Sullivan. 2001. *Assessing the Pasture Soil Resource*. National Center for Appropriate Technology (retrieved November 2007); B. Murphy. 1998, *Greener Pastures on Your Side of the Fence*, 4th ed. Arriba Publishing, Colchester, VT: pp. 72–74.; "Soil sampling." NRCS Fact Sheet MN-NUTR3, July 2002 (retrieved November 2007).

TABLE 3.2 Sources: P. G. Sullivan. 2001. *Assessing the Pasture Soil Resource.* National Center for Appropriate Technology (retrieved November 2007). USDA. 1997. *Maryland Soil Quality Assessment Book* (retrieved November 2007). USDA. 1999. *Ohio Soil Health Card* (retrieved November 2007).

Which approach should be followed? Farmers need to weigh the costs and benefits of the two approaches. Several field crop studies have been conducted that show no crop yield benefit from cation balancing.[2, 3, 4] In addition, the sufficiency level approach generally results in lower application rates than the balanced cation approach.[5] For these reasons, the more conservative and economical approach would be to use a lab that follows the critical level or sufficiency approach.

On the other hand, little or no research has been done on the cation balancing approach in organic systems or on its effect on such attributes as crop quality. Transitioning farmers also may opt for the cation balancing approach because a significant group of experienced organic farmers follow this method. Farmers who choose this approach should be sure to pay attention not only to cation ratios but also to the other soil fertility indicators included on the soil test report, e.g., phosphorus levels, organic matter content, etc.

Whatever soil lab is chosen, farmers should use soil test reports in two ways. First, deficiencies in nutrient levels and pH should be corrected using the methods discussed in the following section. Second, once a series of soil tests have been conducted on a field over time, it can be studied for trends. Do the results show trends toward optimum fertility levels, or are nutrients being depleted or building to excess levels? Farmers should also monitor the organic matter level reported in soil tests.

ORGANIC SOIL MANAGEMENT PRACTICES

As discussed in the section on natural systems above, to achieve long-term sustainability, organic management strives to maximize the use of on-farm resources and naturally occurring processes. In practical terms, this means emphasizing the use of N-fixing legumes and manure to supply soil fertility needs and to build and maintain soil biological and physical health. Only an occasional use of off-farm inputs, for example, lime, will be needed to maintain soil health. This is especially true on organic dairy farms, which have abundant uses for legumes and a steady supply of manure.

SUPPLYING NITROGEN THROUGH BIOLOGICAL N FIXATION: FORAGE LEGUMES

On organic farms, forage legumes are a major source of nitrogen. On organic dairy farms, forage legumes provide the N needed to grow protein-rich pastures and hay crops, which when fed to cows, produce N-rich manure. In crop fields, legume-based hay crops can contribute to the N requirements of heavy feeders like corn and sorghum. Forage legumes also are an excellent source of organic matter, and the products of their decomposition are particularly effective at helping to stick together soil particles to form aggregates, thus contributing to good soil structure. (Forage legumes also provide weed and insect pest management benefits. See Chapter 5, "Organic Crop Management" for details.)

Farmers can choose from a wide variety of forage legumes that are adapted to temperate growing conditions. Forage legumes vary in the amount of N they can provide through biological nitrogen fixation (see Table 3.4). The legume's N-fixing ability needs to be balanced with other requirements including palatability, persistence under grazing, tolerance of suboptimal soil conditions, and even seed cost. Table 5.2 details the growing requirements and special attributes of each of the major forage legume species.

Because legumes are so critical to organic systems, organic farmers should manage them with the following points in mind in order to provide optimal benefits, especially in terms of N fixation.

Presence of Rhizobium.
Forage legume roots need to be in contact with the appropriate *Rhizobium* bacteria for N fixation to occur.

- The best way to insure that the correct bacteria are present in a sufficient amount is to inoculate the legume seed immediately before planting or buy seed that is coated with inoculants. For the correct *Rhizobium* inoculant for individual forage legume species, see Table 5.2. *Rhizobium* inoculants, like all soil amendments, need to be checked to be sure that they are not genetically modified and that they conform to certification requirements—before they are purchased and applied.
- Legumes can be checked when close to or at flowering to see if fixation is occurring. Roots need to be dug out of the soil rather than pulled. The presence of nodules that are pink or reddish when cut open indicates that N fixation is occurring. No nodules or nodules that are not red suggests that conditions are not right for fixation. This can be due to a lack of or improper inoculation, too much plant-available N in the soil (see below), or droughty or waterlogged soil conditions.

Soil Nitrogen Level.
N fixation will not take place if plant-available N is abundant in the soil. This is because there is a cost to the legume plant to obtain biologically fixed N—energy must be supplied by the plant to support the bacteria. Therefore, legumes will obtain N from the soil if it is readily available.

- When possible, avoid adding heavy amounts of manure or compost to areas where forage legumes are growing or are to be planted.
- Growing the legume as an intercrop with small grain or grass crops will tend to favor increased N fixation since the small grain or grass will compete for soil N with the legume.

Soil pH.
Most forage legumes require a near neutral pH (6.5–7) for good growth. They also require medium to high levels of P and K. Moderate applications of manure or compost (see below) should supply these as well as the other macro and micronutrients needed. If legume forage or hay is regularly sold off the farm, nutrient deficiencies are likely to develop.

TABLE 3.4 POTENTIAL NITROGEN CONTRIBUTIONS FROM LEGUMES

LEGUME	POTENTIAL N CONTRIBUTION (LB. N/ACRE)
Alfalfa	150–200
Hairy vetch	80–250
Red clover	70–150
Crimson clover	70–150
Alsike clover	60–120
Field peas	90–150
Sweetclover	100–200
White clover	80–130

Sources: Managing Cover Crops Profitably. 3rd ed. 2007. Sustainable Agriculture Network, Beltsville, MD; M. Sarrantonio. 1994. Northeast Cover Crop Handbook. Rodale Institute, Emmaus, PA; Craig Sheaffer, George Rehm, and Paul Peterson. 2005. "Nitrogen Credit Contributions by Alfalfa to Corn." University of Minnesota Extension Service (retrieved March 2008).

MANURE MANAGEMENT

For organic systems, manure has always been and will continue to be a highly valuable soil amendment. Indeed, its use on dairy farms is an essential part of nutrient cycling. Like forage legumes, manure is not only a source of fertility, but also improves soil structure and stimulates biological activity in the soil.

Manure as a source of macro and micronutrients

Manure is commonly thought of as a source of N, P, and K (Table 3.5), but it is also a good source of calcium, magnesium, and sulfur as well as micronutrients (see Table 3.6). The amount of nutrients actually supplied by manure varies widely, as can be seen in the extreme range in nutrient values reported in Table 3.5. The way manure is managed from farm to farm is a major cause of this variability.

Storing and handling manure to prevent nutrient loss

On organic farms where it is a major source of soil fertility input, manure must be managed to avoid nutrient losses. Of equal consideration are the negative effects poorly managed manure can have on the environment. Through good manure management, farmers can avoid four major avenues of nutrient loss from manure: urine loss, leaching, volatilization, and runoff.

Since significant concentrations of N and K exist in cow urine, failure to conserve this portion of manure can lead to substantial nutrient loss. Further losses of N are possible depending on the handling and spreading of the manure through (1) volatilization, in which ammonium-N is transformed into ammonia gas and lost to the atmosphere and (2) leaching, in which nitrate-N, which forms readily from ammonium–N, moves with water through the soil. The P contained in manure, which is largely contained in the solid portion, can be lost from the system through runoff if the manure is left unincorporated on the soil surface. Box 3.3 lists practices that can help reduce nutrient loss and environmental damage from manure.

Rate of application

As valuable a soil amendment as manure is, its use can lead to nutrient imbalances in crops and serious pollution issues if it is applied improperly. Manure should not be applied to frozen ground. It is also important to determine its application rate based on:

- the nutrient content of the manure,
- previous manure applications to the field (manure will continue to release nutrients over several years),
- the soil test results of the field to which it will be applied, and
- the nutrient requirements of the crop.

The nutrient content of manure can be determined by submitting a sample to a soil testing lab. This can be expensive, but as shown in Table 3.5, manure nutrient content varies widely from farm to farm. Farms should consider having manure tested at least once. As long as the same handling procedures are applied, the results of that test should at least be an approximation of the nutrient content

TABLE 3.5 MAJOR NUTRIENT COMPOSITION OF DAIRY MANURE

	NITROGEN (TOTAL N)	PHOSPHORUS (P_2O_5)	POTASSIUM (K_2O)
Lot-scraped manure (lb./ton)	10	6	9
Range (lb./ton)	3–20	0.6–13	2–20
Liquid manure slurry (lb./1,000 gal.)	22	14	21
Range (lb./1,000 gal.)	8–50	0.2–38	0.7–50

Source: J. P. Zublena, J. C. Barker, and D. P. Wessen. 1997. Dairy Manure as a Fertilizer Source. North Carolina Cooperative Extension Service Publication AG 439–28.

TABLE 3.6 AVERAGE SECONDARY MACRONUTRIENT AND MICRONUTRIENT CONTENT OF DAIRY MANURE

	Ca	Mg	Se	Na	Fe	Mn	B	Zn	Cu	Cl
Lot-scraped manure (lb./ton)	5	2.2	1.7	1.3	0.9	0.1	0.01	0.1	0.02	3.3
Liquid manure (lb./1,000 gal.)	10	4.8	3.1	3.2	1.8	0.2	0.02	0.2	0.05	6.1

Source: J. P. Zublena, J. C. Barker, and D. P. Wessen. 1997. Dairy Manure as a Fertilizer Source. North Carolina Cooperative Extension Service Publication AG 439–28.

of the farm's manure. Resources that can help better "fine-tune" manure applications are listed at the end of this chapter.

THE COMPOST OPTION

Another option for managing manure is to compost it on the farm. There are some major advantages to composting manure. Compost contains lower levels of nutrients, chiefly N, than most manure, but its N is contained in a stabilized form. This means that N will be mineralized from compost slowly, reducing the potential for nitrate leaching and thus groundwater contamination. Compost can therefore be more safely applied than manure in the fall. If well managed, the composting process will destroy most weed seed and pathogens. Manure is an excellent soil conditioner and fertilizer, but the composting process can add further value to the manure by developing a suppressive effect against soilborne diseases. In terms of the time and labor required to construct and manage the compost pile, some of this may be recouped in the greater ease of handling of compost, which can be reduced in volume and weight compared to manure by 50%. The end product of the composting process has a soil-like particle size and consistency allowing for more-even application in the field than manure.

There are some drawbacks to composting. The chief drawback appears to be the additional labor required in making the compost pile or windrow. Although in theory compost can be made using standard farm equipment (e.g., a bucket loader), farmers who start composting and stick with it often purchase a compost-turner. Box 3.4 lists some additional benefits and drawbacks of compost as compared to manure. For insights on why organic dairy farmers

BOX 3.3 REDUCING NUTRIENT LOSS AND ENVIRONMENTAL DAMAGE IN MANURE STORAGE AND HANDLING

- Build outdoor piles on water-impermeable pads with a collection system for run-off.
- Reduce leaching losses by:
  - covering or sheltering the pile from precipitation;
  - building the pile with high steep sides rather than with shallow sides and flat top;
  - applying in spring rather than fall—especially in sandy soils.
- Reduce runoff and N volatilization losses by:
  - incorporating solid manure in the soil as soon as possible after application;
  - injecting liquid manure into the soil.
- Reduce N volatilization loss by avoiding application during hot weather or immediately after liming.

Sources: N. Lampkin. 1994. Organic Farming. Farming Press, Tonbridge,.UK: pp. 86-124; Manure Management. University of Minnesota Extension BU-07401 (retrieved January 2009).

Paul and Maureen Knapp have become committed composters, plus a brief description of the compost process, see the case study of their farm at the end of this chapter.

NOP regulations on manure and compost
The National Organic Program (NOP) has strict standards *if manure or compost is going to be applied to crops that will be consumed by humans*. See NOP regulation §205.203 for the requirements.

PURCHASED INPUTS
As mentioned earlier in the chapter, there is one off-farm input that farms cannot do without. The generally acidic soils of the Northeast need inputs of lime to achieve and maintain a pH of 6–7. This is the pH range in which most pasture plants and other crops grow best and soil biological diversity and activity increases. There is a variety of types of lime available. NOP standards do not allow the use of hydrated lime or by-product lime. Mined limestone is generally acceptable. Dolomitic limestone (calcium-magnesium carbonate) can be used if soil test results show that magnesium levels are low.

BOX 3.4 BENEFITS AND DRAWBACKS OF COMPOST COMPARED TO MANURE

BENEFITS

- Stabilization of N in organic form—functions as a slow-release N fertilizer, reducing the potential for leaching and volatilization
- Low in soluble salts—can be safely applied to growing plants
- Suppression of soil-borne diseases
- Destruction of most weed seeds and pathogens
- Destruction of contaminants and toxins (e.g., antibiotics, pesticides, toxic compounds formed during the decomposition process)
- Improved handling
  - reduced weight and moisture content
  - reduced odors and flies
  - improved spreading due to smaller and more consistent particle size

DRAWBACKS

- Requires time and labor to construct, turn, and monitor the pile or windrow
- Sensitive to weather—cold temperatures slow composting process
- Composting process involves some loss of nutrients
- May require special equipment, e.g., a compost-turner

Sources: R. Rynk, ed. 1992. *On-Farm Composting Handbook*. Natural Resource, Agriculture, and Engineering Service, Ithaca, NY: pp. 3–5; N. Lampkin. 1994. *Organic Farming*. Farming Press, Tonbridge, UK: pp. 86-124; *Manure Management*. University of Minnesota Extension BU-07401 (retrieved January 2009).

Calcitic limestone (calcium carbonate) should be used if Mg levels are high or to raise calcium levels.

Another off-farm input used by some farms is gypsum (calcium sulphate). Gypsum (not a liming agent) is used by some farmers to raise soil Ca levels instead of calcium carbonate if the soil pH is already optimum or high. Farmers have also used gypsum as a soil conditioner. Its application has been shown to inhibit soil crusting, increase water infiltration, reduce erosion, and reduce the strength of subsurface compaction layers. (Inputs of organic matter and use of crops with deep taproots such as alfalfa and sweetclover also have these effects.)

Before purchasing and applying any soil amendment, farmers should check with their certifying agency to make sure it is allowed under certified organic production.

What about other purchased inputs? With the rapid increase in the organic market, there are now hosts of "organic" soil amendments for sale. Farmers need to vet a product thoroughly before deciding to buy and to ask themselves several questions.

- Does the advertising or the salesperson for the product make sensational claims?
- Do other farmers use the product? If so, what have their experiences been?
- What will the product cost to apply, and will the expected benefit justify the cost?
- Finally, do I really need this or can I do just as well without it or by using inputs from my farm?

PASTURE SOIL MANAGEMENT

So far, this chapter has focused on principles, strategies, and practices that apply to all soils on an organic dairy farm. However, pasture soils have several unique soil management challenges, including maintaining adequate legume stands, fertilizing pastures without negatively affecting grazing or excessive loss of nutrients, and avoiding soil compaction.

To provide high-quality forage for the herd, farmers should maintain forage legumes at 50–70% of the sward. This can be challenging especially when forage legumes species tend to be highly palatable; can be outcompeted for light, water, and nutrients by grasses; and are sometimes either biennials or short-lived perennials. In addition to using management intensive grazing (MIG) techniques to help forage legumes persist (see Chapter 4, "Pasture Management"), the soil should be managed to optimize legume growth and persistence—see the earlier section on forage legumes.

In terms of fertilizing pastures, good grazing management is key since it will result in good cow pie distribution. The rule of thumb when spreading manure on pastures is to apply it only to areas that are no longer being grazed but that are actively growing, e.g., to pastures in fall. A better alternative is to compost the manure before application since its nutrients will then be less susceptible to leaching. The use of compost may also help reduce the competitive edge that

N applications give grasses over legumes, given that compost has a lower N content than manure and its N is more slowly available.

Soil compaction in pastures decreases the activity of soil organisms, restricts root growth, and decreases water infiltration, thus increasing the potential for runoff and erosion. These conditions substantially reduce pasture productivity. A major cause of compaction is grazing pastures under wet soil conditions. Farmers can reduce compaction during rainy periods if they graze cows only on well-drained pastures. Poorly drained pastures should be grazed only during dry periods in the summer. A good general strategy to avoid/reduce compaction is the use of a MIG system with high stock density, short periods of stay in the paddock, and extended periods for regrowth, since this enhances root growth. Additional strategies to reduce the potential for compaction include the following.

- Graze only well-established pastures that have no gaps and that have developed extensive root systems to a depth of 6–10 inches.
- Grow a mixture of species that have a diversity of root systems. If soil compaction has occurred, pasture species that develop deep taproots, e.g., chicory, alfalfa, and dandelion, can penetrate and break up compacted soil layers over the course of several years.

FIGURE 3.2
A cover crop of rye and hairy vetch, planted in early September, photo-graphed in mid-October

Winter cover crops are especially important to protect soil from erosion over winter and early spring. The legume in this combination will also supply N to the following crop.

SOIL MANAGEMENT IN TILLED CROPLAND

Tilled cropland exhibits a very different set of soil management challenges from pasture. Tillage and cultivation lay the soil bare, accelerate the decomposition and loss of soil organic matter, and break down soil aggregates.

CROP ROTATION AND COVER CROPS

Strong counterforces to the negative effects of tillage include a well-designed organic crop rotation and the extensive use of cover crops. A crop rotation that includes a sod/hay crop component that is in the ground for several seasons can compensate for the negative effects on soil health from a row crop. (For more information on designing sound crop rotations, see Chapter 5, "Organic Crop Management.") Cover crops, in addition to protecting the soil and adding organic matter, offer such potential benefits as fixing N, scavenging nutrients that would otherwise be lost over the fall, winter, and early spring; suppressing weeds; and increasing the habitat for beneficial species that help control insect pests (see Figure 3.2).

Farmers can grow cover crops whenever there are "bare spots" in the rotation—when no cash or feed crops are being grown for a period of two or more months. A major bare spot, or niche, in northern field-crop systems is over winter. Farmers can plant a winter cover crop or mix of crops after small-grain harvest, e.g., winter rye with or without hairy vetch. Although trickier to establish, cover crops can also be grown after corn or soybean: e.g., rye can be no-till drilled into corn stubble and hairy vetch can be overseeded into soybean at leaf yellowing.

CAREFUL USE OF TILLAGE

In addition to crop rotation and cover crops, the negative effects of tillage can be mitigated by reducing tillage whenever possible. Strategies to decrease tillage include the following.

- *Reduce or eliminate moldboard plowing.* Alfalfa, for example, can be chisel plowed in the fall, which reduces the impact on soil structure and leaves substantial amounts of alfalfa residue on the soil surface over winter to protect against soil erosion. A soil-saver instead of a moldboard plow can also be used after corn.
- *Use no-till planting whenever possible.* For example, spring-planted small grains can be no-till drilled following soybean. (Legumes can still be interseeded in the same operation using a grass-seeding box and chains attached to the drill to incorporate the legume seed lightly.)
- *Reduce tillage associated with weed control.* Use of multiple techniques for weed control, like crop rotation and smother cropping, reduces the need for tillage and cultivation. Knowledge of weed biology coupled with the intensive scouting of fields allows cultivations to be better timed and thus more effective, thereby reducing the need for further cultivation passes. (See Chapter 5, "Organic Crop Management" for more details.)
- *Kill green manures/cover crops without incorporation in the soil.* If such crops can be killed without tillage, not only is the soil left undisturbed but the mulch of plant material left on the soil surface further protects against erosion and may help in weed suppression. A number of methods can be effective in killing green manures/cover crops without soil incorporation including mowing and chopping and use of a roller-crimper (the roller kills the crop by breaking or crimping the cover crop stems).

EVALUATING ORGANIC SOIL MANAGEMENT

As discussed in this chapter, transitioning to organic soil management requires farmers to refocus their attention from soil fertility to soil health and from purchased fertilizers to homegrown soil amendments like legumes, manure, and cover crops. This requires a big shift in perspective and management practices.

Farmers can use the techniques outlined in Table 3.2 to monitor their farm's progress in increasing the soil's physical and biological health. Soil is resilient: within a couple of years after the shift to organic management, farmers should see increased biological activity and diversity and improved soil structure. At the same time, periodic testing of soil fertility will show whether nutrients and pH are approaching optimum levels or whether deficits are occurring or excesses are piling up. For transitioning farmers and those who have recently certified, this two-prong approach to monitoring is essential so that course corrections in management can be made as needed.

RESOURCES

General, Soil Assessment

Brady, N. C. and R. R. Weil. 1999. The Nature and Properties of Soil, 12th ed. Prentice Hall, Upper Saddle River, NJ.

Lampkin, Nicholas. 1994. Organic Farming. Farming Press, Tonbridge, UK.

Magdoff, F. and H. van Es. 2000. Building Soils for Better Crops, 2nd ed. Sustainable Agriculture Network, Beltsville, MD. (Available at: www.sare.org/publications/bsbc/bsbc.pdf).

National Resources Conservation Service website, Soil Quality section: (www.soils.usda.gov/sqi).

Sullivan, P. G. 2001. Assessing the Pasture Soil Resource. ATTRA, National Sustainable Agriculture Information Service. (Available at: www.attra.ncat.org/attra-pub/PDF/assess.pdf).

Soil survey website, webpage of list of surveys by state: http://soils.usda.gov/survey/printed_surveys (on-line county soil maps).

N-Fixing Legumes, Cover Crops

Clark, A. (ed.). 2007. Managing Cover Crops Profitably, 3rd ed. Sustainable Agriculture Network, Beltsville, MD. (Available at: www.sare.org/publications/covercrops/covercrops.pdf)

Sarrantonio, M. 1994. Northeast Cover Crop Handbook. Rodale Institute, Emmaus, PA. (Available at: www.rodaleinstitutestore.org).

Manure Management

Blanchett, K. and M. A. Schmitt. 2007. Manure Management in Minnesota. University of Minnesota Extension, WW-03553. (Available at: www.extension.umn.edu/distribution/cropsystems/ DC3553.html).

Kuepper, G. 2003. Manures for Organic Crop Production. ATTRA, National Sustainable Agriculture Information Service. (Available at: www.attra.ncat.org/attra-pub/PDF/manures.pdf).

Compost

Rynk, R. (ed.). 1992. On-Farm Composting Handbook. Natural Resource, Agriculture, and Engineering Service, Ithaca, NY (available at: www.nraes.org).

Dougherty, M. (ed.). 1999. Field Guide to On-Farm Composting. Natural Resource, Agriculture, and Engineering Service, Ithaca, NY. (Available at: www.nraes.org).

Pasture Soil Management

Bellows, B. 2001. Nutrient Cycling in Pastures. ATTRA, National Sustainable Agriculture Information Service. (Available at: www.attra.ncat.org/attra-pub/PDF/nutrientcycling.pdf).

Reduced Tillage

Kuepper, G. 2001. Pursuing Conservation Tillage Systems for Organic Crop Production. ATTRA, National Sustainable Agriculture Information Service. (Available at: www.attra.ncat.org/attra-pub/PDF/omconservtill.pdf).

NOTES

1 Magdoff F. and H. van Es. 2000. Building Soils for Better Crops, 2nd ed. Sustainable Agriculture Network: Beltsville, MD, p. 10.

2 W. C. Liebhardt. 1981. "The Basic Cation Saturation Ratio Concept and Lime and Potassium Recommendations on Delaware's Coastal Plain Soils," Soil Science Society of America Journal, vol. 45 (3) pp. 544-549.

3 Simson, C.R., R. B. Corey, and M. E. Sumner. 1979. "Effect of Varying Ca:Mg Ratios on Yield and Composition of Corn (Zea mays) and Alfalfa (Medicago sativa)," Communications in Soil Science and Plant Analysis, vol. 10 (1/2) pp. 153-162.

4 E. O. McLean. 1977. "Fertilizer and Lime Recommendations Based on Soil Tests: Good, But Could They Be Better?," Communications in Soil Science and Plant Analysis, vol. 8 (6) pp. 441-464.

5 Magdoff F., and H. van Es. 2000. p. 181.

PAUL AND MAUREEN KNAPP, COBBLESTONE VALLEY FARM

ELIZABETH DYCK AND BETHANY WALLIS

There is always a lot going on at Cobblestone Farm, LLC, a certified organic dairy farm located in central New York. To supplement their dairy enterprise, Maureen and Paul Knapp run an organic U-pick strawberry operation and raise organic poultry. They also practice the art and science of compost making.

Maureen and Paul see compost as the basis of their soil management program. They value compost less as a nutrient source than as a "biological activator" of their soils, which enhances nutrient cycling and productivity on all their fields—from strawberry beds to pasture. The Knapps acknowledge that compost making requires time and labor—which are in short supply on a dairy farm. However, they have worked to make their compost system as efficient as possible. They were able to buy a compost turner with the help of a NY State Ag and Markets Farm Viability grant. Through projects with NRCS, they have built a compost pad and a storage area that can hold manure until they are ready to compost. The Knapps, through study, experimentation, and commitment, have developed a great deal of composting expertise that they are willing to share with others.

Maureen and Paul Knapp.

The Knapps have a diversified cropping system, but its heart is 85 acres of permanent pasture, a mix of orchardgrass and red and white clover. They also have an additional 40 to 50 acres that they can utilize as pasture as needed. The Knapps intensively graze their 90-cow milking herd, rotating pastures every 12 hours. For supplemental forage, they also grow oats, barley, and sorghum, which are either grazed or round baled. The small grains also double as nurse crops when establishing forage legumes. They do purchase a grain concentrate mix consisting mostly of corn with some soy, oats and peas when available, as well as some organic forages when needed, e.g., if their milking herd size increases.

For many years, the Knapp family has had a pick-your-own strawberry operation. The strawberry enterprise helps spread the farm's financial risk and utilize the dairy manure resource. When the Knapps transitioned to organic, they decided to transition the strawberry production as well—but eased into it. They stopped spraying and switched to cultural practices to manage weeds, e.g., rotating through hay and then oats or sorghum before the berry crop. They also took time to educate their customers about the benefits of organic agriculture before marketing and pricing the strawberries specifically as organic.

Maureen and Paul are big proponents of farm planning. It helps clarify their goals and objectives and keeps everyone on the same page. It is also clearly a factor in their ability to successfully integrate the diverse enterprises on their farm.

PRACTICAL ON-FARM COMPOSTING

The Knapps advise that it is vital to get the starting material for composting right. The suite of microorganisms that do the work of composting needs a food supply that has a ratio of carbon to nitrogen between 20:1 and 40:1. The Knapps find that the bedded pack manure from their coverall barn usually works well. In recent years they have also started adding the aquatic weed Eurasian water milfoil, which is removed by the truckload from a nearby infested lake and delivered to them for free. They believe the lakeweed adds minerals to the compost and has stimulated crop growth.

FIGURE 3.3
Aerating the windrow.
The Knapps use a Sandberger compost turner.

A good way to determine if the starting material has the right moisture content and consistency for composting is to start spreading it to form a windrow. Paul notes that if the material is too loose (wet), it will not hold together in a pile. He finds that the manure from his tiestall barn is often too wet and so adds straw or hay—and sometimes sawdust, wood chips, or wood shavings as available. This is a critical step since, while the composting microbes need moisture, they also need oxygen. If the starting material is too wet or does not have enough roughage in it to support pore space, composting will be inhibited.

The Knapps form windrows that are about 8 feet wide and 5 feet high. Windrows need to be small enough to be able to keep aerated while large enough to build up adequate temperatures. Turning the windrow keeps the pile supplied with enough oxygen to support vigorous microbial decomposition. This activity generates temperatures (130–145°F) that accelerate the composting process and are needed to kill most pathogens, fly larvae, and weeds seeds. Turning is also used to keep the pile from overheating (i.e., reaching temperatures >145°F). Using a windrow turner, the Knapps generally turn the pile aggressively both to meet NOP standards and ensure weed seed and pathogen destruction. They try to turn daily during the first week, every other day the second week, and then as needed for the next two weeks. Other composters turn the pile less frequently. When the pile no longer heats up after turning, the active phase of composting is over and curing begins. Maureen says if all goes well, they can have finished compost in 6 weeks after forming the pile. Because of labor "bottlenecks" on the farm, however, the windrow is likely to sit for several months before application.

For more information on composting, see the Resources section on page 61. For NOP composting requirements, see regulation §205.203.

DAVE AND SUSAN HARDY

BETHANY WALLIS AND ELIZABETH DYCK

Dave and Susan Hardy's certified organic dairy farm is nestled into a hillside near Mohawk, New York. They grow pasture and hay for their herd of 75 milking cows—and do it well. Good grazing management and a strong commitment to monitoring and sustaining soil health work together to make their operation a success.

Dave tests his soils every couple of years, paying special attention to cation ratios and sulfur and micronutrient levels. He applies a yearly maintenance application of a certifier-approved NPK fertilizer plus sulfur and boron. When asked for his best advice on soil management for transitioning farmers, he emphasizes maintaining sufficient micronutrient levels (based on soil test analysis) and correctly timing manure application. Summer manure applications are tempting but "make a mess," adversely affecting grazing. He piles manure and hay together during the summer and then waits to spread until the fall. He also cautions against applying manure too heavily to a particular field. Each year he tries to cover about a third of the farm so that pastures receive manure every three years.

Dave Hardy.

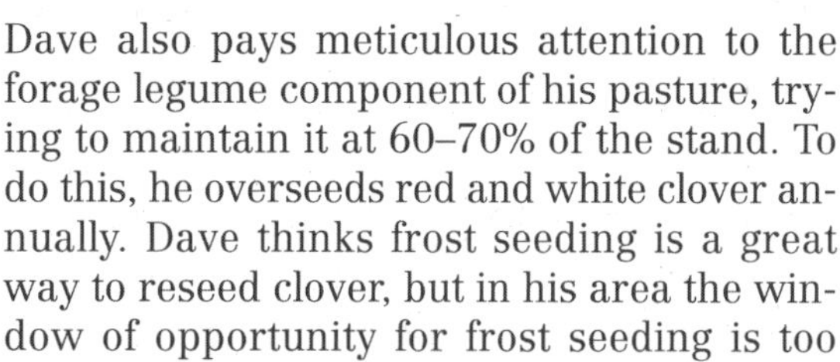

Dave also pays meticulous attention to the forage legume component of his pasture, trying to maintain it at 60–70% of the stand. To do this, he overseeds red and white clover annually. Dave thinks frost seeding is a great way to reseed clover, but in his area the window of opportunity for frost seeding is too short. Instead, just before the cows do their first rotation in the spring, he spreads clover seed using a four-wheeler and a spin seeder. The cows then trample the seed into the soil.

The farm's fence system consists of a permanent high-tensile perimeter with moveable polywire divisions. A permanent watering system provides water to each paddock and to the beginning of the laneway. The Hardys move milkers on a daily basis to new paddocks. The rotation is normally 16 days but can be increased to 22 days or more during dry weather. During the grazing season, the milking herd is fed a diet of 90–95% grass and dry hay to complete the forage portion of the ration. All of the farm's winter feeds are custom-harvested: the Hardys annually store around 1,400 dry round and 500 baleage bales. To maintain a high butterfat component, they supplement with 14 pounds of a 16% protein grain mixture. The cows are also provided free-choice kelp and Redmond salt. The Hardys ship milk throughout the year but only a small amount in March and April. Dave strongly believes that seasonal

Dairy herd grazing at the Hardys' farm.

milk production is a key factor in maintaining herd health because his cows freshen on pasture when it is warm.

The typical vet bill on the farm is less than $700 per year, largely for pregnancy checks and an occasional emergency—typically during calving. The herd's cull rate is less than 5%. Occasionally, there is the odd case of hardware or milk fever and once in a while a hairy wart that needs to be wrapped in copper sulfate, but these are isolated cases that do not spread through the herd. Due to good rotation practices, fecal sampling has shown low parasite loads—eliminating the need for worming. Dave's herd does not experience ketosis problems, and he has treated only one displaced abomasum in the past 13 years. He attributes the herd's high milk quality to good milking procedures, clean cows, pre- and post-dipping, a clean towel for every cow, and yearly milking system analyses. If mastitis occurs in his herd, Dave milks the infected cow into a bucket and gives her a few days to clear up. Most cases, he says, heal themselves.

The Hardys pride themselves on their calf-raising practices. All calves are born on pasture and are then tied in the barn for a day until they are used to being restrained. On the second day, the calves are moved into a hutch outside of the barn on fresh grass where they stay for a couple of days until they are eating well and are acclimated to the hutch. From there, Dave moves them to a hutch rotation where they receive new grass daily. They are fed milk until they are 8–10 weeks old. They are then weaned with a small amount of grain and put into paddocks with group hutches and begin rotational grazing in a fenced system. Calves remain on pasture until the snow begins to fall and are fed supplemental hay as needed.

As Dave guides a visitor around the farm, he repeats the phrase "keep things simple." The Hardy farm is a classic example of how a simple concept—let cows eat grass—can be parlayed though expert management into a successful organic dairy operation.

CHAPTER FOUR

PASTURE MANAGEMENT

SARAH FLACK

This chapter begins with a discussion of the benefits of good quality pasture on organic and transitioning farms, an overview of the range of successful grazing systems used in the Northeast, and some details on the potential pitfalls of a poorly managed pasture system. It provides an overview of all the basic steps needed to set up a new grazing system, fine-tune an existing one, and avoid the most common potential pasture problems seen on organic and transitioning dairy farms.

KEYS POINTS TO THE ORGANIC TRANSITION

- Read and understand the current National Organic Program (NOP) rules for organic pasture. Note the NOP will likely issue a revised organic pasture standard after 2009, which will include more specific rules (§205.2, §205.102, §205.237, and §205.239–§205.240).
- Decide what sort of grazing system you will set up and then calculate the number of acres of pasture your herd will need.
- Make sure that there is enough pasture within walking distance of the barn.
- Do not forget to include your pastures in your crop record-keeping system and to list them as fields on your organic system plan when you apply for certification. Check with your certifier about the specific records needed to comply with the NOP pasture intake requirements.
- Develop your supplemental feeding plan so you can avoid some of the potential pitfalls including inadequate DMI or overfeeding protein.

THE BENEFITS OF GOOD PASTURE MANAGEMENT

An organic dairy farm with a well-designed and properly managed grazing system has many advantages. These include higher feed quality, lower feed costs, improved animal health, excellent animal welfare, and the ability to meet consumer expectations. During the 12-month dairy herd transition, while feeding organic grain but selling nonorganic milk, having an already established and successful grazing system can reduce costs significantly.

A well-managed grazing system provides low-cost, good-quality, high-protein feed. This is particularly helpful for organic and transitioning dairy farmers, for whom grain costs are high and increasing. When feeding well-managed pasture, organic dairy farmers can switch to supplementing with energy feeds such as corn, barley, or molasses. Other cost savings can come from having the cows harvest their own feed and spread their own manure as they graze.

Good grazing management can:

- lower feed costs,
- improve livestock health and lower vet bills,
- reduce cull rates, and
- produce longer-lived animals and make it possible to have additional income from the sale of heifers and cows.

Over time, a well-managed pasture will improve in both quality and yield. Good management can convert weedy or brushy pastures where animals have to search to find good-quality forage into highly productive pastures that can feed more animals high-quality forage, produced and harvested at lower cost. This will also reduce the need to renovate and reseed pastures by encouraging the growth of more productive and palatable perennial pasture plants.

Livestock whose diet is mostly or all pasture produce meat and milk (or more specifically the fats in the meat and milk) that may contain different amounts and types of nutrients than grain-fed livestock. These nutrients can include beta-carotene; vitamins A, E, and D; omega-3 fatty acids; and conjugated linoleic acid (CLA). In addition to potential nutritional differences, many consumers are also attracted to some of the other benefits of grass farming including improved animal welfare and environmental sustainability.

PASTURE PITFALLS IN POORLY MANAGED GRAZING SYSTEMS

While well-managed pasture is an asset to a dairy farm, poorly managed pasture can create serious problems. These can include:

- parasite problems,
- poor animal performance and production due to inadequate dry matter intake (DMI) from pasture or incorrect supplemental feeding,
- reduced milk production,
- poor reproductive performance, and
- animal health problems.

Later sections of this chapter will discuss these problems and how to prevent them in more detail. Chapter 6, "Managing Dairy Nutrition" and Chapter 7, "Organic Dairy Herd Health" also address them.

OVERVIEW OF TYPES OF GRAZING SYSTEMS AND METHODS

The methods of pasture management on organic dairy farms in the Northeast vary from management intensive grazing (MIG) with high-quality pastures where cows are moved to fresh pasture twice or more every 24 hours to large extensively grazed pastures.

When compared to MIG, the large pasture system or extensive grazing requires less daily management in moving fence and water tubs. However, extensive systems require more clipping or occasional pasture renovation over time due to overgrazing damage. Extensive grazing systems, particularly systems where cows continuously graze the same pasture for most of the summer, provide lower total dry matter intake and more-variable feed quality. These types of management generally result in lower milk yields and profitability, but require less day-to-day management.

Management intensive grazing (MIG) is NOT the same as a rotational system because in a MIG system, close attention is paid to how fast plants are growing. The number of days pastures rest after each grazing increases significantly as growth rates slow. One of the key guidelines of a successful grazing system is variable recovery periods.

The information in this chapter will emphasize the MIG system of grazing management. It is also possible, however, to use this information to set up a less intensive system that still provides enough pasture to meet your farm goals. If a less intensive grazing system is the choice, additional acres per cow will be needed since the productivity of the pasture will be lower. Additional land will need to be added to the grazing rotation as plant growth rates slow in summer. The type of pasture system used should be determined by the farm's overall goal and the production objective for the livestock.

A few organic dairy farmers use a type of grazing called holistic planned grazing. This system will also produce high-quality pastures and may be particularly useful to farmers who experience extended drought conditions. This method of grazing requires specific planning, mapping, and recordkeeping to monitor the grazing season. Refer to the Resources section of this chapter for more information.

MANAGEMENT INTENSIVE GRAZING (MIG)

Good grazing management such as MIG will favor the pasture plant species you want, reduce weed problems, and increase the quantity of pasture dry matter produced while improving the nutritional quality of the feed. A high-quality pasture, particularly with good soil management, will produce feed with the highest nutritional value and vitality so that animals are healthy and the meat, milk, and manure produced is of the highest possible benefit to you and your farm.

A good-quality MIG pasture will contain a mix of many plant species, no bare soil, and will have uniformly distributed cow pies from the most recent grazing. The pasture will have patches that were not closely grazed during the last grazing, since cows do not like to eat the grass right next to their manure.

In a continuously grazed or simple rotation system where plants are not allowed to recover fully between grazings, there are more likely to be patches of bare soil, less-desirable grass and clover plants, and increasing weeds. Patches will appear that never grow very tall, and clover and other legumes may be completely absent. These are all symptoms of overgrazing damage. You may also see buildup of dead plant material or thatch on the soil surface and cow pies that have not decomposed quickly.

WHAT YOU NEED TO KNOW TO SET UP A SUCCESSFUL GRAZING SYSTEM

HOW PLANTS GROW

The basis of good grazing management is that pasture plants need time to rest after each grazing in order to photosynthesize and replenish the energy stored in their roots. Continuously grazing animals in the same pasture, or returning them to a pasture before it is fully regrown, does not give plants time to recover and results in overgrazing damage. The resulting weak plants may stop growing or die. These weakened overgrazed plants will not compete well with weed species, will not hold soil well, and will result in bare soil and erosion. Some grasses and clovers will survive by staying very short, never growing tall enough for livestock to graze easily, while livestock will reject other areas that will soon grow up into weeds, brush, or small trees.

If a plant is grazed while it is still growing from carbohydrate reserves rather than from active photosynthesis, it has been overgrazed.

PREVENTING OVERGRAZING DAMAGE

Learning all the ways in which plants can be damaged by overgrazing is essential to managing pastures well. Overgrazing most often results from a few common mistakes, as described in Box 4.1.

BOX 4.1 COMMON SITUATIONS THAT CAUSE OVERGRAZING

To avoid overgrazing, avoid the following situations.

- Taking down all the interior fences in the fall and letting cows "clean up" the pastures
- Having a "rotational" system of 6 or 7 paddocks, with each grazed for 1 or 2 days
- Leaving animals in a pasture for more than 3 days in a row
- Returning animals to the pasture before all of the plants have regrown
- Not adding additional acres into summer grazing rotation when plant growth rates slow

Untoward Acceleration—A Common Rotational Grazing Mistake

Untoward acceleration is a term used by Andre Voisin in his book *Grass Productivity* to describe what happens when paddocks are not rested long enough between grazings. Each grazing of the paddock provides less forage and the regrowth period gets shorter throughout the grazing season until most of the plants are overgrazed and there is little or no feed left.

Winter access to pasture

Allowing livestock access to the pastures during the nongrowing season can also cause damage, particularly during wet or freeze/thaw conditions. Soil compaction and damage to plant root reserves will sometimes be obvious, while in other situations it will result in slower spring growth rates and overall decreased pasture productivity.

Using MIG to prevent overgrazing damage

Using MIG, most dairy farmers give animals a fresh pasture after each milking. It is possible, however, to use larger paddocks and move cows less frequently. Livestock return to the pasture when it has fully recovered by regrowing to a minimum of 8 inches. In the Northeast, this may be as soon as 14 days in early summer when plants are growing rapidly, but it may be 40 days or longer later in the summer. Farmers move animals frequently so that cows do not graze each paddock for more than three consecutive days. Using smaller paddocks and moving the herd more often will provide more consistent high-quality feed and higher overall pasture productivity.

Managing for a dense pasture sward of the correct height results in better animal performance and a more profitable farm.

An important part of MIG is that when pasture growth slows in later summer, the total number of grazing acres will have to increase. If the number of grazing acres is not increased, plants will not get enough rest, and animal dry matter intake will drop, resulting in poor animal and pasture performance. Timing the first cut of hay early enough to allow some areas to grow back tall enough for grazing is the easiest way to provide good-quality pasture when plant growth slows in the summer.

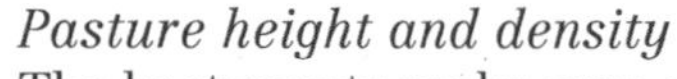

Pasture height and density
The best way to make sure cows are eating enough DMI from pasture is to pay close attention to the size of the bite of pasture they receive. Pasture height and density determines this bite size. If the pasture is too short, then they cannot get enough pasture in each bite to meet their DMI needs, even if you give them a larger area to graze.

Cows take only a certain number of bites each day and graze for only part of each day because they must also spend time resting and ruminating. This is one reason to provide a new pasture, which is tall and dense, after each milking. When you turn cows into a pasture that is tall enough, they can rapidly fill their rumens with high-quality high-protein feed.

FIGURE 4.1
A Holstein reaching with her tongue for a bite of good-quality pasture

Selective grazing
When cows graze, they wrap their tongues around pasture plants and snip them off with their lower teeth and upper dental pad (they do not have upper teeth). See Figure 4.1. They generally first eat tender young plants and the tops of plants (leaves instead of stems). If they are grazing the same pasture for several days or a week, the nutritional quality of what they eat each day will change due to this selective grazing behavior. Using a higher stocking density (smaller paddocks) and moving cows to new pastures more often

will result in more predictable pasture nutrient intake, which can make ration balancing and milk production easier to manage.

Cows will reject some pasture around manure. This natural instinct helps them avoid areas containing parasites. The best way to manage these rejected areas is to improve the biological activity of the soil and insects, which rapidly incorporate manure into the soil. Clipping pastures immediately after grazing may help manage rejected forage for some farms, particularly in the first few years of grazing.

What pregrazing height is best?

The traditional MIG guidelines outline a pregrazing height of 8 inches for dairy cows. Turning cows into pastures shorter than 8 inches will result in lower DMI and some overgrazing damage to plants. There may sometimes be an advantage in allowing pastures to grow even taller than 8 inches before grazing, but once plants become too mature the nutritional quality and palatability of the pasture will drop.

A few graziers in the Northeast are choosing to allow pastures to grow to 12–14 inches before grazing. This practice is particularly common on farms whose goal is to improve pasture productivity and soil organic matter. Graziers managing these taller pregrazing heights must maintain a high stocking density so that grazing and hoof impact benefits are maximized. The challenge with grazing tall is that the digestibility of most pasture plants at these heights is lower and it can be difficult to maintain a high level of milk production when grazing more mature and fibrous pasture plants. Most dairy farms will find that choosing a pasture height that allows the plants to recover fully, but does not give them time to become fibrous or too tall, will yield the highest quality dairy pasture feed.

When to start grazing in the spring

Many farmers begin grazing in the spring when the pastures are under 8 inches tall so that they can begin to shift the dairy ration over to pasture gradually. This also makes it easier to keep up with the grass as it begins to grow rapidly in the spring. Grazing at under the recommended 8-inch pregrazing height for this first grazing in the spring is acceptable as long as the pastures are fully recovered before returning the cows to the paddock. Chapter 6, "Managing Dairy Nutrition" also has some guidelines for spring grazing to help the herd adjust to the new pasture ration.

While the traditional spring grazing recommendation in most MIG systems has been to begin grazing early to prevent pastures from growing too tall, some graziers are now delaying their spring turnout date. The advantages of a later spring turnout include less grazing during the mud season and improved plant vigor due to additional time for root growth. The disadvantage of a later turnout date is that more of the pastures can grow too tall and become overmature. A later grazing start date requires good management and preparation that includes mechanically harvesting or clipping some fields early to allow them to regrow and be added into the grazing rotation. A later spring turnout may also require a high-stocking-density grazing system that can handle the higher pre-grazing heights.

CALCULATING PADDOCK SIZES, DMI, AND ACREAGE NEEDED

This section provides an overview of how to size paddocks and estimate the number of acres of pasture needed for a dairy herd. It is also helpful to attend a grazing school or intensive workshop to learn how to set up a pasture system, estimate dry matter, and gain other important grazing skills that this handbook does not address in detail.

In a MIG system, cows graze pasture rapidly down as short as 2 inches, then plants are permitted to grow back up to at least 8 inches. The density of plants in the pasture will have a large influence on how much available dry matter is in your pastures. If the plants are far apart (low density) and you can easily see soil between the plants, there will be far less dry matter per acre in the pasture. The best way to learn how to estimate dry matter is to attend pasture walks or host a walk on your farm. The density and quality of your pastures will increase as you practice good grazing management.

The amount of pasture dry matter an animal will eat depends on many factors (lactation, growth, animal size, supplemental feed, and pasture height and density). A reasonable estimation of the amount of DMI that a lactating cow can harvest by grazing well-managed pasture is at least 3% of her body weight. This means a 1,000-lb. cow will eat 30 lb. of pasture dry matter per day. *Note that this is just an estimate of pasture DMI and assumes some additional feeding in the barn.* It is always best to use real numbers from your own farm.

Paddock size

The paddock sizes your herd requires will depend on how many animals you have, how long they will be in the paddock, and how much feed is in the pasture (see Figure 4.2). The following example calculates the paddock size needed for a herd of 50 Jerseys on a farm where they receive a fresh paddock after each milking. Refer to Worksheet 4.1 to plug in the example data. After following along with the example, try calculating the required paddock size for your milking herd.

Total number of pasture acres needed

The number of paddocks or total number of acres of pasture needed can be calculated if we know how long the pastures will need to regrow after each grazing. Research done by Dr. Bill Murphy in Vermont produced the following average regrowth periods. The

FIGURE 4.2
Example of a correctly sized paddock with a high stocking density on Morvin Allen's farm in Massachusetts

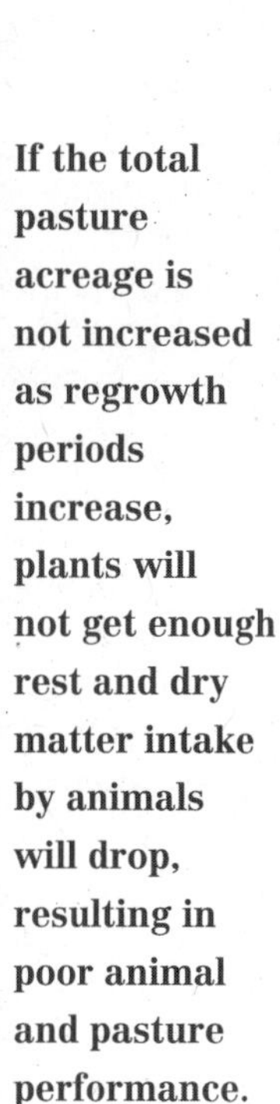

If the total pasture acreage is not increased as regrowth periods increase, plants will not get enough rest and dry matter intake by animals will drop, resulting in poor animal and pasture performance.

regrowth periods on your farm may be shorter or longer. In a dry summer, the regrowth periods may easily reach 60 days or longer.

- 12–15 days in late April to early May
- 18 days by May 31
- 24 days by July 1
- 30 days by August 1
- 36 days by September 1
- 42 days (and longer) by October 1

These numbers are just averages. The actual amount of time needed for complete regrowth will probably be different on your farm. Example 4.2 calculates the total number of pasture acres needed for the example farm we discussed earlier in this section.

For grazing planning on your farm, it is helpful to keep records of how often and for how long you graze each paddock each year. You can keep these records on a copy of your farm map or by using a notebook or worksheet where you write down the date, where the animals grazed, and for how long. These records will allow you to know what the real rest periods are throughout the growing season. Walk through all your pastures at least once each week and record how many inches each paddock has grown back. This can help you plan the order in which to graze your paddocks.

What if you do not have enough land?
Once you calculate the number of acres needed to set up the grazing system, you may realize that your farm does not have enough land suitable for grazing within cow-walking distance of the milk-

EXAMPLE 4.1 USING WORKSHEET 4.1

- Line 1: Enter the estimated weight per cow of **1,000 lb**.
- Line 2: Enter **3%**. Note that on this example farm they plan to supplement with some dry hay and grain in the barn.
- Line 3: Multiply **1,000 lb. × 0.03** to get 30 lb. of DMI from pasture per cow.
- Line 4: Multiply **50 cows × 30 lb. DMI** to get 1,500 lb. of DM needed for the herd per day.
- Line 5: This example farm has decided to use a pregrazing height of 8 inches and has estimated their pasture plant density to be "average." Use the chart in the worksheet to get an average pasture mass or dry matter per acre. According to the chart, the pasture mass is 2,600 lb. Enter **2,600 lb.** on line 5.
- Line 6: This example farm has decided to have the herd graze the pastures to approximately 2 inches. Using this information in the same chart you used above, enter **1,200 lb.** on line 6.
- Line 7: Calculate **2,600 lb. – 1,200 lb. = 1,400 lb**. This is the estimated amount of DM available for the herd to eat per acre.
- Line 8: Divide the amount of DM required (line 4) by the amount available (line 7) to calculate the number of acres needed: **1,500 lb. ÷ 1,400 lb. = 1.07 acres**.
  Note that this is the number of acres needed per day (24 hours) to feed the herd. Since this example farm decided to give them a new paddock after each milking, or two paddocks a day, the half-day paddocks will be **0.54 acres** each (1.07 acres ÷ 2 = 0.54 acres after each milking).

This method of paddock size calculation gives you a good place to start. You will, however, have to adjust the actual paddock size based on several factors, including these.

- How much feed are you giving in the barn? This will affect how much DMI your cows actually need from the pasture.
- Pasture height—the plants must be tall enough. Plants too short will not allow cows to get enough DMI even if you make larger paddocks.
- Weeds, rocks, swamps, and other things in the pasture will need to be factored in.
- Plant density may be lower or higher than estimated.

EXAMPLE 4.2: GRAZING ACREAGE NEEDED

For the 50-cow farm that needs 1.07 acres per day, total acreage needed to graze in the spring will be 16 acres (1.07 × 15 days = 16 acres). When pasture growth slows down to 35 to 40 days, the total needed acreage increases to 38 to 43 acres.

ing facility. In some cases, unless you can obtain additional land, the farm may not be a good candidate for organic certification. Plan ahead to make sure the number of cows and amount of available pasture acreage will allow the design of a good grazing system.

Extending your grazing season

If you have enough land, you may be able to plan your grazing so that you have pasture available into the fall and even winter in some climates. This requires careful planning so that plants are tall enough when growth stops in the fall. Some conditions such as excessive soil moisture or too much snow can make fall stockpiled grazing difficult. For some farms, however, this is a successful way to feed forages. See Chapter 6, "Managing Dairy Nutrition" for more information on stockpiling and nutrition.

» *continued on page 76*

WORKSHEET 4.1 CALCULATING PADDOCK SIZES AND NUMBER OF ACRES NEEDED

Attending pasture walks is a great way to learn how to make theses estimates more accurate!

ESTIMATING FORAGE DRY MATTER INTAKE (DMI)

Average body weight: _____ (Line 1)

Estimated DMI (as % of body weight): _____ (Line 2)

Daily DMI required for single animal (Line 1 × Line 2): _____ (Line 3)

Daily DMI required for herd (Line 3 × number of animals): _____ (Line 4)

ESTIMATING PASTURE MASS (FORAGE DRY MATTER)

HEIGHT (inches)	**AVERAGE DENSITY*** (Pasture lb. DM/acre)	**LOW DENSITY** (Pasture lb. DM/acre)	**HIGH DENSITY** (Pasture lb. DM/acre)
8	2,600	2,200	2,800
6	2,400	2,100	2,600
4	1,800	1,500	2,100
2	1,200	1,000	1,400
1	900	600	1,000

*Pounds of dry matter per acre at each height varies widely with plant density and species.

CALCULATING AVAILABLE DRY MATTER

Available forage dry matter = Pregrazing mass – Postgrazing mass

EXAMPLE:

Pregrazing 6" 2,400

Post Grazing 2" –1,200

1,200 lb. DM/acre

YOUR FARM:

Pregrazing mass: ________ (Line 5)

Post Grazing Mass ________ (Line 6)

________ (Line 7) Available lb. DM/acre

CALCULATING PADDOCK SIZE

Paddock size (in acres per day) = Daily DM required ÷ Available dry matter

Daily DM required _____ (Line 4) ÷ Available DM/acre _____ (Line 7)

= _____ Paddock size in acres/day (Line 8)

(There are 43,560 square feet in an acre, which is a square about 210 feet on each side)

CALCULATING REST PERIOD

Maximum possible rest period = Your total pasture acres ÷ Paddock size

Your pasture acres available _____ ÷ Paddock size in acres per day _____ (Line 8)

= _____ maximum rest period

continued from page 74 »

INFLUENCE OF WEATHER CONDITIONS

During some weather conditions, grazing some or all of the pastures may not be possible. Flooding in the Northeast has forced some graziers to either move livestock to high, dry land or keep them in the barn. Anytime grazing is not possible due to an emergency like this you should notify the certifier. High rainfall periods may result in muddy pastures or lanes that make grazing difficult and may even cause hoof or other health problems. Building well-surfaced lanes, using culverts, laying filter fabric under surface material, and other techniques will make the grazing system much easier to manage in these weather conditions.

As global climate change causes more extreme weather conditions, all Northeast farms will need to prepare for extended dry periods. Planning for possible drought may mean saving extra stored forages that can be fed during a drought. Other techniques being used successfully by farmers who regularly experience summer droughts include grazing annual crops such as millet or sorghum-sudangrass as well as adding significant acres of hay land which is mechanically harvested early in the growing season.

DESIGNING YOUR GRAZING SYSTEM

When setting up a new pasture system as part of transitioning to organic, or improving a new system, it may be useful to use an aerial photograph or survey map. Your organic certifier will require copies of maps for all your fields anyway, so make extra copies. Colored pens are useful to draw the existing or planned lanes, fences, paddocks, and water lines. When subdividing your grazing land, here are a few things to keep in mind.

- Initially, keep your system as flexible as possible so you can move fences and water lines as your herd size, grazing skills, and pasture quality improve.
- You can use permanently fenced paddocks, larger permanently fenced areas subdivided into paddocks with temporary fences, or you can do strip grazing with a front and back fence.
- Put lanes on high, dry ground. You will probably need to do some improvement to and maintenance on muddy wet areas.
- Some areas on your farm will grow faster than others, so whenever possible try to include fast-growing areas in one paddock and slow-growing areas in another.
- Consider topography and keylines: put south-facing slopes in one paddock and north slopes in another.
- You may want to include shade in some pastures, particularly if you have extended hot weather in the summer. In cool weather, however, it is advantageous to minimize shady loafing areas, where animals tend to congregate and concentrate manure.
- Try to provide water in each paddock so animals do not have to walk to find it and drop manure around permanent water tubs.
- Do not forget to check with your certifier about any required

buffers between your pastures and any adjacent nonorganically managed land.

- Put gates in the corner of the pasture closest to the barn.
- Attend a pasture walk and visit other grass-based farms and learn from their experiences.

Read the farm case studies in this chapter for descriptions of real grazing systems that northeast farmers are using.

Electric fencing

Electric fence technology makes MIG grazing management much easier. You will need to install a good-quality, low impedance energizer and ground it correctly. It is also helpful to have high-quality perimeter fencing that can conduct electricity with minimal resistance. High tensile fence using 12.5 gauge wire requires the least maintenance and has the best conductivity over time, but other types of steel wire fence of at least 14 gauge will also make a good perimeter fence. Most people use some kind of portable fencing to subdivide larger areas. There are significant differences in the quality of energizers, portable posts, and polywire, so it pays to do some research before purchasing them. In the Northeast, you will also need lightning protection for your energizer. Check with your organic certifier if you plan to use any type of treated fence posts, as many of these are prohibited in new installations on organic or transitioning farms. See Figure 4.3.

Check with your NRCS office for available cost-share or other programs that may help pay for fence, lanes, water systems, and other costs of setting up the grazing system.

Water systems

Water can be provided in paddocks several ways. If piping water to pastures is not possible, you can use a tank on a wagon parked in the pasture. Some systems pump water from a pond, well, or stream to a tank. If using pipe to provide water in each paddock, use a large-diameter pipe (1- or 1.5-inch) so the water pressure does not drop too much as it travels the length of the farm. Water lines can be buried, or pipe can lay on the ground along a fence

FIGURE 4.3
High tensile fence and interior subdivisions with portable fence posts and polywire on Stonypond Farm in Vermont

line or a lane. Most farmers prefer to keep the pipe aboveground so it is accessible for repairs and can be relocated if needed.

Water tubs can be small and portable as long as the water flow rate is fast enough to keep them full while the herd drinks. Larger, permanent tubs may be easier if the water flow rate is slower or if the herd is very large. It is less expensive to have a few portable tubs than many permanent tubs. Wherever the water source is, use a float valve to control water flow. See Figure 4.4.

FIGURE 4.4
Portable pasture water tub with float valve

Fixed water systems where cows walk to drink from a tub near the barn or near a spring may be your only option when you are getting your grazing system set up, but it will be helpful to begin providing piped water in the paddocks as soon as possible. Drinking from ponds or streams may also be possible but only if well-managed so that water quality is excellent and erosion is prevented. Check with your local Natural Resources Conservation Service (NRCS) office as well as your certifier to see what their guidelines are.

Lanes

Particularly during wet spring and fall weather, having well-built lanes in the right locations will make managing pastures much easier. Dry, well-surfaced lanes can also help prevent hoof injuries and other problems. In wet areas or heavily used lane areas, a road-building filter fabric used under the gravel will create longer-lasting lanes that require much less maintenance. In some areas, you may need culverts and erosion control.

FIGURE 4.5
The main lane from the barn to the pastures on Earl Fournier's farm allows access to paddocks. High tensile fences and water lines are parallel to the lane.

BOX 4.2 GRAZING GUIDELINES SUMMARY

- Learn how overgrazing occurs and then manage so as to avoid it.
- Walk through all your pastures each week and record the height of each pasture.
- Keep records of when you graze each pasture.
- Do not let animals return to a paddock until it has fully regrown (recovery periods will be variable).
- Do not let animals stay in one area for more than three days; 12–24 hours is better.
- Move animals frequently. Moving animals more often can increase dry matter intake and more rapidly improve pasture quality.
- Do not follow a set rotation. Graze according to plant growth rates. If one pasture grows faster than the others do, graze it more often. If you have a pasture that grows slowly, graze other areas and let the plants grow back.
- Lock animals in each paddock so they cannot wander back to the barn.
- When strip grazing, use a back fence to prevent "back grazing," so animals do not overgraze their favorite plants.
- Monitor body condition score and milk urea nitrogen levels and watch manure consistency to see if cows are being overfed protein or not getting enough dry matter intake.

Shade and other methods of heat relief
There are many factors to consider when managing cows during hot weather. These include temperature, humidity, what type and quantity of shade is available, what sort of facilities are on the farm, the type of breed, and how well the cows are adapted to the local climate. Some hot weather strategies include:

- grazing "shade pastures" that are set aside for the hottest summer days,
- keeping cows in the barn (if it is cool) during the hottest days and grazing at night,
- moving cows more frequently to high-quality pasture,
- using sprinklers or portable shade structures (portable shade structures can be moved around to help distribute manure more evenly), and
- providing ample drinking water in each paddock.

Youngstock
The same grazing principles apply to grazing youngstock, but there are some special considerations to keep in mind.

- Training heifers and calves to electric fence early on can make them much easier to manage. A training area in a barnyard with the electric fence inside the barnyard fence works well. It is essential that the fence carry a high voltage when training young animals. Low volts on the fence will result in poorly trained animals that are more likely to run through fences or even injure themselves.

- Youngstock are more susceptible to parasites than adults, so youngstock pastures should be managed to minimize infection.
- Some farmers raise calves on nurse cows while on pasture or raise calves on the mothers. In these situations, the calves graze with adults and an additional wire may be needed in the fence to keep the calves from ducking under a single-strand fence.

"It is no accident that good pastures have perhaps 40 to 50 species. . . . Nor is it an accident that fully half the weeds in the USDA index are also listed in manuals on medicinal plants."
—Gearld Fry and Charles Walters, *Reproduction and Animal Health*

PASTURE PLANTS

DIVERSITY

When animals go into a pasture, they do not just eat. They trample weeds and dead plants into the soil, which adds organic matter. They select plants they want to eat, and they spread manure. Well-managed perennial pastures provide livestock with a salad of many types of plants, providing a variety of nutrients and medicinal compounds. Having a diversity of plants will provide cows with many options to choose from so they can balance their own ration (see Box 4.2). Plant diversity assures that some plants grow even when weather conditions are extreme. Different plants grow differently throughout the growing season.

FIGURE 4.6
Earl Fournier's high-quality pasture contains a mix of grasses and legumes.

GRASSES AND LEGUMES

One group of plants known as the cool season grasses have their highest growth rates during the cool part of the growing season. Some common cool season perennial grasses suitable for grazing include orchardgrass, Kentucky bluegrass, and perennial ryegrass. Another group of grasses that may be grazed is the warm season grasses. These grow fastest during the warmer part of the season.

Sorghum-sudangrass is one of the more commonly grazed warm season grasses. Perennial legumes also show seasonal variations in growth and include white and red clover as well as alfalfa and birdsfoot trefoil. Chapter 5, "Organic Crop Management" includes specific information on these different plant species.

Many dairy farms have land used for both grazing and haying, depending on the time of year. Many of the cool season perennial grasses do well under either good grazing management or hay management. Areas grazed more often tend to have plants better adapted for grazing, such as white clover, Kentucky bluegrass, perennial rye grass, and some fescues. Areas primarily harvested mechanically tend to have species less adapted for grazing such as alfalfa, red clover, timothy, and brome grass. When selecting varieties of plants to grow in pastures, it is important to choose varieties that are suited to the soil conditions, palatable, and locally adapted. Reed canary grass for example is well adapted to wet soils. It can, however, be less palatable than other species of grasses in a pasture setting. Tall fescue grows well in areas with hot, dry summers and can be used for fall/winter stockpiled grazing. However, tall fescue is highly unpalatable when compared to other cool season perennial grasses, so it should be planted only if the local climate makes it necessary.

ANNUAL SPECIES FOR GRAZING

Depending on soil types and the crops grown on the farm, pastures may be entirely permanent perennial fields or they may rotate with some annual crops. Farmers who have access to equipment, labor, and suitable land can rotate grain or corn silage with a pasture grass/legume sod. Farms in areas with extended summer dry periods may grow summer annual crops such as sorghum-sudangrass or millet for grazing. Some farms may benefit from grazing brassica crops or other alternative annual pasture mixes.

Note that there are some risks associated with grazing some annual crops. When grazing sorghum-sudangrass, caution should be used to prevent prussic acid poisoning. To prevent problems, avoid grazing sorghum-sudan during or right after a frost and be sure plant height is at least 24 inches before grazing. Since most dairy farmers use these crops as midsummer drought pasture, they easily avoid these issues. Pearl millet, which can be grazed at 14–24 inches of height, does not cause prussic acid poisoning. Forage brassicas such as turnips, rapeseed, and kale are high yielding and can be grazed, although, there are only a relatively small number of dairy farms in the Northeast doing this. They maintain quality well into freezing temperatures and can be used to extend the grazing season on some farms. Health problems can occur if brassica grazing is not done correctly. These can include bloat, atypical pneumonia, nitrate poisoning, and hypothyroidism. To avoid health disorders, the following two management practices should be followed. First, introduce brassica pastures slowly (over 3 or 4 days) and avoid sudden changes from regular pasture to lush brassica pastures. Second, brassica crops should not be the only pasture or forage in the ration. Supplement with dry hay, silage, or "regular" pasture while grazing brassicas.

Each farm will have to assess the feasibility of growing annual crops according the farm family's goals and the economics of production. A permanent, well-managed grass legume pasture provides significant benefits in building and maintaining soil fertility, and with a good grazing management system there should be few reasons to plow and reseed permanent pastures. Refer to Chapter 5, "Organic Crop Management" for more information on management of these plants.

Dave Johnson, an organic dairy farmer in Pennsylvania, plants oats and turnips together in mid to late April; grazing starts in early August. Grazing is done in 10-foot-wide strips to prevent too much trampling. He is considering using rapeseed instead of turnips next year.

Medicinal plants or weeds

Many plants commonly considered weeds have interesting nutritional or medicinal uses. While you may not want all these plants growing in your pastures, having them available to cows in hedgerows, browse areas, or hospital paddocks may be something to consider. Some of these species include dandelion, plantain, raspberry, yarrow, curly dock, shepherd's purse, chicory, willow, and many others.

Improving pasture

Before plowing and reseeding a grass/legume pasture, be sure that you have not overlooked other options for renovating and improving the field. If you plow and reseed but do not change the grazing management system, the reseeding will only provide a temporary solution to poor pasture quality.

An economical way to improve some areas may include frost seeding clovers; changing the grazing management; mob stocking; using an aerator, Yeomans plow, or chisel plow to improve aeration; or adding compost or some other amendment to improve soil health. When reseeding is necessary as part of a rotation with annual crops or for another reason, choosing the right species is important. Selecting the right species, variety, or mix of species involves a consideration of soil pH, drainage, fertility, climate, crop palatability, weed pressure, harvest method, and the length of time you need the stand.

Seeds

When buying seed, be sure it is not genetically engineered or treated with a fungicide. The use of GMO or fungicide-treated seeds is not allowed on organic farms. GMO seeds are often not clearly labeled, and it is up to the farmer to find out. Organically certified farmers need to buy organic seed whenever it is commercially available.

Frost seeding of clovers and some grasses can be a helpful method to introduce new pasture species to a permanent pasture. Frost seeding generally is less successful for grasses, but works well for white and red clover and involves broadcasting a small amount of seed per acre into the pasture in late fall or early spring. Some grass species such as perennial rye grass have been frost seeded with some success, but require good frost seeding conditions.

The following practices improve the success of frost seeding.

- Use heavy fall grazing to control grass growth and create more open soil.
- Test soils and fix imbalances that hamper legume growth. Pay particular attention to lime, potassium, boron, and phosphorus.
- Seed in late winter or early spring.
- White clover will persist longer than red clover. Consider frost seeding a mix of each.

Shifts in species composition

How you choose to implement your grazing management will shift the plant species in the pasture. Under good grazing management, plants that are grazing-adapted will thrive, while poor-quality pasture plants will die out. White clover will naturally spread by its stolons (stems that travel on the soil surface), grasses will spread by tillering (side shoots), while woody weed species that do not like to be grazed frequently will grow less vigorously and die out.

Mob stocking is a term from New Zealand that describes a system where large groups of cattle or sheep are used to rapidly defoliate pastures or brushy areas repeatedly to kill weed species and encourage pasture plant growth. This method requires a high stocking density and is best done with a group of animals that does not have high-quality feed needs.

PASTURE SOIL HEALTH

Pastures with properly mineralized soils, high organic matter, and biological activity produce higher-quality feed and result in better animal performance and fewer herd health problems. Farms with healthy soils growing more nutritious, higher-quality pasture can produce higher milk yields while feeding fewer concentrates. Managing internal parasites, which can be a challenge with youngstock on organic dairy farms, is reported by some farmers to be easier after improving soil health. They also report experiencing fewer metabolic and other herd health problems, longer-lived animals, and more replacements reaching productive years.

An important part of organic dairy farming is creating a farm where microorganisms in both the soil and the livestock's digestive system are abundant and healthy. These organisms play an essential role in the process of cycling nutrients and energy from soil, air, and sun to plant, animal, manure, and back to the soil. For more information on soils, manure management, and soil testing, see Chapter 3, "Organic Soil Management."

Note these pasture-specific soil fertility issues.

- A good grazing system will provide an even distribution of manure in pastures.

- Application of compost instead of raw manure to pastures can help eliminate rejected forage in the next grazing cycle. Fall application generally results in less rejected forage.
- Healthy soil biology incorporates manure into pasture and meadow soils as rapidly as possible to minimize nutrient loss. You can monitor the rate at which earthworms, dung beetles, or other beneficial soil life incorporate cow manure into the soil.

Earl Fournier, who converted his confinement dairy to a successful grass-based organic dairy farm, found that it is a common mistake among transitioning farmers to pasture milkers on land they cannot do something else with. He encourages farmers to "use your prime meadows for grazing!"

SETTING UP A NEW GRAZING SYSTEM

Converting from a confinement (nongrazing) system to a grazing system is a major change. Ideally, the transition from confinement to grazing should be done before the start of the organic transition of the dairy herd in order to give the animals, land, and farmer all time to adapt. In order to be successful, shifting a farm and herd to a pasture-based feeding system requires careful planning and preparation. Cows will need time to adapt to the new diet and management system, the farmer will need time to gain new skills, and the pastures will likely need work and time to become good quality pasture.

IMPORTANT POINTS TO CONSIDER FOR NEW GRAZING SYSTEMS

Converting cropland to pasture

Start by first confirming there is enough land within walking distance of the milking area to provide enough pasture for the cows. (See Calculating Paddock Sizes, DMI, and Acreage Needed on page 72). Be sure to include enough land for the heifers and dry cows. Land currently in hay/legume crops can be converted to pasture without reseeding. If fields are pure grass, you can add white or red clover by frost seeding. Fields that have been in annual crops will need to be seeded to perennial grass/legume mixes well adapted to grazing.

Introducing cows to the new system

If the cows have never been exposed to electric fencing, be sure to train them to the fence before turning them on pasture. Use a well-grounded, high-quality energizer and set up the wire just inside the barnyard fence so that when cows touch it they back up instead of running forward. Once they are respectful of the fence, they are ready to go out to pasture. Do not put hungry cows on pasture the first time. Instead, begin the grazing season by grazing for just a few hours a day, letting their rumens adjust to the new diet. In a new pasture system, it is important to monitor pregrazing height, plant density, and paddock size regularly to be sure the cows are getting enough dry matter intake. Working with a grazing mentor can be very helpful at this critical stage.

Converting from one system to another

If you are converting from a rotational system with a few paddocks or a continuous grazing system, the good news is that even small changes in how you move the cows can result in noticeable improvement. The bad news is that sometimes the overgrazing damage (weeds, plants that will not grow tall in some areas, poor quality feed) can take time to repair. Good grazing management may need to be combined with soil tests, compost spreading, and frost seeding to improve some areas. Start by learning more about how plants grow, why recovery periods for pastures are variable, and what a fully regrown and recovered pasture looks like. You may decide to redesign your system with new fences and lanes and a water system that provides water tubs in each paddock, or you may decide instead to intensify your grazing management system gradually. If you decide to subdivide the existing paddocks and add in new ones, the increased rest period for the pastures will allow stressed, overgrazed plants to recover (see Table 4.1).

TABLE 4.1 COMPARING ROTATIONAL, CONTINUOUS, AND MIG GRAZING SYSTEMS

CONTINUOUS GRAZING	ROTATIONAL GRAZING	MIG GRAZING
Cows are in the same paddock for the whole grazing season.	Cows are rotated around several pastures, usually on a set rotation. Recovery periods are not varied as plant growth rates slow.	Cows are moved to a new paddock only when it has fully regrown. They are moved at least every 3 days. Recovery periods are variable.
Cows will graze selectively, making it difficult to balance the ration. Pastures will generally provide enough feed in spring, but later in the summer pasture will be too short, or too over-mature to provide enough dairy quality feed.	Cows may have adequate DMI and pasture quality in the spring, but as plant growth rates slow, the pastures will be too short or plants will over-mature and provide less dairy quality feed.	Cows will have adequate DMI and pasture quality throughout the grazing season.
Pasture quality and quantity will significantly decline as the season progresses.	As cows rotate back into pastures that are not fully regrown the quantity and quality of feed will decline.	Cows only rotate back into pastures that are fully recovered. Additional acres will be added into the grazing rotation as growth rates slow.
Pasture quality will decrease each year due to overgrazing damage, increased weeds, and rejected forage. Clipping and reseeding may be needed.	Pasture quality will decrease each year due to overgrazing damage, increased weeds, and rejected forage. Clipping and reseeding may be needed.	Pasture quality will improve over time.

BOX 4.3 FORAGING BEHAVIOR AND DIET SELECTION: MORE THAN THE LUCK OF THE BITE!

DR. DARRELL L. EMMICK

It is commonly believed that modern dairy cows have been fed in confinement for so long they possess little if any ability to evaluate the nutritional adequacy of their own diets. However, this view has little merit. It is important to recognize that modern confinement-based feeding strategies have been in existence only about 50 years. Herbivores, which include dairy cows, have been using their senses of sight, smell, touch, taste, and postingestive feedback mechanisms to locate, discriminate, and select diets for millions of years. It is, therefore, not likely that dairy cows have lost this ability in the few short years humans have been feeding them in confinement. In fact, because all foods are a combination of nutrients and toxins, and even nutrients become toxic when eaten in excess, if we were successful in eliminating the cow's ability to evaluate foods, she would likely not live long enough to reach her first lactation.

WHAT IS BEHAVIOR?

Behavior is anything an organism does that we can measure. It can be fairly simple, such as a reflexive response to a single stimulus i.e., things like the knee jerk response, eye blinking, or jumping at the sound of a loud noise, or it can encompass more complex activities, such as foraging and diet selection. Researchers argued for years which complex behaviors were more important, genetic inheritance or learning through environmental experience. However, today we know that behavior of individual animals is a unique combination of both. Modern dairy cows behave the way they do partly because of their evolutionary histories, their unique genetic endowments, and the result of social and individual learning that takes place throughout their lifetime.

MECHANISMS CONTROLLING DIET SELECTION AND FORAGING BEHAVIOR

Animals eat what they eat because of the interactions of two interrelated systems. One system is called the cognitive or voluntary system. This system uses the senses of sight, smell, touch, and taste along with information learned from mothers, other members of the herd or flock, and past trial and error encounters to determine what to eat or not to eat. Experiences early in life and guided by mothers are extremely important in determining what any animal will choose to eat, even years later. If mom eats it and baby watches, chances are baby will eat what mom eats. The other system at work is called the affective or involuntary system. This is a subconscious process that operates without any effort on the part of the animal and links the taste of a food with the food's postingestive (after it is eaten) consequences relative to the requirements of the animal.

Postingestive feedback is essentially an information system that operates within animals at the subconscious level. This system utilizes chemical, osmotic, and mechanical receptors within the gut to evaluate the unique chemical content of each food ingested relative to the particular animal's nutrient requirements. This information is fed back to the brain where decisions about the food are made. Simplistically, if an animal eats a particular food and, shortly after, feels sick, discomfort, or in some other way "not good," the taste of this food will be paired with the discomfort and the animal will likely shy away from or become averted to consuming this food. On the other hand, if a food is consumed and the animal feels satiated i.e., no ill effects or feels "good," the animal will generally pair the flavor of this food with the feeling of satiety and thus develop a preference for the food.

TAKE HOME MESSAGE

While the voluntary and involuntary systems function as two separate systems, they are integrated through the senses of sight, smell, taste, and postingestive feedback. Animals use the involuntary system to evaluate the postingestive consequences of consuming a food; they use the voluntary system to change their behavior toward the food depending on whether the postingestive feedback was positive or negative. Through this interactive exchange of information, animals constantly monitor the foods they consume and alter their diets in response to their own ever-changing nutritional requirements, changes in the foods they eat, and variations in the foraging environment.

As we continue to look for ways to improve milk production from pastured dairy cows, it is important to recognize that dairy cows are not the machines they are often made out to be. Machines do not care what a food looks, smells, tastes, or feels like. Cows do. When dairy cows express their concerns, it is usually as a decrease in intake and a loss in milk production. Thus, we need to pay better attention to what our cows are telling us about our management and the foods we feed to them. Behavior is a function of its consequences, and the more we accommodate the needs of our cows, rather than dictate what they are going to get, the more contented and productive they will be.

Dr. Darrell L. Emmick is the State Grazing Land Management Specialist with the USDA-NRCS in New York state. His work over the past 25 years includes the research, development, and promotion of cost-effective grass-based livestock production systems for use by dairy and livestock producers in the northeast region of the United States.

RATION BALANCING AND PASTURES

The two most common nutritional pitfalls during the grazing season are: feeding too high a protein concentrate, and inadequate DMI due to pastures being too short.

Protein

As soon as cows are grazing daily in the spring, the grain and other supplemental feed in the barn must be changed to a lower- protein mix. Adjusting the total mixed ration (TMR) to take out high-protein feeds such as haylage can also help improve the grazing behavior and performance of the cows. If the protein content of the supplemental feed is not reduced, cows will lose condition and drop in milk production. If this imbalance continues, it may result in lower fertility and other health problems. Testing pasture forage quality and milk urea nitrogen (MUN) can help you convince your nutritionist to adjust your ration. If your nutritionist is hesitant to adjust your rations for the grazing season, you may need to change nutritionists to work with someone who has experience with rations for pastured cows. See Chapter 6, "Managing Dairy Nutrition" for more information on MUN, BCS (body condition scoring), and observing manure to determine if your ration is well balanced.

BOX 4.4 COMMON GRAZING NUTRITION PITFALLS

The two most common nutritional pitfalls during the grazing season are:

- feeding too high a protein concentrate, and
- inadequate DMI due to pastures being too short.

Dry matter intake (DMI)

Under a continuous or rotational system, there may be enough high-quality pasture during the early part of the grazing season, but as the season continues, the quality and quantity available to the cows will decline. This can result in much lower total DMI, and if not provided with additional forages in the barn, cows that will perform poorly. Understanding grazing behavior can be a helpful tool in maximizing DMI and animal performance. Cows prefer to graze at dawn and dusk, so turning them into a fresh paddock after the afternoon milking can encourage more dry matter intake from pasture. The type and amount of what is fed in the barn also will have a significant effect on how well they graze the pastures. Box 4.3 provides additional information.

Testing pastures for forage nutritional quality

Testing the forage quality of your pastures can be very helpful, particularly in the first few years of learning how to graze. When taking the samples, be sure to observe what the cows are actually grazing before taking the sample. Use your hand (not clippers) to take the sample and try to mimic how the cows are grazing as you collect the forage.

CONVERTING TO LOW OR ZERO GRAIN FEEDING

Due to market demand for these products, there is increasing interest in producing milk and meat from animals fed little or no grain. Interest in grass-based systems also stems from the rising cost of grains and the interest in reducing petroleum-based farm inputs. However, transitioning to no-grain feeding is challenging, and if done too fast, with the wrong genetics, or anything less than the best-quality forages, it can result in poor animal performance and low profits. A poorly planned transition to no-grain can cause cows to lose condition, not breed back, and significantly drop in milk production. If a farm is interested in the goal of zero-grain feeding, it is important to be clear about how that will fit into the overall farm and family goal. Consider the following questions.

- Will the decrease in total milk production still provide enough cash flow to cover all farm and labor costs?
- Is the quality of the winter stored forage and summer pasture excellent and consistent?
- Does the farm manager have grazing skills and knowledge?
- Is there a system to allow sufficient nutrient and mineral supplementation?
- Is there enough market demand and is the price for products high enough?

Moving toward zero-grain feeding requires beginning by rearing youngstock and selecting suitable breeds or bloodlines that thrive on an all-forage diet. Growing high-quality stored forages and a high level of understanding and skill with grazing management is essential. A clear understanding of the economic incentives (or lack of them) to cutting out the grain is important. At this time, the only farms paid a premium for no-grain feeding have a local value-added market. Approach zero-grain feeding with an open mind and allow yourself plenty of time and flexibility to feed grain as needed during the transition to no-grain. See the Chapter 6 case study on Rob and Pam Moore's farm for an example of a successful zero-grain feeding farm.

PASTURE AND INTERNAL PARASITE MANAGEMENT

Pastures can become a source of internal parasite infection, particularly for youngstock. Preventing internal parasite problems is essential on organic farms because treatment options are limited. After cows shed the parasites in manure on pasture, the parasites will live and continue to be infective for up to a year (longer in some areas). Animals that shed the most eggs are usually mothers after birthing or youngstock. With this understanding, it is possible to manage pastures to keep the most susceptible animals (youngstock) out of the most infective areas. This done by grazing them in areas that were not grazed within the past year. Sheep and goats have (mostly) different parasites than cows, so in some cases multispecies grazing can be part of the strategy to create clean pastures. Other ways to decrease parasite loads in a pasture include harvesting hay, fallowing an area, and rotating with an annual crop. For more information on parasites, see Box 4.3 as well as Chapter 7, "Organic Dairy Herd Health."

PASTURE REQUIREMENTS OF MILK BUYERS AND ORGANIC STANDARDS

NATIONAL ORGANIC PROGRAM PASTURE STANDARDS

Pasture is required by organic standards. In addition to the National Organic Program (NOP) pasture standard, individual milk buyers may have a more specific pasture requirement. The current NOP rule defines pasture as, "Land used for livestock grazing that is managed to provide feed value and maintain or improve soil, water, and vegetative resources." Pastures need to be managed to "maintain or improve the physical, chemical, and biological condition of soil and minimize soil erosion." However, lack of enforcement and clarity in the rule has allowed a small number of farms and certifiers to continue to operate and certify organic dairy farms where the cows do not have access to real pasture. At the time of writing this chapter, the NOP has released the long awaited proposed rule on pasture, which has been requested as part of the solution to this issue. The proposed rule will be released after 2009 and may more clearly define pasture by requiring a minimum DMI from pasture. Check the current NOP rules on pasture, §205.2, §205.102, §205.237, and §205.239–§205.240 for the current pasture requirements. See also Table 4.2.

BOX 4.5 PARASITE MANAGEMENT AND PREVENTION GUIDELINES

- Pay close attention to vitamins and minerals in the ration.
- Keep livestock in good condition.
- Minimize infection by using feeders that animals cannot get manure into and avoiding ground-feeding.
- Include high-tannin plants in hedgerows and pastures to reduce parasite loads (birdsfoot trefoil, chicory, brambles, and many woody species).
- Farms with a lower stocking rate (total number of animals on the farm) generally have lower parasite loads.
- Since most of the infective parasites are in the bottom 2 inches of pasture, manage pastures to prevent animals from grazing the pastures too short. This is particularly important in wet weather and on more heavily infected pastures.
- Read the section in this chapter: What pre-grazing height is best? (p.71).
- Some farms report that improving soil fertility and using compost seems to help decrease parasite problems.

CERTIFYING AGENCY

Your certifier may have specific recordkeeping requirements or other guidelines for pasture management. They may also have guidelines about animals having access to streams or ponds in pastures. If you have any existing erosion problems in pastures, lanes, or stream crossings, you will need to develop a plan to fix them and be sure that any new grazing areas will not create additional problems.

IMPROVING YOUR PASTURES

Pastures do not suddenly become poor quality. It is a gradual decline over time, usually due to overgrazing damage, which encourages weed growth and reduces the more-productive desirable grazing species. A knowledge of soil fertility and a good understanding of grazing management techniques, along with trained observational skills, are needed to develop this vital part of the organic system.

TABLE 4.2
MEASURING 30% AVERAGE DAILY DMI PER ANIMAL FROM PASTURE

Examples of some farm records that will show your certifier or milk buyer that you are able to provide at least 30% of the average daily DMI from pasture include the following:

- What date is spring turnout for your adults?
- What date is spring turnout for your youngstock?
- What date do you bring in your adult animals at the end of the growing season?
- What date do you bring in your youngstock at the end of the growing season?

These records demonstrate the amount of time livestock spend on pasture during the year and can easily be kept on the barn calendar or in a herd notebook.

The next records are one way to show how much daily DMI comes from pasture compared to the amount that comes from stored forages or concentrates. This is a calculation of the average ration per cow in the nongrazing season (winter) and the average ration per cow in the grazing season. When you compare the two, it is easy to see how much less DM is being fed in the barn during the grazing season and then calculate what percent of the daily DMI is from pasture. You can find the percent dry matter for each of the stored feeds from your forage test results or grain slips. If that information is not available you can use average percent dry matter numbers.

Nongrazing (winter) Feed Ration

Avg. lb. of hay consumed _____lb. × _____ % dry matter = _______ lb. dry matter

Avg. lb. of baleage consumed _____lb. × _____ % dry matter = _______ lb. dry matter

Avg. lb. of corn silage consumed _____lb. × _____ % dry matter = _______ lb. dry matter

Avg. lb. of grain consumed _____lb. × _____ % dry matter = _______ lb. dry matter

*Total (Average) Lb. Dry Matter per Cow per Day in Winter =*_______________

Grazing Season Feed Ration (all feed other than pasture)

Avg. lb. of hay consumed _____lb. × _____ % dry matter = _______ lb. dry matter

Avg. lb. of baleage consumed _____lb. × _____ % dry matter = _______ lb. dry matter

Avg. lb. corn silage consumed _____lb. × _____ % dry matter = _______ lb. dry matter

Avg. lb. of grain consumed _____lb. × _____ % dry matter = _______ lb. dry matter

Total (Average) Lb. Dry Matter per Cow per Day in Grazing Season =_____________

DM in winter ration – DM in grazing season ration = Estimated Pasture Dry Matter Intake

_______________ – ____________________ = ____________________________

Assessing pasture quality: Assessing if your pasture quality has improved over time is a good way to measure the success of your grazing system. There are several useful systems of pasture condition or quality scoring, which involve scoring the same pastures each year to measure how the production of quality feed changes. It is also helpful to do this condition scoring several times each season and to then repeat them the next year, on the same dates, in the same pastures so you can track changes over time. Some useful questions to ask yourself as you begin to assess your current pastures include the following. See also Box 4.6 and Table 4.3.

- Are you capturing all the sunlight that lands on your farm, or are there areas of bare soil and nonproductive plants in your pastures?
- Are soils fertile and biologically active so that plant health and the conversion of sunlight are maximized?
- Does water cycle through the farm so it is available to the plants and cows when they need it and does not create erosion?
- Is the livestock grazing system improving your farm's ability to convert sunlight into high-quality forages?
- Are livestock harvesting forages efficiently and producing enough milk to meet your farm goals?

BOX 4.6 GRAZING RECORDKEEPING

- Check with your certifier to see what grazing records are required.
- Keep records of the stored feeds (grain and forages) you supplement your cows' diet with to show your certifier or milk processor what percentage of their daily DMI is from pasture.
- Keep records of when you begin grazing in the spring and when you run out of pasture in the fall.
- Records of when you graze each pasture will allow you to track farm regrowth periods so your pasture planning for next year can be more accurate.
- If your records allow you to track daily milk production, you can start to see patterns in animal performance and its relation to your grazing system.
- Consider using one of the pasture quality or condition scoring systems to track changes in pasture quality over time.
- There are very helpful methods to measure overall "cover" of pasture dry matter on the farm. One of these is the "grazing wedge" calculator found on the University of Missouri Extension webpage: (http://plantsci.missouri.edu/grazingwedge/).

TABLE 4.3 PASTURE QUALITY AND CONDITION SCORE CHART

DIRECTIONS: Walk through each pasture area on your farm and use this chart to assess the quality or score of each category. Rank each pasture area as poor, fair, good, or very good in each category. Include comments and descriptions of what you observe to help you track changes over time in your pastures. *Consider monitoring soil health at the same time you check your pasture quality using Table 3.2.* Date: ____________________

CATEGORY	PASTURE AREA	SCORE *(poor, fair, good, very good)*	NOTES AND COMMENTS
Plant diversity: How many different species of plants are in the pasture? A poor pasture will have few legumes and a small number of grass species. A good-quality pasture will contain two or more legume species and many grass species.			
Plant density: Poor density will have soil visible between the plants, whereas a good-quality pasture will be higher density with a more complete cover.			
Palatability of plants: Good-quality pasture will contain most or all plants that are the species and at the stage of maturity cows prefer to graze. A poor-quality pasture will contain nonpalatable weed species and other plants the cows will not eat.			
Plant growth rate: Ideally, pasture plants will grow vigorously throughout the growing season. There will normally be slower growth of the perennial species in midsummer. Pastures that are more drought prone, have poor soil fertility, or are overgrazed will be less vigorous and will show more seasonal variation.			

RESOURCES

Books

Emmick, Darrell, Karen Hoffman Sullivan, and Robert Declue. 2000. *Prescribed Grazing and Feeding Management for Lactating Dairy Cows.* USDA NRCS, NY State GLCI, Syracuse NY [(607)–756–0851].

Murphy, Bill. 1998. *Greener Pastures on Your Side of the Fence.* Arriba Publications, Colchester, VT. [(800) 639–4178].

Robinson, Jo, 2004. *Pasture Perfect.* Vashon Island Press, Vashon WA., [(866) 453–8489].

Smith, Burt. 1998. *Moving 'Em: A Guide to Low Stress Animal Handling.* Graziers Hui, Kamuela, Hawaii.

Undersander, Dan, Michael Casler, and Dennis Cosgrove. 1996. *Identifying Pasture Grasses.* Publication #A3637. Cooperative Extension Publications, Madison, WI. [Available at (603) 262–2655 and online at http://learningstore.uwex.edu/pdf/A3637.pdf].

Zartman, D. L. 1994. *Intensive Grazing Seasonal Dairying: The Mahoning County Dairy Program, 1987–1991.* OARDC Research Bulletin 1190, Wooster, OH: pp.1–49.

Magazines

Acres USA: The Voice of Eco-Agriculture. [(800) 355–5313; www.acresusa.com/magazines/magazine.htm].

Graze: by Graziers for Graziers. [(608) 455–3311, www.grazeonline.com].

Stockman Grass Farmer. [(800) 748–9808, www.stockmangrassfarmer.net].

ATTRA Publications

This organization has many excellent free publications on topics including: Grass-based and Seasonal Dairying, Financial Tips and Resources for Grass Farmers, Integrated Parasite Management for Livestock, Introduction to Paddock Design and Fencing Water Systems for Controlled Grazing, Matching Livestock and Forage Resources in Controlled Grazing, Meeting the Nutritional Needs of Livestock with Pasture, Nutrient Cycling in Pastures, Rotational Grazing, and Sustainable Pasture Management.
[http://www.attra.ncat.org; (800) 346–9140].

Websites

Behavioral Education for Human, Animal, Vegetation, & Ecosystem Management (www.behave.net). This website contains information on why animals choose to eat certain plants and not others.

Eatwild (www.eatwild.com). This website is written by Jo Robinson and contains information on the nutritional content of grass-fed milk and meat.

Grazinghandbook.com (http://www.chaps2000.com/grazing.asp).This website is a comprehensive online grazing handbook developed by Dr. Llewellyn L. Manske, Range Scientist at the North Dakota State University Dickinson Research Extension Center.

Holistic Management International (www.holisticmanagement.org) contains information on Alan Savory's method of holistic planned grazing.

Northeast Grazing Guide (www.umaine.edu/grazingguide). This website is part of the Northeast pasture consortium and links to several sources of grazing information.

Pasture Based Dairies in Missouri (http://agebb.missouri.edu/dairy/grazing) contains information on calculating overall farm cover or the "grazing wedge."

BRENT AND REGINA BEIDLER

SARAH FLACK AND BETHANY WALLIS

In the beautiful green hills just around the corner from Vermont Technical College sits Brent and Regina Beidler's organic dairy farm, which overlooks the White River Valley in Randolph Center, Vermont. The Beidlers purchased their first farm in 1998. They immediately began their organic transition and certified in March 2000 since the land was managed organically the year before they purchased it. At that time, Organic Valley had already begun picking up milk in their area, and they became the sixth farm added to the pickup route.

Brent, Erin and Regina Beidler. Photo by Carrie Branovan.

Numerous influences have shaped Brent's management style and decision to manage an organic dairy. When Brent was young, the well on his uncle's dairy farm was contaminated with nitrates from agricultural runoff from his own conventional farming practices and those of surrounding farms. Alarmed, he transitioned his herd to grazing to restore the land's health and eventually became an organic dairy. Brent was in his early twenties when his uncle transitioned to organic and became interested in these alternative methods. A pasture management course taught by University of Vermont Professor Bill Murphy also influenced Brent's interest in organic management. Brent and Regina had both worked on a number of conventional farms so they were familiar with conventional dairy management and its effect on herd health and farming lifestyle. They wanted something else for their own farm.

Brent and Regina began building their own herd while still managing another farmer's conventional dairy. They started with 15 heifers ready to freshen and bought 30 additional cows from the farm they purchased. Their current herd size of 40 animals has been a good fit for the 105 tillable acres on the farm, the majority of which is pasture. They grow a diversity of crops including summer annuals, an acre of wheat and camelina, 5 acres of BMR sorghum-sudan, and 5–10 acres of Japanese millet. Brent recently increased his organic land base with an additional 20 acres of forages. The Beidlers' feeding program consists of pasture, dry hay, and 6–10 pounds per cow (depending on pasture quality) of 10% protein grain during the summer months. In the winter, they increase to 12 pounds of 16% protein grain per cow. To determine their cows' ration, the Beidlers routinely test their forages and watch their animals' manure, body condition, and production levels. They supplement the diet with kelp and Redmond salt to keep the herd's diet balanced. Their nutritional goal is to maintain good

Beidler Farm's dairy herd grazing

herd health and avoid stressing their animals. Their biggest nutrition challenge is maintaining consistent high-quality feed.

Brent and Regina Beidler's herd grazes most of their 145 acres by the end of the summer. The permanent pastures are divided into 5-acre paddocks, which are strip grazed so the cows get fresh pasture after every milking. In spring, they begin grazing the herd on 15 acres, which they double to 30 acres in June. As plant growth rates slow, they continue to add land, so by the end of the summer they graze 100 or more acres.

Their fencing is high tensile perimeter with a mix of steel wire and polywire for the interior subdivisions. Some of the improvements they have made in the past several years include improved lanes, frost seeding, higher-quality grasses, and a piped water system so cows can graze longer each day. This has helped them improve pasture quality, access to paddocks, and has allowed them to reduce the amount of grain fed in the barn. Another change they have made is to begin their grazing season later in the spring, giving plants more time to grow before the first grazing.

In addition to permanent pasture and hay land they graze after the first and second cuts, they also grow a crop of sorghum- sudangrass or millet to graze in dry years when they are short on other pasture during the summer. Brent recently switched to millet because he can grow it with less nitrogen input and does not have to worry about the prussic acid problems associated with sorghum-sudangrass.

The Beidlers have this grazing advice for new farmers.

- Put water in every pasture.

- Spread compost on your pastures late in the season.
- Use some plowing and annual crops in your rotation if possible.

In addition to forage sampling, the Beidlers conduct soil sampling on all of their fields every three years. They are excited about the new Cornell University system for soil management and work with Heather Darby, UVM extension, to balance their soils. Currently, they do not apply many soil amendments, but instead rely on lime, manure, and compost to increase soil fertility. Brent and Regina have identified that their soils are low in nitrogen and are planting more legumes to address this nutrient issue. They are concerned about soil compaction from harvesting and plant Japanese millet and rye to loosen soils with their root structures. The Beidlers have a semisolid stored manure system and do some composting. Brent advises transitioning farmers to "expect to feed more to your cows and the soils and remember that you cannot just give your soils a shot of nutrients midseason like you may have done under conventional management." Organic dairying requires more holistic planning and an integrated management approach.

The Beidlers chose organic management to maintain good herd health. Brent admits, however, that it was difficult for him to change his mindset from getting more than 100 pounds of milk from a cow to wanting only 50 pounds of milk production. They initially struggled to maintain good herd health during their transition, which was attributed to taking bad advice from feed salesmen as well as getting acclimated to a new management system and establishing a good grazing program. At first they battled with metabolic issues, but now experience fewer herd health problems overall.

The herd still experiences occasional mastitis. They utilize herbs, whey products, and homemade tinctures to treat their herd. The Beidlers use California Mastitis Tests to identify affected quarters and occasionally self-test with Dairy Herd Improvement (DHI). Brent believes that knowledge is power when managing for low SCC counts and they have consistently achieved SCC counts between 150,000 and 200,000 over the last five years. Their milking procedures include pre- and post-dipping with an iodine product, using a clean dry towel on every animal, and changing inflations every eight weeks. Brent does not shy from three-teated cows; he dries off quarters if they have a persistent problem. They view the dry period as a time of rest for the animal and an opportunity to clear up infections and build up their natural immunity. When drying off a cow, they wait to milk her out until five days have passed, even if she drips a little. When they have questions about herd health, the Beidlers rely on other farmers, the Organic Valley regional newsletter, and their vet. They believe that overall, their animals are more robust and resilient under organic management.

Financial viability as a whole has been the Beidlers' biggest accomplishment. Brent states, "Coming to the organic milk industry has been empowering for us because instead of producing something that we as a society have too much of, our organic milk is valued and consumers are interested in and happy with the way we produce it. Our processors value milk quality and they compensate us for it. For the same milk quality that would get $0.35 in conventional quality premiums, we receive $2.00 from the organic processors, so we feel our hard work is recognized." Their community and other farmers in their information network energize them, something they did not feel in the conventional market where many farmers get their information from sales representatives.

Regina believes there is dignity in the organic market as well, "We have some control over our own market and pay price; people in general are profitable enough to be doing well and pay down their debts." Brent and Regina state, "We don't feel like we are a needle at the bottom of the haystack or that the only option we have to become financially sustainable is to become more efficient, which is confining because it's never enough. Under many conventional systems, the only way to improve is to get more out of your cows, do it cheaper, or get bigger. Organic allows us to be happy with our cows the way they are."

The Beidlers' advice for transitioning and newly transitioned farmers is to be patient. There is an expectation that you will immediately become a well-managed organic system and things are going to be terrific, but it often takes a while to acclimate to a new management system—both for the farmers and for the cows! Brent believes that many of the organic certification requirements are good for your farm—even if you are conventional. He sees the organic grazing, recordkeeping, and nutrient management requirements mirroring the direction consumers would like dairies to go.

EARL FOURNIER

SARAH FLACK AND BETHANY WALLIS

When Earl Fournier and his family began looking at organic management, they were the quintessential small conventional dairy farm. They milked around 150 cows, used *bovine somatotropin* (bST), fed up to 30 pounds of grain a day, and averaged 27,000 pounds of milk per cow. The Fourniers were well respected in the Vermont farming community, so when Earl transitioned to organic management, the local farmers started paying attention to organics.

Earl's son, David, was interested in continuing the dairy, and Earl knew the farm needed to change to remain profitable for the next generation. He began looking for an environmentally and financially sustainable method for carrying on the family farm. After working with a planning group and running input spreadsheets on production and grain costs, Earl and David found that organic marketing would provide this higher net farm profit, even if their annual production fell as low as 13,000 pounds per cow. Since the Fournier farm already maintained a low debt-to-asset ratio, the farm was a safe candidate for the financial investment needed to transition to organic, and they presented their proposal to their lending agency with little resistance.

David and Earl Fournier.

The Fourniers began their transition in December 2003, and Earl admits that it was a "steep, steep learning curve." Becoming accustomed to less milk production and getting the cows to graze were his biggest challenges. Earl and David now intensively graze 80 milking cows on approximately 100 acres. Together they manage the farm full-time and employ three part-time workers and Earl's youngest son, William. They store forages, haylage, baleage, and dry hay from an additional 165 acres (both owned and rented).

The Fourniers feed a summer ration of dry hay, baleage, and 14 pounds of grain. Earl would like to move away from feeding this partial total mixed ration (TMR) and replace it by feeding grain and dry hay separately in the summer. During the winter months, the herd receives haylage, dry hay, and up to 15 pounds of grain per day. They supplement the feed year-round with trace minerals and 2 ounces of kelp. Earl's forages typically produce good protein levels, between 18 and 22 percent. He believes the TMR should contain only around 16.5% protein, unless you are pushing for high milk production.

Earl developed his grazing system on some of his highest-quality cropland when he converted from a conventional confinement dairy to a grass-based organic system. With the help of NRCS cost sharing programs, he was able to invest in grazing infrastructure, which includes gravel lanes, high tensile fences, and water piped to every paddock. His pastureland is a mix of perennial ryegrass, orchard grass, Kentucky bluegrass, festolium, red clover, and white clover. He grazes his hay fields and seeds them with a mix of brome, timothy, meadow fescue, orchard grass, alfalfa, and red clover. He has also tried grazing some brown midrib (BMR) sorghum-sudangrass.

In the spring, Earl grazes his herd on 35 acres of pasture and increases that acreage to 80 acres of pasture by the end of the grazing season for the milking group, plus another 20 acres for heifers and dry cows. He also sends 25 heifers to a certified organic custom-grazing farm during the summer, and keeps his dry cows and some other heifers in a separate grazing rotation near the barn.

The pasture is fenced into 1.2-acre paddocks that Earl subdivides with polywire depending on herd size and amount of feed in the pasture. His strip grazing method allows him to give the cows fresh pasture three times a day, encouraging good grazing.

The Fourniers turn cows into the pasture when the forage is about 10 inches or taller. They clip the pastures after grazing several times each year to reduce rejected forage and mature grass. Earl calculated that he spends half the time and half the cost feeding the cows in the summer compared to the winter, and he figures it is easier to make more money on grass even if you are making less milk. He also feels it is better for the cows.

Earl's advice to new graziers is to "use your prime meadows for grazing." A common mistake transitioning farmers make is to think they should pasture the land they cannot use for something else. He remembers his first year of grazing as frustrating because the cows were not making as much milk, but now his cows graze well and he is happy with the milk production level. If Earl were to change the way he managed his transition, he would have immediately cut the haylage out of the TMR to help the cows start grazing more efficiently in the spring. Once he switched from haylage to dry hay and baleage, the system worked much better.

Earl's approach to soil management has also changed under organic production, and he now aims not to leave any bare land. The Fourniers heavily incorporate cover crops into their crop rotation and plant them in the spring and summer. Earl examines his soil's mineral balances and applies lime, sul-po-mag, boron, and zinc when needed. He suggests, "Spreading manure lightly allows the plants to better absorb the nutrients." He currently spreads a light application of manure from his liquid manure lagoon three times a year.

Under conventional management, Earl's herd suffered from lameness and breeding difficulty. A typical milking cow would last only

Dairy herd grazing at Fournier Farm.

2–2.5 lactations due to high cull rates. Herd health has improved, but Earl has found that "even now, cows are genetically bred for high production, not to maintain themselves." The Fourniers crossbreed to improve their herd health through hybrid vigor. They use DHIA to monitor mastitis and identify problem animals for culling. Earl uses herbal therapies to treat herd health problems, and believes that although alternative treatments take more time and effort, they are well worth it. He would still like to see more support from the local veterinarians, but acknowledges that their involvement in and knowledge of organics has improved. Some herd health resources the Fourniers rely on are publications by Hubert Karreman and Paul Dettloff, NOFA Vermont meetings, the Northeast Organic Producers Association (NODPA) newsletter, and other farmers.

Earl's advice to farmers considering transitioning to organic dairy farming is that "when it's all said and done, organic management and sustainable agriculture can be done. The management level of a farm determines how successful the farm will be, but the practices do work and you can make a living." He encourages producers considering organic to secure a market prior to transitioning. Earl also recommends "looking at your crop needs, because they are very important. Depending on your forage needs, dry matter intake from forages may increase under organic management due to feeding less grain. Additional forages may be difficult and expensive to come by and quality may be poor. If you are in a situation where you barely have enough feed now, you most likely will need additional feed under organic management."

After transitioning to organic, Earl says, "The future looks brighter than before." He feels confident that transitioning to organic dairy will provide an environmentally and economically sustainable future for his farm. The consistent organic pay price has been good for his financial planning, and he feels less stressed knowing what his income and pay price will be at the beginning of the year. Earl admits, "Now I feel I have some control, and the responsibility for success is more on me than on the government or anyone else."

CHAPTER FIVE

ORGANIC CROP MANAGEMENT

ELIZABETH DYCK

The intent of this chapter is to provide information to transitioning and certified organic dairy farmers on various cropping options and their management. While pasture is the key crop on organic dairy farms, annual grain and forage crops as well as cover crops can also provide important benefits for some operations. This chapter includes guidelines for developing a crop plan that will satisfy organic feed requirements as well as strategies to avoid crop yield loss during the transition years. Crop rotation design and the management of specific crops are also covered. Finally, practical approaches to managing pests organically, including weeds in cropland and pasture, are discussed.

KEY POINTS FOR TRANSITIONING FARMERS

- Read and understand NOP rules for organic crop management (sections §205.202–§205.206 and §205.600–§205.602).
- Develop an organic crop plan that includes pasture and feed crops as needed.
- Consider options to reduce the risk of yield slumps during transition.
- Develop a sound crop rotation.
- Become familiar with the diverse array of grain, forage, and cover crops that can be used in the rotation.
- Gain knowledge of organic weed and pest management concepts and practices.

GROWING CROPS OTHER THAN PERMANENT PASTURE

The benefits of pasture-based dairy systems are rightly emphasized in nearly every chapter in this handbook. Well-managed pastures provide high-quality, lower-cost feed, improved animal health and welfare, and excellent protection and maintenance of the soil. However, crops other than pasture and hay can play useful roles on organic dairy farms, including:

- grain crops to reduce or eliminate the need for purchased feed,
- annual forage crops to supplement feed during slumps in pasture production, and
- cash crops to provide supplementary income to the farm.

However, growing crops other than pasture does come with costs, including:

- intensified farm management to deal with increased crop diversity,
- the need for specialized equipment and storage facilities,

- labor bottlenecks when cropping activities conflict with livestock and pasture management needs, and
- an increased risk of soil erosion and degradation on land that is not under permanent cover.

So what is the best approach for cropping in an organic dairy system? The case studies of successful organic dairy operations in this handbook show that while these farms are all pasture based, they vary from zero-grain systems to those with extensive acreage of grain and other crops. Therefore, the answer to the above question is: it depends. The approach to cropping depends on a number of factors, including farmer experience, the type and amount of land available, feed costs, milk prices, and the availability of labor and equipment. Because many of these factors change over time, a farm's approach to cropping is likely to change as well. As farmers transition and gain more experience with management intensive grazing (MIG), they may convert much of their tilled cropland to pasture. On the other hand, experienced organic dairy farmers may bring some pasture back into annual crop production in response to increasing costs for purchased grain or to take advantage of a developing market for a cash crop.

MEETING FEED NEEDS THROUGH AN ORGANIC CROP PLAN

Planning for all aspects of the transition is critical, but the experience of transitioning farmers tell us that developing an organic crop plan for the transition and beyond is crucial, for the following reasons.

- During the transition year, 100% of the feed used for the herd must be organically grown, either produced on the farm itself on land that is certified or in the third year of transition, or purchased as certified organic feed.
- Purchased certified organic feed is expensive and often hard to source.
- There is a risk of lower crop productivity during the transition as farmers shift from conventional to organic methods.

Therefore, if farmers have not planned ahead on how to meet their organic feed needs, transitioning or newly certified organic dairy farms may be hit with unexpected feed shortages and no choice but to make emergency purchases of expensive organic feed.

COMPONENTS OF THE ORGANIC CROP PLAN

Pasture

As the chapters on pasture management and animal nutrition in this handbook detail, successful organic dairy farming is built around maximizing the use of pasture. Developing the pasture component of the crop plan is thus the first priority of transitioning farmers. Worksheet 4.1 can be used to determine the number of acres needed for high-quality pasture production. Once the amount of land needed for pasture is determined, a next step is to assess other feed crop needs and the farm's capacity to grow or source them.

Feed crops

Farmers who under conventional management have been purchasing some or all of their feed can use Worksheet 5.1 to estimate the cost of continuing this strategy under organic management. However, for many farmers, the option of purchasing certified organic feed—the cost of which is likely to remain substantially higher than conventional—is simply not affordable. It is highly recommended that farms become as self-sufficient as possible in supplying their feed needs and devise sustainable feeding strategies.

Farmers who want to grow their own feed need to determine if sufficient and suitable land is available on the farm. If there is additional acreage available beyond what is needed for high-

WORKSHEET 5.1 ASSESSING THE COST OF PURCHASED ORGANIC FEED

A. CALCULATING FEED AMOUNTS:
(If you know the amount of each purchased feed you use or will need per year, skip to Step B.)

FEED 1: ____________________

Pounds fed/cow/day _______ × # cows _______ × 365 = _______ lb. grain/year

_______ lb. grain/year ÷ 2,000 lb. = _______ tons/year

FEED 2: ____________________

Pounds fed/cow/day _______ × # cows _______ × 365 = _______ lb. grain/year

_______ lb. grain/year ÷ 2,000 lb. = _______ tons/year

FEED 3: ____________________

Pounds fed/cow/day _______ × # cows _______ × 365 = _______ lb. grain/year

_______ lb. grain/year ÷ 2,000 lb. = _______ tons/year

B. CALCULATING ORGANIC FEED COST:
Use the table below to estimate your feed costs. To get current organic prices of feed, check with a supplier of organic feed (see Appendix D) or check prices on the New Farm Organic Price Report (www.newfarm.org/opx).

ORGANIC FEED	PRICE/TON	ORGANIC FEED	PRICE/TON
48% soybean meal		Corn meal	
Roasted soybeans		Oats	
Sunflower meal		Wheat	
Field peas		Barley	

FEED 1: __________ TONS/YEAR × __________ PRICE/TON = $ __________

FEED 2: __________ TONS/YEAR × __________ PRICE/TON = $ __________

FEED 3: __________ TONS/YEAR × __________ PRICE/TON = $ __________

TOTAL/YEAR = $ __________

quality pasture production, this land needs to be assessed for crop production.

Assessing farm fields

An important step in developing an organic crop plan is to assess each farm field to determine its best use under organic management and optimize its productivity. This also helps to plan the buffer zones needed for organic certification. Buffer zones are areas (e.g., fallow land, windbreaks, roads, ditches, etc.) that separate land in transition or certified land from adjacent noncertified land. Transitioning farmers need to check with their certifiers for the specific requirements for buffer zones.

Worksheet 5.2 can help determine what cropping options are best for the land. For example, a fertile, well-drained field may be suitable for either pasture or annual field crops, but easy accessibility to the milking facility may make it more useful as pasture. A field bordered by a neighbor's land on which GMO crops are planted would not be a good fit for growing corn because of increased chance of contamination through cross pollination. A poorly drained field is likely better suited for hay than pasture (to avoid serious soil compaction). In some cases, although total farm acreage appeared adequate, field assessments may show that there is not enough suitable land available for feed crops given pasture needs.

Other feed crop options

If cropland is not available on the farm, an option is to rent land that is or can be certified. When renting land, farmers need to follow NOP guidelines in documenting land use and inputs applied. A word of caution: land may be available that has not had prohibited inputs for 36 or more months and in some cases may therefore be immediately certifiable—but this does not mean that it is ready to grow organic crops! Abandoned or fallow land often has marginal fertility and can harbor huge weed-seed banks. Assessing the condition of land (see Worksheet 5.2) before renting it and attempting organic crop production can avoid costly cropping disasters.

If land is not available to rent for growing feed crops, farmers may be able to contract with a neighboring organic farm to grow feed. This option allows farmers to customize purchased feed to fit the needs of their operations and save on transport cost. Another option is to alter the feed ration. Replacing higher-priced grains like soybean with lower-priced alternatives like small grain/pea mixtures could make purchased feed more affordable. Alternatively, it may be possible to feed less grain. This may require rethinking milk production goals, or it may also be possible to compensate by feeding higher-quality forages. Before altering feeding strategy, farmers should read Chapter 6, "Managing Dairy Nutrition" and consult a nutritionist.

Avoiding yield slumps during the transition

Some farmers see significant yield declines in field crops during transition while others see very little change in yield. There are a

WORKSHEET 5.2 ASSESSING FARM FIELDS

Complete this exercise for each existing field on the farm.

Field ______________________ Owned or rented _______ Acres _______ Shape _______

Soil type(s) ____________ Slope and orientation _____________ Drainage _____________

Cropping history	YEAR	CROP GROWN	AMENDMENT(S) APPLIED
	______	______________	______________________________
	______	______________	______________________________
	______	______________	______________________________

Overall fertility based on past performance: ____

Date of last soil test ________ Summary of soil test results: ___________________________

____

Weed pressure _____________ Close enough to milking facility to allow grazing? _________

Distance from manure storage ____________ Adequate access for farm equipment?________

Danger of GMO contamination, spray drift, or runoff from neighboring adjacent field? ______

Buffer zones needed in ft^2: (calculate as certifier mandated buffer × length of field border)

N ______________ S _______________ E _______________ W _______________

Acreage remaining for crop production ________________ (Total acreage – [N + S + E + W])

Crop(s) to be grown: ____

____

____

number of strategies that transitioning growers can use to reduce the risk of yield loss.

- *Transition the farm's cropland first,* followed by the dairy herd. This strategy is especially important if the farmer is new to MIG and/or growing feed crops.
- *Transition cropland gradually*, on a field-by-field basis, rather than all at once. This approach requires additional recordkeeping to document that organic and nonorganic crops and agricultural products are kept separate on the farm, but it also allows farmers to gain experience with organic cropping techniques gradually and to get fields under sound organic soil and weed management individually or a few at a time.

- *Use a low-risk crop sequence during transition.* Corn (which requires high fertility and intensive cultivation for weeds) is actually the riskiest crop to grow during the transition, while hay/sod crops pose the least risk. Soybeans and small grains pose an intermediate risk. A strategy suggested by longtime organic growers is to transition a field by putting it into a hay/sod crop for several years before attempting an annual crop like corn. Two or more years of a hay/sod crop accumulates N in the soil, adds organic matter, stimulates soil biological activity, and suppresses many weeds (see also the sections on crop rotation and weed management below).

Figure 5.1 illustrates the results of a 3-year experiment contrasting different crop sequences during organic transition. Yield of corn following two years of a sod crop (oats/alfalfa-alfalfa) is nearly equivalent to the conventional corn yield. However, corn yield after a crop sequence of wheat/red clover-soybean shows a yield loss of 42% compared to conventionally grown corn. There are two primary factors for this yield loss: (1) reduced fertility given that only one year of a forage legume is grown and (2) increased weed pressure because of two row crops grown back-to-back.

Cash crops
An optional third component of the organic crop plan includes food and feed crops to sell. Many successful organic farmers spread the risk of market slumps or weather-related disasters by engaging in a diversity of farm enterprises. During the transition, the focus of the organic crop production plan should be pasture and other feed crops for the farm's herd. Once farmers have gained experience with organic pasture and crop management, however, growing certified organic cash crops may be a profitable option. This may especially be true for farmers who have grown specialty cash crops conventionally. Paul and Maureen Knapp's family dairy farm in up-

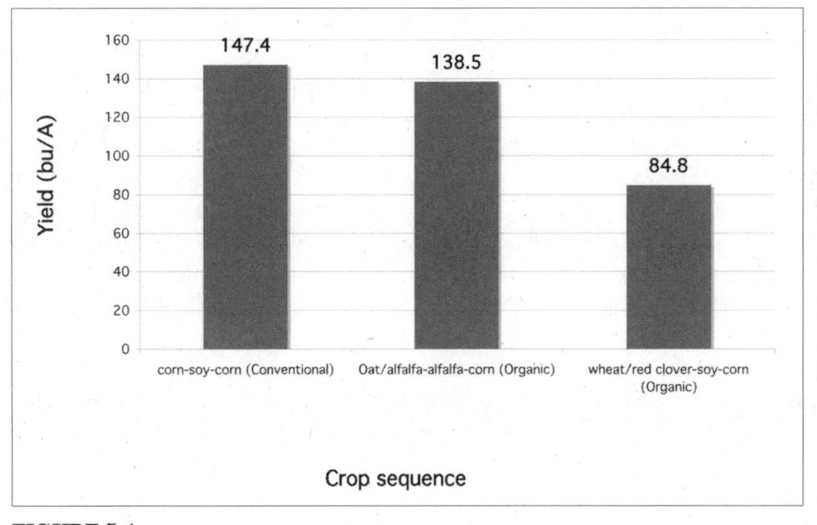

FIGURE 5.1
Corn yield during transition

state New York, for example, had a long-standing pick-your-own strawberry enterprise. When they transitioned to organic dairy, they successfully transitioned their strawberry production as well. (See the Knapp case study in Chapter 3 for more details.) Before including a cash crop in the organic crop production plan, farmers need to first research the market for the crop, how it is grown organically, and what records must be maintained for certification.

DEVELOPING A SOUND CROP ROTATION

On an organic farm, crops cannot be grown successfully without a sound crop rotation. This is because rotation is a major tool for both maintaining soil productivity and controlling weeds, disease, and insect pests.

Crop rotation also reduces the risk of weather or market-related disasters by maintaining a diversity of crops on the farm. For these reasons, the National Organic Program (NOP) standards *require* a sound crop rotation for farms seeking organic certification (see Box 5.1).

There is no "one size fits all" rotation for organic farms. Each farm has unique circumstances and requirements, and the rotation can be adapted to meet them. It is also rare to find a farm that has not changed its rotation over time to adapt to changes in production costs and markets. That being said, there are general principles for crop rotation that need to be followed to ensure that the rotation works to sustain soil productivity and fights against weed buildup, disease, and insect pests.

Table 5.1 lists the general requirements that a crop rotation should meet to be sustainable. At first glance, this may look like a bewildering number of criteria that no one rotation could possibly fill. The good news is that including certain crop types in the rotation—specifically, forage legumes and small grains/grasses—satisfies several requirements at once. Cover crops can also supply multiple benefits. Even row crops, such as corn and soybeans, which have net negative effects on soil quality because of wide row spacing and intensive cultivation requirements, can also work to break up both weed and pest cycles (See Organic Pest Management on page 119).

BOX 5.1 NOP CROP ROTATION RULE

NOP crop rotation practice standard (§205.205)

The producer must implement a crop rotation including but not limited to sod, cover crops, green manure crops, and catch crops that provide the following functions that are applicable to the operation:

a. maintain or improve soil organic matter content,
b. provide for pest management in annual and perennial crops,
c. manage deficient or excess plant nutrients, and
d. provide erosion control.

EXAMPLES OF CROP ROTATIONS

In addition to keeping basic principles in mind, it is useful to look at examples of crop rotations, including those used extensively in the past. Examples of historical crop rotations used on dairy farms in the Northeast include:

corn—oats—hay
corn—oats—winter wheat—hay.

In both of these rotations, a legume (often mixed with timothy) was seeded with or into a small grain. The hay or sod phase of the rotation lasted one to several years. Note that the hay, a soil-fertility-enhancing crop, was followed by corn, a heavy feeder.

TABLE 5.1 PRINCIPLES FOR BUILDING A SOUND ROTATION

PRINCIPLE	EXAMPLES/BENEFITS
Include a nitrogen-fixing forage legume sod crop.	A dense, healthy stand of legumes like alfalfa and red clover can • provide high-quality hay • supply 150 lbs. or more N per acre to a following crop • add organic matter to the soil • improve soil structure • suppress weeds and some insect pests and diseases.
Strive for crop diversity.	• Alternate grass and broadleaf crops (which tend to have different diseases and insect pests) for a baseline defense against pest and disease buildup. • Include "break" crops, that is, crops that are disrupters of disease and pest lifecycles for many other crops, e.g., buckwheat, oats, and sorghum-sudan. • Plant intercrops to avoid crop failure to pest/disease attack or even poor growing conditions (e.g., plant a mixture of forage such as red clover, alfalfa, and sweetclover, rather than one of these legumes by itself). • Include crops with different planting dates, including early spring, late spring, summer, and fall, to – disrupt weed lifecycles: e.g., use a fall-planted crop to suppress spring-emerging weeds, – provide protection for the soil over winter months and scavenge nutrients in fall and early spring, and – reduce labor bottlenecks by spreading crop management tasks throughout the growing season.
Include crops that contribute large amounts of shoot and root residue.	• Grasses, including the small grains, not only leave behind large amounts of straw or aboveground residue but also dense fibrous root systems, thereby making a major contribution to soil organic matter. • Forage legumes continuously slough off leaf, stem, and root material.
Include deep-rooted crops.	Crops that develop deep taproots, e.g., alfalfa and sweetclover, can • break up compacted layers of soil and • take up nutrients from deep in the soil profile.
Alternate crops that have different methods of weed control.	Weeds thrive when the same type of crop is grown over and over in the rotation. When a mix of row crops (e.g., corn), smother crops (e.g., oats or buckwheat), and sod crops (alfalfa or red clover) are grown, weeds cannot build up because they are subjected to a wide variety of control measures.
Include drought-tolerant crops.	By including crops in the rotation that require less water (e.g., barley, sorghum, or sorghum-sudan) the risk of crop loss to drought can be reduced.
Include as many cover crops as possible.	Cover crops, such as buckwheat, oats, rye, sorghum-sudan, Japanese millet, hairy vetch, clovers (planted alone or as an intercrop) can • protect the soil between main crops • add N (if a legume) • add organic matter • suppress weeds and reduce disease/pest incidence.

An example of a modern organic dairy farm rotation is that of Kevin Engelbert, Engelbert Farms, in the southern tier of New York. His rotation is: corn—soybean—corn—hay. The hay in this case is a mix of alfalfa, red clover, and orchardgrass, which is grown for 3–6 years before being plowed down. Kevin Engelbert uses this mix to optimize stand establishment and forage production in the face of variable growing conditions: orchardgrass can survive and fill in under wet conditions while alfalfa is tolerant of drought. Red clover is included because his cows love it, it is more tolerant of lower pH and fertility than alfalfa, and it can be frost seeded every couple of years to keep up stand quality and density. The three years of row cropping is made possible by the well-managed multiyear hay crop—and his use of multiple weed control techniques (see the case study at the end of this chapter on page 131).

None of these three rotations follow all of the principles listed in Table 5.1. Crop rotations have to be economically feasible. But all three do rely on the basic rotation principles of crop diversity and the inclusion of a hay or sod crop that are key to maintaining good soil quality and managing weeds and pests.

Worksheet 5.3 is designed to help farmers plan an environmentally and economically sound organic crop rotation by incorporating the principles in Table 5.1.

CROP OPTIONS AND THEIR MANAGEMENT

Because of the requirement for a diverse crop rotation and the advantages of growing their own grain and forage crops, transitioning and newly certified organic dairy farmers are likely to find themselves growing at least one or more new crops. It is critical to investigate the growing requirements of these new crops and review the basic requirements of familiar crops. The crop's preferred temperature and soil conditions are especially important factors to consider because organic growers do not use such inputs as seed coatings, synthetic fertilizers, pesticides, and herbicides that can mask poor planting and growing conditions. To establish and maintain healthy crop stands, planting and other management operations need to occur at the optimal time for the crop and the soil.

Timing of harvest is especially critical to produce high-quality forage, which can help reduce or eliminate the need for expensive purchased organic grain (see the section Different Feeding Systems and Methods on page 144.). For first-cut hay, harvesting when grass is in the late boot to early head stage should provide a good yield of high-quality hay. For highest quality, later harvests should be made when the legume is in the late bud stage.

It is also important to give cover crops, which have critical roles in managing soil, weeds, and pests, the same careful attention as feed or cash crops. Cover crops need to be managed to establish a dense, uniform stand that grows vigorously until crop kill or die back.

WORKSHEET 5.3
DEVELOPING A SUSTAINABLE ORGANIC CROP ROTATION

Indicate below your current crop rotation. Fill in only as many blanks as there are years in your current rotation.

________	________	________	________	________	________
YEAR 1	YEAR 2	YEAR 3	YEAR 4	YEAR 5	YEAR 6

Indicate below how you will change your current rotation to meet organic standards, meet your livestock feed needs, and develop a sustainable organic rotation. Fill in only as many columns as there are years in your planned rotation. (There is no set standard for the length of a sound rotation. However, rotations of fewer than 3 years are not likely to be sustainable under organic management). Refer to Tables 5.1 and 5.2 for guidelines and crop possibilities.

	YEAR 1	YEAR 2	YEAR 3	YEAR 4	YEAR 5	YEAR 6
SPRING						
SUMMER						
FALL/WINTER						
SCORE						

To evaluate the sustainability of your rotation, assign points to each crop as follows:

1 point for a legume sod or hay crop (score 1 point for every year the hay crop is grown)
1 point for every small grain crop grown until maturity
1 point for every legume intercrop within a small grain
0.5 point for every short-duration cover crop (e.g., buckwheat, winter rye, sorghum-sudan)
–1 point for every row crop
–0.5 point for each season when the soil is bare in the rotation

Example: Evaluate the historical 4-year corn—oat—winter wheat—hay rotation. In year 1, the score would be –1.5 (row crop plus bare soil in winter); in year 2, +1 (small grain); in year 3, +2 (small grain and legume interseed); in year 4, +1. Total score: +2.5

Total score: If your score is 1 or more, you are likely developing a sound organic crop rotation. If the score is less than 1, work to enhance the rotation with more cover crops, small grains, legume, and other forage crops.

A critical aspect of growing organic crops is sourcing organic seed. The National Organic Program standards require the use of organically certified seed unless such seed is not commercially available for the crop or variety to be grown (NOP Rule §205.204). Growers must document their attempts to source organic seed and should check with the certifying agency for specific documentation requirements. If organic seed is not available, untreated seed can be used if it is not genetically modified. Transitioning farmers should get into the habit of requesting a letter from the seed company stating that the seed is nonGMO *before* buying the seed.

Table 5.2 provides information on the organic management of specific crops, including grain, forage, pasture, and cover crop species.

TABLE 5.2 CHARACTERISTICS AND MANAGEMENT OF SPECIFIC CROPS

CROP USAGE CODES: GZ: GRAZE GR: GRAIN F: FORAGE CC: COVER CROP

BRASSICAS, FORAGE GZ, F

Highly digestible, high-yielding, fast-growing crops that can provide forage in summer and into late fall. Crude protein levels range from 15–25% in tops, 8–15% in roots. Can be rotationally grazed or harvested for green chop or silage. Should not comprise more than 2/3 of forage intake due to low fiber content and sulfur compounds that can cause disorders such as anemia and micronutrient deficiencies. Supplement with dry hay or grass pasture and/or seed with small grains such as oats (at 1–2 bu./A). Introduce grazing herd to brassica pastures over 3–4 days. Strip grazing is recommended to avoid trampling. Avoid feeding brassicas for 2 hours before milking to avoid off-flavors. Must be managed to avoid nitrate poisoning: (1) avoid building up high levels of N in the soil; (2) do not graze or harvest during or immediately after stress conditions (e.g., prolonged drought and high temperatures, periods of cool, cloudy weather, hail, and frost) as these conditions can raise nitrate levels in forage plants.

Requirements: Well-drained soils, pH 5.3–6.8, moderate to high fertility. Not drought tolerant; need good rainfall to establish.

Planting: Plant in spring for grazing/harvest in Aug.–Sept. Plant July to mid-Aug. for grazing/harvest into late fall. Seed turnips or swedes at 1.5–3 lb./A. Seed kale, rape, or Tyfon at 3.5–4 lb./A. Seed no deeper than ½". Drill or broadcast followed by cultipacking.

BUCKWHEAT CC, GR

Quick emergence and rapid growth shade out both annual weed seedlings and tough perennials, including quackgrass and Canada thistle. Grow back-to-back cover crops of buckwheat to deal with severe weed problems. Can be used as break crop for diseases of both grass and broadleaf crops. Excellent honey crop; attracts beneficial insects. Requires 10–12 weeks if grown for grain. Grain has a lower feeding value than small grains or corn, but can be used as a partial substitute. Use as no more than 30% of the feed ration to avoid sunlight-sensitivity in light-skinned animals.

Requirements: Grows in a wide variety of well-drained soils, but does not tolerate "wet feet;" pH 5.4–6.5. Light feeder; high levels of N will cause lodging. Frost sensitive; will germinate at 45°F, but prefers soil temps above 55°F.

Planting: Because high temperature impairs flowering, for grain production plant after last frost date in late spring/early summer or plant in midsummer (late June to mid-July). A cover crop can be planted any time after last spring frost to four weeks before fall frost. Plant 40–60 lb./A for grain; 60–100 lb./A for cover crop. Use higher number if broadcasting rather than drilling. Plant 0.5"–1.5" deep.

CHICORY, FORAGE GZ

A perennial forage whose leaves have greater digestibility and higher mineral content than alfalfa. Produces abundant forage in spring and summer if well managed. A deep taproot makes it drought tolerant and can reduce compaction in pasture soils. Chicory can persist from 5–7 years, but is more likely to require reseeding after several years. After producing lush leaf growth in spring, flower stems emerge that must be checked for good continued productivity. To prevent bolting, do not allow stems to exceed a pregrazing height of 6" and graze down to 1.5"–2" stubble.

Requirements: Prefers moderately to well-drained soils; pH 5.5 and above; requires moderate to high fertility.

Planting: Can be frost seeded in late winter into existing pasture. If preparing a new planting of chicory alone or intercropped with a legume, seed in spring. An early August planting is also possible, but chicory is less likely to survive winter if planted at later dates. If broadcasting, cultipack before and after seeding. Plant at 3–4 lb./A (with or without legumes), 0.25"–0.5" deep.

CORN GR, F

Traditional mainstay of rotations on both field crop and dairy farms. Organically grown seed of both hybrid and open pollinated (OP) corn is available. Hybrids are higher yielding by 15–20%, while OP varieties generally have higher protein content, can be saved for seed, and can be adapted to local growing conditions. Unadvisable to grow corn during transition unless it is preceded by a legume sod/hay crop. Corn requires early season tine-weeding or rotary hoeing plus inter-row cultivation. Heavy nutrient uptake, required cultivation, and wide row spacing that favors erosion result in negative effects on soil quality. Corn must be grown in rotation with soil-building crops. Can be grazed, but there are many other forage options that are better in terms of diversifying the crop rotation and for soil management.

Requirements: Grows best on well-drained, loamy soil; pH 5.6–7.5. Requires heavy fertility, especially N: a crop that produces 100 bu. of grain and 3.3 tons of stover/A is estimated to remove 153 lb. of N, 28 lb. of P, and 108 lb. of K from the soil.

Planting: To reduce rot and insect attack of seed as well as weed pressure, plant after the soil temperature remains over 50–55°F for several days—often means waiting 2 or more weeks after the start of conventional corn planting. Choose corn with a relative maturity rating that allows for the later planting date needed. Farmers' planting rates for hybrids vary from 24,000–30,000 seeds/A. For OP varieties, 18,000–22,000 seeds/A. Seed 1.5"–2.5" deep.

FIELD/FORAGE PEA GR, F, GZ

A cool-season legume whose seed contains from 21–25% protein. Can be fed directly to livestock without processing. An excellent intercrop with oats and/or barley. Makes excellent hay or silage and can be grazed. If grazing, plant with a small grain to keep peas upright and provide a more balanced forage. Strip grazing is recommended. Begin grazing when pods are forming and introduce animals gradually. A multipurpose use for field pea is to seed it in spring along with oats and alfalfa. The oats and peas can be harvested off for forage in early to midsummer, leaving the alfalfa to establish. For a full season green manure that both smothers weeds and accumulates N and organic matter, spring-seed peas with oats and hairy vetch.

Requirements: Adapted to cool temperatures and well-drained clay or clay-loam soils; shallow root systems make them susceptible to drought on lighter soils; pH of 6.5–7.5; moderate fertility. Use a pea-vetch inoculant before planting.

Planting: Plant in early spring for grain or forage. Peas can also be seeded with a small grain as a fall cover crop. Recommended seeding rates for pea sole crops vary widely, from 70 to more than 200 lb./A. Sow small-seeded types at the lower end of the range and large-seeded types more heavily. With a small grain, sow 25–50 lb./A. Seed 1"–3" deep: it is important that the seed contact moisture. Seeds left on the surface in spring or fall are unlikely to establish, although frost seeding into small grains is possible.

FORAGE GRASSES, COOL SEASON GZ, F

Along with forage legumes, cool season perennial grasses are key components of pasture in temperate areas. They typically follow a seasonal pattern of growth: high productivity in spring/early summer, slowed growth in midseason, and a recovery in fall.

Requirements: While often grown on marginal land, these grasses are most productive on moderately to well-drained, loamy soils with a pH of 6–7. Specific grasses may be more sensitive to or tolerant of marginal growing conditions (see descriptions below).

Planting: Plant in early spring or late summer (before the end of August in southern Pennsylvania, earlier further north). Prepare a firm seedbed and seed 0.25–0.5" deep. See also individual descriptions below.

KENTUCKY BLUEGRASS

A sod-forming, low-growing grass that is winter hardy and well adapted to grazing. Not tolerant of drought or high summer temperatures. Highly compatible with white, red, and alsike clovers and birds-

Continued on next page »

Continued ≫

foot trefoil. Can be frost seeded into existing pastures to increase stand density. If planting as a new pasture, best to seed in late summer at 10–14 lb./A (4–8 lb./A with legumes), no deeper than 0.25".

ORCHARDGRASS

A tall-growing bunchgrass that can be used for pasture, hay, green chop, or silage. Moderately tolerant of droughty, wet, and acidic soils and summer heat; moderately winter hardy. Will persist in pastures if a tall pregrazing height is used and if grazed to leave 3"–4" stubble. Although more competitive with a legume intercrop than timothy or Kentucky bluegrass, works well with red clover, and alfalfa if managed to avoid shading out the legume. Seed at 8–12 lb./A; 2–6 lb./A when seeding in combination with a legume. For hay, best quality and yield is obtained by harvesting in the boot stage.

REED CANARYGRASS

A tall, sod-forming grass that can be used for pasture, hay, and silage. Well-adapted to poorly drained soils, but also tolerant to drought and to acidic and alkaline soil conditions; winter hardy. More productive than other cool season grasses during the "summer slump." Palatability can be lower than that of other cool season grasses. Use low alkaloid cultivars and prevent from becoming overly mature by maintaining the grass below 12" in spring and early summer. Harvest before heading to get best-quality hay in the first crop. Plant at 14 lb./A (6–8 lb./A in mixture with a legume).

RYEGRASS, PERENNIAL

A bunch-type grass of early to medium maturity that makes a high-yielding and nutritious forage. Not as winter hardy as other cool season grasses and not drought tolerant. Can be frost seeded. Plant at 15–20 lb./A alone, 4–8 lb./A in a mixture. Diploid (doubled chromosome) and tetraploid (quadrupled chromosome) types are available. Diploid types tend to exhibit greater growth and persistence after establishment. Tetraploid types have larger leaves, more open growth (better suited for legume intercrops), and higher digestibility.

SMOOTH BROMEGRASS

A tall, sod-forming grass that puts down deep roots and masses of rhizomes and can be used for pasture and hay. Shows good drought tolerance and is moderately tolerant to wet, acidic, or alkaline soils. Needs a longer regrowth period after grazing or cutting. Harvest the spring crop when seedheads emerge. Because of its root characteristics, smooth bromegrass is often used for erosion control on roadsides and steep slopes. To establish on eroded areas, boost fertility with composted manure and protect seed and seedlings from water flow. Seed at 12–16 lb./A (6–8 lb./A in combination with a legume).

TALL FESCUE

A deep-rooted, sod-forming grass that can be used for hay, silage, or pasture. It is tolerant of acidic soils and continues to grow during hot, dry weather. Much of the existing acreage is infected with an endophyte fungus that produces alkaloids that reduce palatability and can cause poor animal performance. Risk of adverse effects can be reduced by growing endophyte-infested fescue in a diverse stand with other grasses and legumes and by restricting its use to spring and fall when animals may be less affected by the endophyte's alkaloids. When establishing a new stand, plant an endophyte-free variety. Good fall regrowth and maintenance of forage quality under freezing temperatures make tall fescue a good choice for stockpiling for late season grazing. Seed at 12 lb./A (8–10 lb./A in combination with a legume).

TIMOTHY

A shallow-rooted bunch grass that intercrops well with legumes to produce high-quality hay. Extremely winter hardy, but not drought or heat tolerant. Although less well adapted to grazing than hay production, it is often used in a pasture mix. Allowing a longer regrowth period may be necessary to enhance persistence. Also, avoid grazing from early to mid-fall to allow for restocking of energy reserves. Plant from 8–10 lb./A (2–6 lb./A in mixture).

FORAGE/GREEN MANURE LEGUMES GZ, F, CC

These plants are essential components of high-quality pastures, hay crops, and sustainable crop rotations. See the Chapter 3, "Organic Soil Management" for best management practices for N fixation in forage legumes. Forage legumes make excellent intercrops with forage grasses and small grains.
Requirements: Forage legumes grow best in moderately to well-drained, fertile soil with pH of 6.5–7. Specific legume species may be more sensitive to or tolerant of marginal growing conditions (see individual descriptions below).
Planting: If seeding a forage/hay crop, plant in spring or late summer (late July/early August in extreme northern areas, mid- to late August further south). In spring plantings, seed into a well-prepared seed bed and interseed a grass/small grain nurse crop to control weed pressure. (Oats make an excellent nurse crop seeded at 2–3 bu./A.) Planting can be a one-pass operation with the small grain/grass in the drill box and the legume in the grass box. Chains dragged behind the drill are enough to incorporate the legume seed. In late summer plantings, nurse crops may be too competitive for moisture; instead, plant in fields with low weed pressure. To optimize N fixation, forage legume seed should be treated with the appropriate inoculant immediately before planting. Some inoculants currently on the market are genetically modified. Be sure to check with (and obtain documentation from) the supplier before applying an inoculant to make sure it is GMO-free. See also individual descriptions below.

ALFALFA

Produces large amounts of high-quality forage as a hay crop. One of the best N-fixers, reliably supplying 150 lb. N to a subsequent corn crop. Deep taproot not only helps it tolerate droughty conditions but breaks up subsoil compaction. Not suitable for waterlogged areas or acidic soil. Most varieties cannot withstand frequent or close grazing. Under management intensive grazing, harvest first for hay or haylage, then allow 25–30 days for recovery before grazing. Not as palatable as red or white clover to livestock and can cause bloat. If alfalfa hay is routinely sold off the farm, nutrient depletion is likely. High-quality composted manure from off the farm will need to be applied to restore nutrients. Because of high seed cost, alfalfa is usually maintained in the rotation for at least 2 years. Seed at 12–15 lb./A. Use inoculant for alfalfa/sweetclovers.

ALSIKE CLOVER

A short-lived perennial clover that is well adapted to poorly drained soils. Tolerant of acidic soils but not of drought. Has a reclining growth habit but can be held more erect by interseeded grasses. Can be toxic to horses. Seed at 4–10 lb./A; use the higher rate when overseeding. Not as well-adapted to frost-seeding as white or red clovers. Use inoculant for red and white true clovers.

BIRDSFOOT TREFOIL

Tolerates a wide range of soil types and soil conditions, including waterlogged, saline, and acidic soils (although performs best on fertile, well-drained soils of >pH 6.2). Has a nutritive value equal to or greater than alfalfa, but is less productive than alfalfa under optimal soil conditions. Does not cause bloat. Has slower early growth than alfalfa or red clover. Best mixed with less competitive forage grasses, e.g., Kentucky bluegrass. Excellent for stockpiling in the second half of summer and grazing in fall after the first killing frost. Adequate soil moisture is critical, and a firm seedbed helps to ensure a good stand. Plant at 4–10 lb./A, no deeper than 0.25". Use trefoil/lupine inoculant. Can also be frost seeded.

HAIRY VETCH

A winter annual that can be grown in a wide range of soils and is tolerant of acidic/alkaline soils. Winter hardy even in northern areas. Grows rapidly in the spring, and if allowed to grow through late spring, can produce large amounts of biomass and N. Forms a weed-suppressive mat, especially when grown with rye as a cover crop. Can be killed through tillage or, after it has flowered, by using a roller/crimper (or even driving over it). Has a high percentage of hard seeds (up to 20%) that may cause weed problems especially in later small grain crops. Because of vining growth habit, should

Continued on next page »

not be grown as an interseed in small grain crops grown for seed. If grazing, wait until plants are at least 6" tall. Grazing below the lowest leaf axil will result in slow regrowth. For hay, cut at ¾ bloom and process quickly: leaves dry much faster than stems. For late summer or fall planting, seed 30–40 days before a killing frost. Can also be planted in early spring. Seed at 30–40 lb./A as a sole crop, 20–30 lb./A if intercropping with rye. Seed is larger than that of other forage legumes: drill 0.5"–1" deep or broadcast and lightly disk in. Can be overseeded into standing crops, but establishment can be variable due to low moisture conditions and shading effects. Use pea-vetch inoculant.

RED CLOVER

An excellent N -fixer that is better adapted to wetter and more acidic soils than alfalfa. Not drought tolerant. A short-lived perennial that rarely lasts more than 2–3 years. To maintain in a pasture, frost seed or overseed in the spring. Will also reseed itself if allowed to go to seed. Can cause bloat. Shade tolerant, grows extremely well in an intercrop with a small grain or grass and has been successfully established in between corn rows at last cultivation if moisture conditions permit. Two types are available: Medium Red and Mammoth Red. Mammoth Red provides a higher yield of hay in a first cutting but has slow regrowth. Medium Red, or multicut, is the better option when planning to grow the crop for more than one year or as a hay crop. Red clover is easier to kill than alfalfa in late fall or early spring because of its shallower root system. Seed at 10–15 lb./A. Use an inoculant for red and white true clovers.

WHITE CLOVER

A shallow-rooted, low-growing forage legume well-adapted to grazing. Tolerates wet but not droughty soil conditions; shade tolerant. Highly tolerant of foot and machine traffic. Spreads via creeping stems and can reseed itself under favorable conditions. To optimize persistence in pasture, graze at intervals frequent enough to keep taller-growing grasses from shading out the clover. Varieties are characterized by their size or length of their leaf stems: Large white or "Ladino" clover is the tallest; common or white Dutch clover intermediate; and "wild" white the shortest. As part of a pasture mix, seed at 4 lb./A, no deeper than 0.25". Can be frost-seeded into an existing pasture. Use 7–14 lb./A when establishing a sole crop or overseeding under adverse conditions. Use an inoculant for red and white true clovers.

YELLOW SWEETCLOVER

A biennial legume that is a good N-fixer and excellent soil conditioner. Deep taproot can break up soil compaction and bring up nutrients into the crop root zone. If grown over two seasons, can contribute large amounts of organic matter to the soil as a plowdown. Drought tolerant but not tolerant of acidic conditions. Best used as a green manure/cover crop but can be used as forage if cured properly: moldy hay or silage can be toxic to cattle. Best planted in the spring. Can quickly reach heights of 6–8' before flowering in early summer—one or two spring mowings may help to make incorporation more manageable. Seed at 12–15 lb./A. Use inoculant for alfalfa/sweetclovers.

MILLETS GZ, F

Several of the small-seeded annual grasses known as millets can be used as supplemental pasture or hay crops. These grow quickly under warm temperatures, and some species are well-adapted to droughty conditions. Plantings can be made from when the soil temperature rises above 60°F until 60–70 days before frost. Millets can tolerate marginal growing conditions, including acidic soils, but perform best under good fertility. When used as pasture, strip grazing is recommended. To graze for regrowth or for multicut hay, maintain a stubble height of 6"–8". Millets do not cause prussic acid poisoning, but must be managed to avoid nitrate poisoning (see Brassicas on page 112).

PEARL MILLET

The millet best suited for grazing. Drought tolerant, well-adapted to lighter, sandy soils. Used primarily in the Southeast but a short-season hybrid can perform well in the Northeast. For silage, harvest at seedhead emergence or as soon as frost occurs. Plant 0.5"–1" deep at 15–20 lb./A broadcast or 8–10 lb./A drilled.

FOXTAIL MILLET

Ready for harvest from 50–60 days after planting—can be used as an emergency crop after crop failure. Grows best in loamy soils of moderate to high fertility; not tolerant to severe or prolonged droughty conditions. Can be grazed, but better suited for hay: harvest at late boot to early bloom stage—avoid mature seedhead bristles, which can irritate jaws and eyes of cattle. Plant 0.5"–1" deep at 25–30 lb./A broadcast or 15–20 lb./A drilled.

JAPANESE MILLET CC

A fast-growing grass that tolerates wet soil conditions and can tolerate droughty periods once established. With moderate to high N fertility, can provide several cuts of high-quality forage. Also an excellent smother crop for weed suppression. It is a cultivated form of barnyard grass: on cropland, do not allow seedheads to form. Plant 0.5"–1" deep at 25–30 lb./A broadcast or 15–20 lb./A drilled.

SMALL GRAINS GR, F, GZ, CC

Important food and feed crops that are also highly prized in organic systems for their soil-building roles: straw builds organic matter while fibrous roots scavenge nutrients and help create good soil structure. Also form excellent intercrops with forage legumes. Fall-planted small grains protect the soil over winter months and help fight weed buildup.

Requirements: Prefer well-drained, medium- to heavy-textured soils; pH 6–7.

Planting: In spring, should be planted as early as the ground can be worked—delayed planting into mid- or late spring will cause substantial yield reductions. For winter grain planting dates, see individual crop listings below. For fall and early spring grazing, plant winter small grains in summer/early fall. For summer grazing, plant from early to midspring. Stored grain should be at 14% moisture.

BARLEY

Has a feed value of about 95% of that of corn. Grows well as an intercrop with field pea or any of the forage legumes. Awnless types can be useful for forage. Because barley and wheat are susceptible to several of the same diseases, barley should not precede or follow wheat in the rotation. Winter barley is not as winter hardy as rye or wheat, but may be grown in areas from New York south, with reliability increasing southward. To improve winter survival, plant winter barley 7–10 days before optimum winter wheat planting date. Requires moderate fertility: excessive N will cause lodging and high fertility encourages weeds. Seed at 2.5–3 bu./A (120–144 lb./A); depth: 1"–2".

OATS

Grain protein content is12–13% (up to 15% in hulless oats). An excellent break crop for diseases, a good scavenger for nutrients, and an outstanding nurse crop for spring-planted forage legumes. Also grows well planted in a mixture with peas, peas/alfalfa, or peas/hairy vetch. An excellent choice for fall planting if a winter-killed cover crop is desired. Oats require cooler temperatures and more water than other small grains to produce good yields. Requires moderate fertility. Plant 3 bu./A or 96 lb./A for grain and forage. For fall cover crops, plant from 3–4 bu./A; depth: 1.5"–2".

SPELT

A hulled wheat that has a niche in the organic health food market. May be better tolerated than common wheat by those with wheat allergies. Taller and leafier than common wheat and thus likely to be more competitive with weeds. Susceptible to the same diseases as wheat, although resistant to common bunt and ergot. More tolerant than common wheat to poorly drained soil or lower fertility conditions, although it responds well to moderate to high levels of N. Seed at 112–200 lb./A for hulled seed. Use the highest rate when broadcasting. Use the lower rate for hulless seed. Plant at 1.25"–2" deep for hulled seed, 1" deep for hulless. Hulled seed may be better resistant to cold, wet conditions, but requires a drill with adjustable openings large enough to accommodate the large seed size. Planting dates are the same as for winter and spring wheats.

Continued on next page »

Continued »

TRITICALE

A cross between wheat and rye that provides some of the disease resistance and stress tolerance of rye with the higher yield and nutritional quality of wheat. Has potential for high yields of both grain and straw. Both spring and winter types of triticale are available. Follow the planting recommendations for wheat.

WHEAT

Requires heavier N fertility than other small grains. Plant after a forage legume, soybeans, or another crop where residual N is available. Planting date for winter wheat will vary by location and is somewhat elastic. Ideally, plant to allow development of 3–4 leaves before freeze-up. Earlier planting dates can increase winter injury and disease problems. Later planting dates (e.g., early to mid-November in the Northeast) are possible but increase the risk of winter kill and yield loss. Plant at 2–3 bu./A (120–180 lb./A) at a depth of 1"–2".

WINTER RYE

More tolerant than other small grains of low fertility and acidic or sandy soil. Because of extensive fibrous root system, winter hardiness, and competitiveness with weeds, typically grown as a cover and soil-improving crop. Has an allelopathic effect: crops seeded after rye kill in the spring may have significantly reduced weed incidence from 2–5 weeks after crop planting, although supplementary weed control later in the season may be necessary. Avoid planting a small-seeded crop (such as flax) after rye. For a fall and early spring pasture crop, plant rye in late summer/early fall. To avoid off-flavor, remove cows from rye several hours before milking. For grain and cover crops, plant so that rye will put on at least 4"–6" of top growth before freeze-up. No-till drilling rye into corn stubble is an option to improve the catch of a late planting. Grain infected with the fungus ergot can cause abortions in breeding stock. To minimize risk of ergot, use clean seed and do not plant rye after other cereals or forage grasses in the rotation. Requires moderate fertility; excessive N will cause lodging. Seed at 2.5–3 bu./A (140–168 lb./A) at a depth of 1"–2".

SORGHUM, SUDANGRASS, SORGHUM-SUDANGRASS GZ, F, CC

Fast-growing, warm-season (drought-tolerant) annual grasses that can provide large amounts of biomass during mid- to late summer when grown with good fertility on productive soils. Can be grown for silage, hay, green chop, or grazing. Forage sorghum is essentially one-cut whereas sudangrass and sorghum-sudan may be cut multiple times. Make excellent break crops for disruption of lifecycles of nematodes and soilborne pests. Contain an allelopathic compound that can suppress weed and other plant growth. Sudangrass and sorghum-sudangrass make excellent cover crops to improve soil quality. Can cause prussic acid poisoning (see below) and nitrate poisoning (see Brassicas on page 112) in cattle.

Requirements: Can grow on lighter, droughty soils more successfully than corn. Fertility requirements for sorghum and sorghum-sudangrass are similar to those for corn; sudangrass requires at least moderate fertility, pH range of 5.8–7.5.

Planting: Plant after soil temperature has reached 65°F, usually from mid-May to early June. Cover crops may be planted throughout the warm season up to 7 weeks before frost. Sorghum: 15–20 lb./A drilled, 10–15 lb./A in 18"–20" rows; depth: 0.75"–1.25". Sudangrass or sorghum-sudan for pasture, hay, or green chop: 30–40 lb./A broadcast, 20–25 lb./A drilled. For a weed-suppressive cover crop, plant 35–50 lb./A (use the higher rate for broadcast); depth: 0.5" (heavy soils) to 1.5" (lighter soils).

Grazing/harvesting management: Ready for grazing or green chop when height is over 18"–20" for sudangrass and 24"–30" for sorghum-sudangrass (see prussic Acid poisoning section below). Retain a 6"–8" stubble for regrowth. Do not allow green chop to heat in the wagon as this will increase prussic acid levels. For silage, harvest between 60–72% moisture, usually between the medium dough and hard dough stages. For hay, cut when forage is about 30" high. For a soil-improving cover crop, mow at 3'–4' feet to increase rooting depth to break up compaction layers.

Avoid prussic acid poisoning: These crops contain dhurrin, a compound that can release prussic acid in amounts highly toxic to livestock. Sorghum contains the most, sudangrass the least, and sorghum-sudangrass intermediate amounts. Because of high dhurrin, grazing sorghum is not recommended. Prussic acid is rarely a problem in silage because it dissipates during the harvesting and ensiling processes and in moving it for feeding. Cured hay is rarely hazardous. To avoid poisoning during grazing, follow the recommended pregrazing heights given above, do not graze during or immediately after a drought, do not allow grazing after a frost until the dead plants are dry (5–6 days) or until regrowth has reached the recommended pregrazing height.

SOYBEAN

A warm-season pulse crop grown for both food and feed. Has both high protein content (~40%) and relatively high oil content (~20%). Usually requires early season tine-weeding or rotary hoeing plus inter-row cultivation. Susceptible to a variety of diseases plus crop residues left after harvest do not build organic matter. Therefore, rotate with break crops to prevent soil-borne diseases and with soil-improving crops to maintain soil health.

Requirements: Prefers well-drained, loamy soil; pH range: 6-7.5; Needs moderate amounts of P and K. N is not required. Use soybean inoculant.

Planting: Should be planted when the soil has thoroughly warmed and one or more weed flushes have occurred—often 2 or more weeks after conventional soybean planting. Choose varieties with relative maturity ratings that allow for the later planting date needed. A high density plant population (170,000-230,000 seeds/A) can reduce weed competition.

Sources: Rayburn, E.B., ed., *Forage Utilization for Pasture-Based Livestock Production*, 2007, Nat. Res. Agric., and Eng. Ser. (NRAES), Ithaca, NY; Canadian Organic Growers, *Organic Field Crop Handbook*, 2nd ed., Ottawa, Ont., Canada; Sustainable Agriculture Network, 1998, *Managing Cover Crops Profitably*; Univs. of Wisconsin, Minnesota, and Iowa State Extension, 2004, *Alfalfa Management Guide*; Penn State Forage series (www.forages.psu.edu); Barnes, Miller, and Nelson, eds., 1995, *Forages: An Introduction to Grassland Agriculture*, 5th ed., Iowa State Univ. Press, Ames, IA ; Sarrantonio, M., 1994, *Northeast Cover Crop Handbook*, Rodale Inst., Emmaus, PA; Martin, Leonard, and Stamp, 1976, *Principles of Field Crop Production*, 3rd ed., Macmillan, New York, NY; Univs. of Wisconsin and Minnesota Extension Services, 1992, *Alternative Field Crop Manual*; Univ. of Minnesota Extension Service, 1999, *Minnesota Soybean Field Book*; Abdel-Aal and Wood, eds., 2005, *Specialty Grains for Food and Feed*, Am. Assoc. of Cereal Chemists, St. Paul, MN.

ORGANIC PEST MANAGEMENT

Organic farming involves a very different approach to managing weeds, insect pests, and diseases than conventional agriculture.[1] By keeping the following points in mind, transitioning farmers can avoid falling into the "substitution trap"—substituting organic inputs for conventional pesticides—that inevitably results in higher costs and growing pest problems.

Pest management in an organic system is based on the following.

- It emphasizes the prevention of pest buildup rather than control methods once pest outbreaks have occurred. Organic sprays and other pest management inputs that are acceptable to certifiers may be useful in certain emergency circumstances, but the bulk of pest control efforts are best focused on creating a healthy environment for crop growth.
- It accepts that pests cannot, and should not, be eradicated. Pests have vital roles in the farming system. For example, weeds can be excellent sources of minerals for livestock (see

Chapter 4, "Pasture Management"), often provide habitat and food for beneficials (species that kill or suppress pests), and protect against soil erosion. Insect pests and disease-causing bacteria, fungi, and nematodes themselves serve as food for beneficial species and aid in the breakdown of plant tissue—an essential step for nutrient recycling.

- It combines many techniques, rather than relying on a single "magic bullet." Take the case of quackgrass, for example, which can be useful as a pasture plant but is a highly competitive weed on cropland. No one strategy is likely to control it, but midsummer tillage used in combination with smother cropping, interseeding, and crop rotation can knock quackgrass back on its heels.

WEED MANAGEMENT IN FIELD CROPS

Farmers often list weed control in crops as a major worry or even barrier to transition to organic management. This is understandable because the primary method of conventional weed control is synthetic herbicides, which cannot be used on organic farms. The good news is that organic farms can have excellent weed control—comparable to that on well-managed conventional farms.

Organic farmers who are successful at weed control use many weed control techniques in a coordinated way. The following section discusses the tools they use, starting with three essentials: sound soil management, knowledge of how weeds grow, and a weed-suppressive crop rotation.

Soil management for weed control

Sound soil management is the baseline defense against weeds.

- Soil that is at optimal fertility results in healthy, vigorous crops that compete well with weeds. Excess nutrients, especially N, cause many weeds to thrive, e.g., lambsquarters and pigweed, and can cause crop damage through lodging or increased pest attack.
- Good soil structure not only promotes strong, competitive crop root systems but is also essential for effective cultivation. Roots of weed seedlings are more easily dislodged from soil during tine weeding or rotary hoeing if it is in good tilth rather than compacted or cloddy. Also, well-managed soil will "flow" when it is cultivated, more effectively burying weeds. On the other hand, compacted soil can favor small-seeded weed emergence and growth. In Figure 5.2 numerous foxtails have emerged in the wheel track rows of a soybean field but very few in the uncompacted row (center). By staying out of the field when the soil is too wet, farmers not only avoid soil compaction but also avoid enhancing the emergence and growth of small-seeded weeds as well.

Knowledge of how weeds grow

By becoming knowledgeable about problem weeds, transitioning

FIGURE 5.2
Soil compaction benefits weeds

and organic farmers can more effectively use management techniques to prevent weed survival and growth. Critical information about problem weeds includes the following:

Correct identification: Many different weed species are likely to be present on a farm, and it is easy either to misidentify a weed or to clump together similar-looking weeds that actually have very different growth habits. Fortunately, there are good, cheap weed identification guides that can help farmers quickly and accurately identify both weed seedlings and mature plants (see the Resources section at the end of the chapter).

The lifecycle of weeds and how they survive and spread: Weeds, like crops, can have lifecycles that are annual (completed in one year), biennial (completed in two years), or perennial (the weed usually lives for more than two years). A weed's lifecycle sheds light on how it survives and reproduces itself. Annual weeds reproduce through seed. Biennials overwinter, but they too must set seed in the second year to reproduce. The worst perennial weeds not only set seed but also are able to spread vegetatively, e.g., with rhizomes, root buds, etc. Table 5.3 identifies the lifecycles of common weeds.

When weeds emerge and the growing conditions that favor them. Many weeds tend to emerge at certain times of the year. For example, shepherd's-purse and some mustards emerge in late summer or fall while pigweeds emerge in mid- to late spring. Emergence does not follow calendar dates but depends on growing conditions, so periods of greatest emergence will vary from year to year. Unfortunately, many weeds can continue to emerge, if in reduced numbers, throughout the growing season, e.g., yellow mustard and velvetleaf. Nevertheless, information on the time of greatest weed emergence can help in timing field preparation and crop planting to destroy as many emerging and emerged plants of problem weeds as possible. (For times of emergence of many common weeds, see Table 5.3.)

Weed-suppressive crop rotations
To optimize weed suppression, crop rotations should include diverse crop types, including row crops, smother or cover crops, and sod crops. Each crop type provides specific opportunities for weed control.

TABLE 5.3 LIFECYCLES OF COMMON WEEDS
(with time of emergence of annual weeds)

ANNUALS			
Barnyardgrass (midspring–midsummer)	Eastern black nightshade (late spring–summer)	Horseweed (late summer or spring)	Velvetleaf (late spring–midsummer)
Cocklebur (midspring–summer)	Fall panicum (late spring–midsummer)	Jimsonweed (late spring–summer)	Wild mustard (late summer–spring)
Common ragweed (early/midspring–June)	Field pennycress (late summer–early spring)	Lambsquarters (early spring–early summer)	Wild radish (late summer–early spring)
Corncockle (autumn)	Foxtails (midspring–midsummer)	Pennsylvania smartweed (early spring–early summer)	Witchgrass (late spring–midsummer)
Crabgrass (late spring–late summer)	Giant ragweed (early spring–late spring)	Pigweed (midspring–summer)	
BIENNIALS			
Bull thistle	Common mullein	Poison hemlock	Wild carrot
Common burdock	Musk thistle	Teasel	White campion (or cockle)*
PERENNIALS			
Bindweeds	Common milkweed	Goldenrods	Quackgrass
Buttercups	Dandelion	Horsenettle	Yellow nutsedge
Canada thistle	Docks	Smooth bedstraw	

*Can also grow as an annual or short-lived perennial.

Sources: Richard H. Uva et al. 1997. Weeds of the Northeast. Comstock Publishing, Cornell University Press, Ithaca, NY. Relative Emergence Sequence for Weeds of Corn and Soybean. 1997. Iowa State University Extension, SA-11.

Row crops: (e.g., corn, soybean, dry bean, sunflower)
- Wide row spacing permits cultivation to kill weeds through midseason.
- Relatively late planting dates allows a series of preplant tillage operations that can first stimulate flushes of spring-emerging weeds and then kill them.

Smother/cover crops: (e.g., small grain/legume intercrops, buckwheat, sorghum-sudangrass, Japanese millet, forage brassicas)
- Compete with weeds for light, nutrients, and other resources because of high-density planting or narrow row spacing.
- Allow variation in tillage and planting date from row crops that disrupt weed lifecycles, including:
  - early spring planting → late summer tillage (e.g., spring small grains)
  - tillage throughout spring → early to midsummer planting (e.g., buckwheat, millets, and sorghum-sudan)
  - late summer tillage → fall planting (e.g., winter small grains).

Sod/hay crops: (e.g., alfalfa, red clover, forage legume mixes, legume/grass mixes)
- Compete with weeds through a high-density planting.
- Cause the death of the majority of buried grass weed seeds *if the sod crop is maintained for two or more years* (many broad-leaved weed seeds are less significantly affected).
- Prevent seed set of annual and biennial weeds and reduce root reserves of perennials through periodic mowing or grazing.
- Disrupt the root systems of such perennials as Canada thistle and quackgrass when the soil is inverted at plowdown.
- Create a habitat for organisms that attack or eat weed seeds, e.g., carabid beetles.
- Function as a relatively slow-release nitrogen source for subsequent crops in the rotation, which has sometimes been shown to favor crop over weed growth.

By working through a diverse rotation (for example, corn—soybean—small grain/legume—legume), weeds are subjected to multiple stresses and an unstable environment that prevents their buildup. On the other hand, crop rotations that grow multiple years of row crops under organic conditions are soon infested with a mix of spring and summer annuals and often perennial weeds as well. With good reason, the 1938 *Soils and Men* called crop rotation "the most effective means yet devised for keeping land free of weeds."[2]

Additional weed management tools
Mechanical cultivation: As long as it does not become the only tool farmers reach for, mechanical weed control is a very effective (and satisfying) method of destroying weeds. Cultivation is a skill developed over years of practice and experimentation. However, transitioning farmers with no cultivation experience can greatly benefit from talking to experienced cultivators and actually observing these farmers in action. The following are some recommendations de-

FIGURE 5.3
Some type of early-season cultivation tool is needed for organic row crops, for example, a flex-tine weeder.

rived from consultation with experienced cultivators and scientific experiments.

- In row crops such as corn and soybean, early season cultivation (from time of planting to inter-row cultivation) is important to reduce weed numbers and prevent yield loss to weeds. Inter-row cultivation is also critical. Therefore, transitioning farmers who are growing these crops need to have access to two pieces of equipment: some type of early season cultivator (e.g., flex tine weeder, rotary hoe, or even spike tooth harrow; see Figure 5.3) and an inter-row cultivator.
- There are several different types of cultivation equipment that can be highly successful in controlling weeds—that is, there is no one "indispensable" piece of cultivation equipment. Instead, farmers develop personal preferences as they gain skill and insights into the cultivation process.
- The timing of cultivation is critical. After crop planting, farmers should be looking for germinating weeds in the top two inches of the soil—their roots look like white threads. When these roots are visible in moderate to high numbers, the field should be cultivated. (Some experienced cultivators recommend cultivating at a set number of days after planting—see the Engelbert case study at the end of this chapter). Similarly, in a weedy field, inter-row cultivation should be done as soon as the crop is tall enough to withstand burying.
- Another critical factor is the adjustment of cultivating equipment to provide maximum effectiveness against weeds under field conditions. For inter-row cultivators, sweep placement and depth need to be adjusted each season at minimum and often need several adjustments during the season to allow for changing soil conditions and weed growth. Additional adjustments in terms of experimenting with sweep types, sweep

number, or use of shields can further increase the effectiveness of the cultivation pass.

Use of clean seed: Crop seed from a reputable commercial seed house has been tested (as required by state seed laws) for the total percentage of weed seeds present and for designated noxious (highly undesirable) weeds. The weed-seed test information is noted on the seed bag label. However, many growers use their own seed, purchase it, or swap seeds with neighboring farms. This may be even more likely for transitioning or organic growers since certified organic seed is often still difficult to find and is expensive. The problem with using farm-produced crop seed is that unless it has been well cleaned, it may contain a high proportion of weed seeds that are then planted along with the crop. To avoid this situation, growers should only use their own seed or that from another farm if it has been well cleaned. All farm-produced crop seed should be carefully visually inspected for presence of weed seeds before planting.

Planting dates and soil temperature: Planting dates of warm season row crops like corn and soybeans can be dramatically different in organic as compared to conventional systems. Transitioning and organic growers need to wait until the soil has adequately warmed—to 50–55°F at a depth of 2 inches for at least several days. This higher soil temperature spurs crop emergence and early growth, making the crop more competitive with weeds. Waiting for a higher soil temperature also means that many spring weeds that germinate at lower soil temperatures, e.g., wild mustard and lambsquarters, will have already emerged and can be killed with preplant tillage. Following this rule of thumb means that transitioning and organic growers may be planting their corn and soybeans two weeks or more after their conventional neighbors. It may be tough to sit and wait while other growers are planting, but the rewards in terms of weed control make it worth it. To accommodate the later planting date, it will likely be necessary to grow an earlier maturing variety.

Seeding rate: A good rule of thumb is to seed crops at the maximum rate, which optimizes crop competition with weeds. An exception to this rule is corn, which, because it requires high fertility, is best planted at a more moderate rate. See Table 5.2 for suggested seeding rates for specific crops.

Cover crops and crop mixtures: Short-term, fast-growing cover crops like buckwheat or sorghum-sudan are an excellent way to knock back weeds in problem fields or even in weedy patches or margins of fields. These crops can suppress even noxious weeds like Canada thistle. Crop mixtures that form a dense, living mat over the soil (e.g., rye/hairy vetch, oat/pea/alfalfa or oat/pea/hairy vetch) are not only excellent for soil improvement but are especially good at smothering weeds.

Allelopathic crops: Certain crops have the potential to suppress weeds because they release chemicals into the soil that inhibit germination and growth of small-seeded weeds—an effect termed allelopathy. All small grains, sorghums, and some forage legumes have shown this allelopathic effect, but it has been most widely used for weed control with winter rye. Because this effect does not always occur and is not likely to persist through the cropping season, allelopathy should be used in conjunction with other weed control tools.

Soil amendments: Some longtime organic growers maintain that there is a weed management benefit to keeping calcium and magnesium ratios in balance (see Chapter 3, "Organic Soil Management"). One theory is that the calcium will loosen up tight, compacted soils that are thought to favor certain weeds such as annual grasses. Unlike the other weed management tools in this section, there does not appear to be a scientific basis for this approach. Studies suggest that many of our most troublesome weeds are capable of adapting to a wide range of soil conditions, including varying calcium and magnesium levels. Surface compaction in soil that occurs after running equipment when soil is not fit can increase the emergence of small-seeded weeds. However, it is not clear that calcium addition can significantly reduce equipment-related soil surface compaction. Transitioning farmers interested in using lime or gypsum for weed control should consider using it in combination with other weed management strategies.

Hand weeding: Hand weeding or rogueing can play a critical role in preventing serious weed infestations. If occasional patches of weeds still exist after last cultivation, it is highly beneficial to walk the rows and pull them—thus preventing further weed seed inputs into the soil.

A last resort: In cases where weeds get seriously out of control and no more weed control methods are available, the farmer needs to weigh the available options. If a serious amount of weed seed will be shed by allowing the crop to go to harvest, it may be better for the long term to kill the weeds before they go to seed. It may be possible to salvage the crop by harvesting it for forage rather than grain, or it may be wisest to turn the crop under as a green manure and try another crop later in the season or the following year.

WEED MANAGEMENT IN PASTURES

Many weeds can provide nutritious forage, especially in their vegetative stages. Dandelion, for example, is very close in value in protein and total digestible nutrients to alfalfa. Weedy grasses, including quackgrass, have about 75% of the quality of forage legumes. However, pastures will drop in yield and forage quality if weedy species crowd out forage grasses and legumes. In addition, some weeds are poisonous (e.g., cow cockle, jimsonweed, milkweed).

The most important tool for controlling weeds in pastures is good grazing management. Avoiding overgrazing and maintaining a dense forage stand with no bare spots keeps weeds from getting a foothold. Sound soil practices and providing adequate regrowth periods promote the vigorous growth of both shoots and roots of forages to give them a competitive advantage over weeds. In addition to good grazing management, the following practices can help to minimize weed problems in pastures.

- Scout for weed invasions. Weekly walks through pastures to check on forage growth also allow scouting for weeds. Invasion by a new weed can often be stopped if a single weed or small patch is rogued out or clipped before seedheads can form.
- Maintain diversity in pasture species. A pasture mix that contains multiple species—for example, two or more forage grasses, two or more forage legumes, plus chicory—resists weed invasion better than a mix of just two to three species.
- Keep areas along fencerows, field margins, and roadways mowed to reduce the risk of weed invasion.
- Mow and clip pastures as needed.
  - Mow annual weeds at a height of 8–10" or earlier if flowering. An additional mowing or mowings later in the season may be needed to prevent weeds that have regrown from setting seed.
  - Mow biennial weeds after they have bolted (formed a flower stalk) but before seed has set. If there are relatively few biennial plants in the pasture, these can be killed by cutting them off with a sharp spade at least one inch below the soil surface (below the crown region where bud tissue is located).
  - Suppress perennial weeds by mowing or clipping 3–4 times during the season to weaken root reserves.
- Graze with other animal species to control biennials and perennials. Sheep prefer broadleaved plants to grasses and shrubs and have successfully helped control such weeds as leafy spurge. Goats readily consume brush and spiny weeds such as multiflora rose and thistles.

INSECT PEST AND DISEASE MANAGEMENT: BASIC TOOLS

Just as in weed management, organic farmers rely on sound soil management as an essential tool in insect pest and disease management. Crops that have been planted under optimal temperature and soil conditions and that are supplied with adequate nutrients are much better able to withstand pest attack than those grown under stress. Moreover, a steady and diverse supply of organic matter promotes diverse and abundant populations of soil organisms that can reduce soilborne pest organisms by competing with them for resources, preying on them, or releasing compounds that inhibit their growth.

Pest biology

Knowledge of pest biology is also key in preventing or minimizing pest damage. For example, seedcorn maggot, which attacks corn and soybean seed, is often controlled in conventional systems by seed treatment with an insecticide. However, damage from seedcorn maggot can be avoided or reduced by using information on its lifecycle and feeding habits. Seedcorn maggots are attracted to fields that have high organic matter inputs from manure or green manure plowdowns. Adults lay eggs at or near the soil surface that take a few days to hatch into larvae. Crop damage is caused by these larvae (maggots) that feed underground on the seed or emerging cotyledons. The larvae generally take from a week to 10 days to complete their growth stage. Waiting two to three weeks after organic matter incorporation to plant crop seed helps to avoid the larval stage of the insect. Delaying planting until the soil has warmed up also limits damage, because under these conditions crop germination and emergence will proceed rapidly. Although not all insect pests and diseases can be avoided, information on insect pest and disease biology usually provides growers with good information on the conditions that favor their buildup and cultural practices that help prevent their outbreak or minimize their damage.

Plant diversity

Creating and maintaining plant diversity on the farm is another major tool for insect pest and disease control. Table 5.1 provides examples of how two types of crop diversity, crop rotation and intercropping, can help prevent crop loss to insect pests and diseases. Another means of diversifying the landscape is strip plantings. Plants such as dill, coriander, buckwheat, tansy, and alyssum have been shown to attract beneficial insects that are natural enemies of insect pests. Research suggests that field-scale planting of such species (as strips within a field or along its edges) can help manage crop insect pests. However, careful design is needed. For example, plant species need to be chosen that flower at the appropriate time, attract the specific beneficial insects needed to control major pests, do not attract other crop pest species, and are not a danger of becoming weeds. Strip plantings of crops can also help control insect pests. Strips of alfalfa, for example, provide habitat for hundreds of species, including many beneficials. Hedges that house insect-eating birds are another kind of strip that helps to protect the farm from pest attack.

INSECT PEST AND DISEASE: ADDITIONAL TOOLS

Clean seed

Certain disease pathogens, e.g., soybean mosaic virus, wheat scab, and oat smut can be spread through infected seed. Improperly cleaned seed can also spread disease, as in the case of soybean seed that contains soil contaminated with soybean cyst nematode. Farmers should use well-cleaned, high-quality seed from fields or sources known to be disease free.

Use of resistant varieties

When possible, growers should make use of resistant varieties that have been developed with traditional breeding techniques (that is, without the use of genetic modification techniques).

Sanitation

Washing field equipment before moving between fields can avoid the spread of pathogens throughout the farm. Some disease pathogens (e.g., those causing wheat scab) and insect pests (e.g., European corn borer) can survive on crop residue. Burying crop residue in the soil can reduce the survival of such pests. Clean cultivation to remove all weedy growth two or more weeks before crop planting can also be effective in managing certain pests, e.g., cutworms in corn. However, the use of clean cultivation for pest control should be balanced against its negative impact on soil quality and used only sparingly and in conjunction with soil-enhancing strategies. For example, plowdown of main crop residues should be followed with planting a cover crop.

Manipulation of planting/harvesting
As noted above in the example of seedcorn maggot, field operations can be timed to avoid peak populations of crop pests or to increase beneficial species. Harvest strategies can also reduce the spread and severity of diseases. For example, in harvesting forage legumes, mowing should occur after the morning dew has dried and the youngest stands (which are less likely to have disease) should be mowed first. When stands are hit by leaf spots, they should be mowed a few days earlier than usual, which helps with leaf retention and reduces disease inoculum.

Purchased inputs
A variety of commercial pest management products are now available to organic growers, including natural enemies reared and shipped for mass release, botanical pesticides derived from plants (for example, pyrethrum and various formulations from the neem tree), and biopesticides—e.g., products containing the fungus *Beauveria bassiana*. However, application of these products is usually prohibitively expensive in field crops and their use can have unintended consequences—including the destruction of beneficial species and the development of resistance to the input by the pest species. In addition, growers must check with their certifiers before using any of these pest management products.

EVALUATING AND IMPROVING YOUR CROP PRODUCTION PLAN

Developing an effective organic crop production plan—and the management strategies to effectively execute that plan—takes time, effort, and inevitably involves some trial and error. Plus, as noted at the beginning of this chapter, things change: milk and crop prices and feed availability can fluctuate and the weather (and maybe even the climate) is not predictable. For these reasons, it is best to assume that there will be changes to the crop production plan and to develop strategies to evaluate its effectiveness.

Major problems, such as large shortfalls in feed, require a complete rethinking of the crop plan and an analysis of what basic organic management principles and strategies may have been neglected. However, to avoid developing major problems and increase farm productivity and sustainability, farm fields should be evaluated on a yearly basis. Fortunately, the crop records required for organic certification on a field-by-field basis—including cropping history, yield, and soil amendments—make an excellent basis for evaluating the crop production plan. In addition, crop fields should be walked at least once during the growing season to note any problems with stand establishment, crop growth or quality, weeds, or crop damage from insect pests and diseases. Of course, successes—good stands, vigorously growing crops, clean fields—should also be noted! By keeping these yearly records on a field-by-field basis, farmers can over time determine the best management practices for individual fields and crops.

RESOURCES

Field Crops, Crop Rotation

Canadian Organic Growers. 2005. Gaining Ground: Making a Successful Transition to Organic Farming. Canadian Organic Growers, Ottawa, Ontario (Available at: www.cog.ca).

Lampkin, Nicholas. 1994. Organic Farming. Farming Press, Tonbridge, UK.

Leighty, C. E. 1938. "Crop Rotation," in Soils and Men: Yearbook of Agriculture 1938. USDA, Government Printing Office, Washington, DC., pp. 406–430.

Macey, A. ed.. 2001. Organic Field Crop Handbook, 2nd ed. Canadian Organic Growers, Ottawa, Ontario (Available at: www.cog.ca/ofch.htm).

Forages

Rayburn, E. B. ed. 2007. Forage Utilization for Pasture-Based Livestock Production. Natural Resource, Agriculture, and Engineering Service, Ithaca, NY. (Available at: www.nraes.org).

Penn State Forages website: (www.forages.psu.edu).

Cover Crops

Clark, A. (ed.). 2007. Managing Cover Crops Profitably, 3rd ed. Sustainable Agriculture Network, Beltsville, MD (Available at: www.sare.org/publications/covercrops/covercrops.pdf).

Sarrantonio, M. 1994. Northeast Cover Crop Handbook. Rodale Institute, Emmaus, PA (Available at www.rodaleinstitutestore.org).

Weed Management

Bowman, G. 2001. Steel in the Field: A Farmer's Guide to Weed Management Tools. Sustainable Agriculture Network, Beltsville, MD (Available at: www.sare.org/publications/steel/steel.pdf).

Muenscher, W. C. 1936. Weeds. MacMillan, New York, NY.

Uva, R. H., J. C. Neal, and J .M. DiTomaso. 1997. Weeds of the Northeast. Cornell University Press, Ithaca, NY.

Chomas, A. J., J. J. Kells, and J. Boyd Carey. 2001. Common Weed Seedlings of the North Central States. North Central Regional Extension Publication No. 607. Michigan State University Extension, East Lansing, MI (Available at: field-crop.msu.edu/documents/Ncr607.pdf).

Insect Pest and Disease Management

ATTRA (National Sustainable Agriculture Information Service) website, pest management and field crop sections (attra.ncat.org)

A Whole-Farm Approach to Managing Pests. Sustainable Agriculture Network, Beltsville, MD (Available at: www.sare.org/publications/farmpest.htm).

NOTES

1 Portions of this section have been adapted from E. Dyck, "Pest management in organic cropping systems" in J. Padgham, ed., 2006, Organic Dairy Farming, Orang-utan Press, Gays Mills, WI, pp. 109-117.

2 C. E. Leighty 1938, in Soils and Men: Yearbook of Agriculture 1938. USDA, Government Printing Office, Washington, DC.

KEVIN AND LISA ENGELBERT, ENGELBERT FARMS

ELIZABETH DYCK AND BETHANY WALLIS

Nestled along the Susquehanna River sits the first NOFA-NY certified organic dairy in New York State, Engelbert Farms. Kevin's father was a Cornell University alumnus who farmed with the newest technology. He was the first in the area to use pesticides and fertilizers, and produced record yields. With this chemical-based technology, the farm was rotating herbicides, not crops. By the late 1970s, crops were getting harder to grow. Weeds like nutsedge could no longer be controlled. Plowing produced slabs of dirt that were difficult to break up. Kevin recalls that in 1980, while plowing over 200 acres, he spotted only six earthworms. At the same time, the herd began experiencing severe health problems. The vet bill reached $1,000 per month for weekly herd checks to treat cystic cows, infected uteruses, and sore feet.

Lisa and Kevin Engelbert.

"You didn't have to be a rocket scientist to see the connection," he explains, between their heavy use of chemicals and declining soil and herd health. Kevin decided to try a radically different approach. The farm's manure storage pit allowed him to make annual applications of manure, which he credits as a major factor in restoring soil health. He also began experimenting with growing crops without sprays, for example, by using oats as a nurse crop for alfalfa. After 1981—when chemicals were stopped all together—the herd's health began to improve. Weekly herd checks became monthly, quarterly, yearly and now are on an "as needed" basis. The vet has been out to the farm only 6 times in the last 10 years for difficult births. Kevin believes that if you develop good soil health, everything else will follow.

The Engelberts rotationally graze 75 acres to provide the 125-cow milking herd with fresh pasture twice a day. Kevin has put some of the farm's best cropland into pasture, explaining that these "are still our highest producing fields; only now, we let the cows harvest it instead of us." The farm also produces all its own feed, including baleage, dry hay, and high-moisture corn. The Engelberts grow approximately 200 acres of corn, 300 acres of alfalfa/clover/orchardgrass hay, and 150 acres of soybeans in rotation, keeping the rest of the farm in permanent hay.

Kevin credits their feed self-sufficiency to the farm's rich river bottomland that is suitable for row crops, but his management of the land is critical to sustaining its productivity. Instead of "buying in" fertility, he optimizes the farm's own resources by (1) keeping the density of N-fixing legumes in his hay fields and pastures high through frost seeding red clover and (2) using the herd's manure as a major nutrient source for both corn and hayfields. He also uses

Kevin Engelbert cultivating soybeans.

multiple tools to manage weeds. Some of his tools are mechanical: a rollover plow (which prepares the ground without back furrows or dead furrows that interfere with cultivation and harbor weeds), a Kovar tine weeder, and front- and rear-mount cultivators (used in combination). He also uses other cultural methods, including a diversified crop rotation, crop mixes, and delayed planting of row crops until conditions favor vigorous crop growth.

The Engelbert family runs the farm, with Kevin and his two oldest sons, Kris and Joe, working full-time, and Lisa and their youngest son, John, who is in college, working part-time. This workforce has allowed the farm to diversify: they now raise 50 beef cows and 12 to 15 hogs a year. Any excess in feed crops also contributes to farm income since they sell it as certified organic feed.

A unique aspect of the Engelberts' farm is that they only use their barns for shelter during extreme conditions. They milk the herd in a parlor and keep them outside throughout the year. Kevin believes this is a more humane practice since it keeps the animals in their natural environment. While the Engelberts milk year-round, they do not freshen animals during December–February, allowing their animals to remain outdoors. Newborn calves go into a hutch, then are moved to pasture where they are fed with a mob feeder. At weaning (around 3 months), they go into the barn to individual stalls to prevent them from suckling each another. After 2 months in the barn, they go back outside, usually to a heifer barn with access to the outdoors.

In relation to herd health, Kevin comments, "Ninety percent of your problems will come from 10% of your cows. Those are the 10% you will need to weed out. You don't want those genetics; they are not adapting to the organic system or your operation. Closed herds are the key to organic agriculture. You'll eventually build up a herd immune to anything your system is susceptible to." If needed, Kevin and Lisa do treat their cows with garlic and provide free-choice kelp so that animals can balance themselves. When asked about additional remedies, Kevin warns, "These are still, even though they are not antibiotics, band-aids trying to fix a symptom. It is still an indication that there is something wrong in the operation and that changes need to be made."

The Engelberts became certified in 1983. After so many years, organic management is second nature to them. Still, Kevin recalls that, when they transitioned, there were few to turn to for advice. His father, although originally highly skeptical of organic farming, helped Kevin learn the art and science of cultivation. Now Kevin and his family are part of a growing community of experienced, successful organic dairy farmers who can serve as models for transitioning growers. Kevin's father, who became a huge promoter of organic agriculture, once paid Kevin the highest compliment by saying, "I wish more farmers farmed like he did."

THE ENGELBERT FARM'S WEED TOOLBOX

- Rotate between row and hay crops (see Developing a Sound Crop Rotation on page 108).
- Keep a dense, competitive stand in hayfields by liming and manuring as needed and frost seeding red clover.
- Grow small grain nurse crops (oats, barley) as needed to suppress weeds during hay crop establishment.
- Mow hayfields in a timely fashion to prevent weed seed set.
- In row crops:
    - Completely flip over the soil in spring moldboard plowing (to disrupt perennial root systems).
    - Go through two cycles of field preparation to allow flushes of weeds to germinate.
    - Prepare a level seedbed without clods or ruts (in which weed seeds can hide).
    - Plant immediately after the last field preparation.
    - Plant when soil conditions favor vigorous crop emergence and growth ("when the trees have fully leafed out").
    - Blind harrow 5–6 days after planting (before crop plants emerge) in the same direction as the field was plowed.
    - After crop rows appear (wait for soybean seedlings to straighten out of the crook stage) harrow again, going in the opposite direction from plowing.
    - Cultivate at the 3–4 leaf stage of the crop.

CHAPTER SIX

MANAGING DAIRY NUTRITION

KAREN HOFFMAN

This chapter addresses organic nutrition concerns for beginning dairy farmers, experienced conventional dairy farmers considering making the transition to organic, and newly transitioned organic dairy farmers interested in better managing their herd's nutrition. Sections of the chapter for beginning farmers are so noted, and if you are more experienced please feel free to skip those sections.

KEY ORGANIC TRANSITION POINTS

- Read and understand NOP rules for dairy nutrition (sections §205.201–§205.206, §205.237–§205.239, §205.600, §205.603, and §205.604).
- During the one-year herd transition you must do the following.
  - Feed all transitioning animals either 100% certified organic feed and/or your *own* third-year organic transition crops.
  - Be certain all components of your herd's ration are certified organic.
- Your rations should be balanced for a pasture-based diet during the growing season.
- Forage testing is key on organic farms.
- Feeding a high forage diet with high-quality forages is important.
- Basic ration-balancing knowledge can be helpful in maintaining milk production, animal health, and profitability.
- Monitoring milk production and health indicators are important ways to determine if your feeding program is on target or needs to be changed.

Organic dairy production is different from conventional dairy in a number of ways, including the management of your nutrition program. You may decide to feed less grain and manage the herd at a lower production level to reduce stress on the cows and minimize herd health interventions. You may change the way you manage your herd nutrition through your transition and after you certify organic including greater forage, increasing or improving the use of pasture, and including different forages or feed grains in your ration or crop management. Feeding and nutrition are dynamic areas of your farm because many factors constantly change, and you will find that the forages and grains you feed may change over the long term.

Organic certification requires that all components of your herd's ration must be certified organic. During your one-year transition, you are required to meet this ration requirement and feed all your transitioning animals either 100% certified organic feed and/or your own third-year transition crops for one full year before the herd is eligible to certify organic.

ASSESSING YOUR CURRENT HERD FOR ORGANIC MANAGEMENT

USING A HOLISTIC APPROACH

If you are like most dairy farmers, always busy with the day-to-day tasks and chores of running a dairy farm, you may not find extra time to step back and evaluate your farm to see how the daily tasks and chores integrate to make a successful business. Now that you are either considering or undergoing a transition to organic, it is more important to look at your farm in a more holistic way. This is similar to using a systems approach, but each system is part of the whole farm.

In your nutrition program, a holistic approach evaluates all areas of your farm and their effect on nutrition. For example, your forage and crop programs affect how you feed your cows, and soil management affects the quality of the forages you grow. Thus, when we discuss nutrition in an organic framework, we must look at it from soils to crops to animals and have an awareness and understanding of where and how these systems overlap. Organic nutrition is more than simply balancing a ration based on forage quality. It is also more complex than balancing a ration for a certain level of milk production, as you need to consider animal health as well. You will need to look at all systems in your farm management and gain an understanding of areas that need improvement, make changes, and evaluate the results of those changes on a regular basis.

ANIMAL CHARACTERISTICS

This book covers soils, crops, and herd health in other chapters, so to begin the discussion of your current herd's nutrition program we will begin by assessing the animals themselves.

There are a number of factors that determine the nutritional requirements of the animals on your farm. These include body weight, stage of production, level of production, and genetics. Over many decades of research, a science-based set of requirements has been developed that can be used as a guideline for balancing rations.

In general, animals bred for higher production or a larger mature body weight will have higher nutritional requirements based on the above factors. If your current herd consists of large animals with high production histories or potential, you will need to do an excellent job of managing forage quality and supplemental feeding under organic management to meet their nutrition requirements.

MANAGEMENT CONSIDERATIONS

Under conventional management, you may have emphasized high milk production for various reasons. If this was one of your management goals, you most likely used a variety of management tools to help you achieve high milk production including special feeds or feed ingredients, additives, hormones, or feeding large amounts of grain.

Transitioning to organic means you may need to change most, if not all, of the common conventional dairy management tools relied on. For example note the following.

- Synthetic hormones and certain feed additives are prohibited.
- Some feeds and feed additives are not available in an organic form.
- Some organic feeds are not commonly used in conventional feeding.
- Many organic grains can cost two to three times more than conventional grains.

BOX 6.1 ADJUST PRODUCTION EXPECTATIONS

When Earl Fournier and his family of Swanton, Vermont, first considered organic management, they were the quintessential small conventional dairy farm. They milked around 150 cows, used bovine somatotropin (bST), fed up to 30 pounds of grain a day, and averaged 27,000 pounds of milk per cow. After some careful planning and running input spreadsheets on production and grain costs, the Fourniers decided organic marketing would provide a higher net farm profit, even with annual production falling as low as 13,000 pounds per cow. They began their transition in December 2003, and Earl admits that it was a "steep, steep learning curve." Becoming accustomed to less milk production and getting the cows to graze were his biggest challenges. Earl and his son David now intensively graze 80 milking cows on 100 acres and feed a summer ration of dry hay, baleage, and 14 pounds of grain. Earl hopes to move away from feeding this partial total mixed ration (TMR) and replace it with feeding grain and dry hay individually. During the winter months, the herd receives haylage, dry hay, and up to 15 pounds of grain per day. (See the full Fournier case study on page 99.)

These management changes are significant enough that "pushing for production" is going to be a challenge. If your cows have always been pushed, they will go through an adjustment as they become accustomed to new feeds, less grain, and a new management program. Organic milk production is likely to be lower than conventional. The advantage is that cows will experience less stress when not pushed for high milk output and as a result have a stronger immune system. Creating stronger immune systems and therefore healthier cows with greater longevity is essential to making organic management work.

If you hope to maintain your conventional level of milk production under organic management, you will need to grow or purchase the highest-quality forages. Well-managed pasture during the growing season is also necessary. If you can maximize the intake of nutrient-dense forages and manage a ration balanced for your production goal, the possibility of high production exists.

THE ORGANIC FEEDING PROGRAM

The majority of organic farmers feed their herds a ration comprised primarily of hay crops, such as pasture, dry hay, haylage, or baleage.

Many of them have eliminated corn silage from the ration, as it is difficult to grow organically. Furthermore, fermented feeds like corn silage increase rumen acidity, which may cause herd health problems or reduce immunity if they are fed at a high rate or if fermentation is poor. Some organic dairy farmers have also eliminated corn silage to reduce labor and create a simpler system. However, those who continue to feed corn silage do so for the energy benefits it can provide, as well as adding diversity to their crop rotation.

Organic dairy production requires that all animals more than 6 months of age graze pasture during the growing season. Pasture should be treated as any other crop with a high level of management. Anything that looks like a swamp, brush-lot, exercise area, or mountainside is not the kind of pasture you should have for organic production. On the other hand, pasture that contains lush green grass and legume combinations, is 6 to 8 inches tall, has high plant density, and moves animals from paddock to paddock on a regular basis is the kind of pasture you should have. Grazing requires a different feeding strategy, which will be explained later in this chapter. To learn more about pasture management, see Chapter 4.

CULLING CRITERIA
Under conventional management, it is common to base voluntary culling on production traits rather than functional traits.

Production culling traits include:
- low production,
- low components,
- high somatic cell counts, and
- mastitis.

Functional culling traits include:
- poor feet and legs,
- low udders, and
- poor teat placement.

Since organic production relies on having healthy animals, you may find that some cows have traits that do not work well in the new system. For example, if you have a number of low-uddered cows that tend to get mastitis easily, you may decide to cull them from the herd. Likewise, if cows have difficulty walking to pasture, you may cull them based on poor feet and legs. On the other hand, if you have cows that produce less milk, but are otherwise healthy and functional, you may decide to keep them in the herd.

BREED CHARACTERISTICS
Every dairy farmer has a favorite breed for different reasons such as temperament, level of milk production or components, or because of the facility design constraints. Regardless of breed, most of our modern cattle genetics select for high milk production and large mature body size. The rationale has been that bigger animals make more milk. Unfortunately, big, high-producing animals re-

quire more nutrients for basic maintenance, and even more for high milk production. This is why grain-feeding rates are typically higher in diets of conventionally managed herds—it is the only way to meet these higher nutritional requirements.

In an organic system, a smaller, lower-producing cow selected for a broader range of genetic traits may be advantageous. She is more likely to be able to maintain herself, as well as produce an acceptable amount of milk, on a higher forage diet. That is not to say that big, high-producing Holsteins will not do well under organic management, because there are many who do. It is another factor to consider before transitioning, however.

Some farmers contend that certain breeds are better grazers than others. It is true that smaller breeds, such as Jerseys, are more heat tolerant and require less nutrients (and therefore less grass) for maintenance. However, it is also true that Holsteins take in more feed per bite of pasture, and thus fill up faster. For all breeds, grazing is a learned behavior, and animals born and raised on concrete are not going to know how to graze. Thus, within any breed, there will be both good, efficient grazers as well as slow, inefficient grazers. Based on this information, you may want to consider incorporating a different breed into the herd either through cross-breeding or purchasing. Pick the breed or combination of breeds that you want on your farm, and do the best you can with them.

OVERVIEW OF NUTRITIONAL CONCEPTS

The concepts of nutrition for organic animals are not significantly different from conventional animals, but some of the nutritional challenges are. Organic animals, whether in the barn in winter or on pasture in summer, still require protein, energy, fiber, minerals, and vitamins. They still require a certain level of dry matter intake based on body weight, growth, or production, and should receive the nutrients needed to meet their requirements.

If you are new to the dairy industry, the basic nutrient information in Box 6.2 will be extremely helpful for an understanding of basic nutrition terminology. If you are experienced and comfortable with nutritional concepts, skip to the section below, Alternative Feeds.

ALTERNATIVE FEEDS

Organic molasses: Some organic dairy farmers feed molasses as their only energy source, and it is gaining in popularity. Although there is not much published research on feeding high levels of molasses, some farmers have found that it works well. Molasses has a broad range of 14 types of sugars including glucose, sucrose, fructans, xylotol, etc. The rumen bacteria can quickly utilize these sugars because they do not need to break them down through fermentation. According to farmers who have conducted on-farm research, 1 pound of molasses has the energy equivalent of 3 to 4 pounds of corn meal. This energy value for molasses is not reflected in the NE_L value found in most dairy nutrition references due to the conventional methods that are designed to estimate and express energy content. On the other hand, some farmers have tried feeding molasses as the only energy source without much success, so be prepared to return to grain if the cows do not respond well.

In addition to high energy values, molasses contains a broad range of minerals. Molasses has a high potassium content, which may be a concern for dry cows moving into postcalving time. It also contains calcium, magnesium, copper, iron, manganese, and zinc. The minerals in molasses are highly concentrated, so a relatively small amount may meet the animal's requirements. It is low in phosphorus, so you may need to supplement additional P.

Kelp meal: This ingredient provides a broad range of vitamins and minerals and is used by many organic dairy producers. Many farmers feed kelp free choice and animals usually eat it readily. Initially, they consume large quantities but later drop their intake once they adjust to eating it and their body has met any nutrient or mineral needs that were previously deficient. The minerals in kelp are chelated, making them more available to the animal. It also has a high iodine content, which may cause concern for nutritionists not familiar with kelp. This, however, does not seem to cause any problems.

Fats: Cows can also derive energy from fat sources. Roasted soybeans, whole cottonseed, canola seeds, and whole sunflower seeds are good sources of natural fat. Most of the conventional sources of fat, such as tallow and commercial by-pass fat, are not approved for organic use.

IMPORTANCE OF FIBER

Fiber is critically important to maintain a healthy rumen and in turn the health of the animal. Ruminants evolved with a rumen to digest fiber and have formed a long-term symbiotic relationship with the bacteria and other microorganisms within the rumen. Any time there is a shortage of fiber in the diet, the delicate balance of rumen bugs to cow health can easily be undone through too much acid formation, not enough buffering via cud chewing, and unhealthy shifts in the types of rumen bugs. Fiber digestion also allows lactating cows to produce higher butterfat levels, adding to the farm's bottom line.

Pushing for extremely high levels of milk production has sharply increased the amount of grain fed to cows, disrupting rumen health and compromising this basic principle. In organics, you need to optimize fiber in the diet to maintain healthy cows, a fundamental tenet to success, so cows stay in the herd longer. Forages are the main fiber source for dairy animals. While pasture is a source of good-quality fiber, it is fairly low in fiber, and you may need to include other ingredients like rolled barley or oats (hulls on), soyhulls, whole cottonseed, wheat midds, and citrus or beet pulp to increase fiber in the diet. Although these *Continued on page 142 »*

BOX 6.2 BASIC DAIRY NUTRITION

PROTEIN

This is a primary nutrient needed for growth and production. Proteins are various combinations of amino acids and peptides and are one of the major sources of nitrogen. There are different types of protein classified by how the ruminant animal utilizes them. The following list includes terms and simple definitions.

Crude protein: This is the total protein in a feed or forage, including all the different protein fractions. It is calculated by determining the amount of nitrogen (N) in the feed and multiplying it by 6.25, because there are 16 grams of nitrogen in 100 grams of protein (100/16 = 6.25). The term "crude" reflects that it also includes some nonprotein nitrogen (NPN), so it is usually slightly inaccurate. Crude protein = amount of N in feed × 6.25.

Bypass protein: Bypass protein is also called undegradable protein or rumen undegraded protein (RUP). This protein cannot be broken down to basic peptides or amino acids by microorganisms in the rumen. It bypasses digestion in the rumen, but the animal uses it once it reaches the enzymes and acids in the lower gut or abomasum. In general, RUP should make up 40% of the crude protein portion of rations for milking cows.

Degradable protein: This is also known as rumen degradable protein RDP), and is the opposite of bypass protein. Rumen bugs break it down into peptides and amino acids and also use some of it to reproduce. When digested feed moves out of the rumen, some bugs go along as well. Without their ability to reproduce, the rumen would quickly lose function. The degradable protein is incorporated into the rumen bugs for the animal to use in the abomasum and intestines (lower gut) as well.

Soluble protein: This is the most confusing term in protein nutrition, as it is part of the degradable protein. Soluble protein is easily dissolved and quickly used in the rumen by the rumen bugs. It includes some nonprotein nitrogen that moves into the liquid and includes sources like urea, naturally found in some feeds. Many fermented feeds have a high soluble protein level because fermentation can create or transform proteins into a soluble form. Rations for milking cows should target 30% soluble protein in the diet.

ENERGY

This nutrient is essential for growth and development. All living animals require five to six times more energy on a daily basis than they do protein. Energy is not easily measured. It is estimated from other components of the feed such as fiber levels and digestibility. Common terms are based upon the net energy system, and other terms include total digestible energy (TDN), metabolizable energy (ME), and digestible energy (DE), but are not common in practical on-farm use.

Net energy: The Net Energy system (NE_M, NE_G, NE_L) describes the amount of energy the animal uses (or nets) for different functions. The NE_L requirement tells us how much of the energy consumed is needed or utilized for lactation. The NE_L value for milking cows includes maintenance and growth requirements.

Total digestible nutrients: TDN is not used to balance energy requirements, but is a method to compare feed quality, as it is a measure of feed energy content. A high TDN value indicates that the nutrients in the feed are highly digestible, enabling the animal to derive a higher energy value from the feed. A low TDN value indicates low-energy feed that the animal will not utilize as well.

Relative feed value: This is also used to compare feed quality, but not specifically energy. A relative feed value (RFV) of 100 is the standard and reflects the quality of alfalfa hay with 41% ADF and 53% NDF (see below for definitions). The higher the RFV, the better the forage.

FIBER

The fiber fraction of plants has three main compounds, and they are all part of the plant cell wall that surrounds the fluid part of each cell. These fiber compounds are cellulose, hemicellulose, and lignin, all intertwined and complexed together in the cell wall. Cellulose is a straight chain of sugar molecules, easily broken down by rumen bugs. Hemicellulose is also a chain of sugar molecules but has branches along the chain. Lignin is a mostly indigestible compound—it goes through the animal virtually unchanged. Pectin and other organic acids are also part of the cell wall but are not considered to be a component of the fiber.

Acid detergent fiber (ADF): ADF is the cellulose and lignin, as determined by a chemical procedure that washes away all the other plant components, leaving these fiber components intact. The ADF value estimates the energy content of feeds—the lower the ADF value, the higher the energy.

Neutral detergent fiber (NDF): NDF is cellulose, lignin, and hemicellulose. The NDF value predicts intake. If NDF is low, animals are able to consume more feed.

NUTRIENT SOURCES

Protein can be found in both forages and grains. See Table 6.3 for typical protein hay crop forages and their protein levels. Relying on high-quality forage helps to reduce or eliminate the need to purchase expensive organic protein sources.

Grain sources of protein include soybean meal and soybean by-products such as extruded, roasted, or expelled soy, distillers and brewers grains, canola meal, flaxseed meal, sunflower meal, field peas, corn gluten meal or feed, and peanut meal. Many of these may not be available certified organic in your area, so check their availability first. Organic protein grain sources are usually expensive. For this reason, growing your own high-protein forage supply is worth considering to minimize the amount you need to purchase.

Forages are also sources of energy, and are included in Table 6.3. Corn silage provides energy through both the grain and stover, as the rumen bacteria can easily digest the fiber in the stover. Ruminal digestion of fiber produces chemical by-products called volatile fatty acids (VFAs)-acetic, propionic, and butyric acids. The animal derives energy from VFAs, and pasture and high-quality hay crops provide more energy to the animal via the highly digestible fiber. Some alternative forage crops such as sorghum-sudangrass, small grain silages, and brassicas are high in energy due to low fiber values and higher fiber digestibility.

Grain energy sources (corn, barley, oats, wheat, and other cereal grains like spelt) have high carbohydrate levels. These grains are typically less expensive than protein sources, but are still expensive and suggest that high-quality forages are important for profitability. These grains have high levels of sugars and starches, frequently referred to as either nonfiber carbohydrates (NFC) or nonstructural carbohydrates (NSC).

VITAMINS AND MINERALS

Farmers often overlook this nutrient class because traditionally they have used a standardized premix from a feed company to supplement vitamins and minerals. Oftentimes, these premixes provide more minerals than needed, and conversely, animals sometimes become mineral deficient if not enough attention is paid to the mineral program. In an organic dairy, minerals are important to maintain healthy animals. A mineral imbalance, either in excess or in deficit, will weaken the immune system, contributing to other health problems.

Major minerals: Requirements include calcium, phosphorus, magnesium, potassium, sodium, chlorine, and sulfur. They all play key roles in various metabolic processes, as well as in the structure and function of different tissues and body fluids.

Trace minerals: Trace minerals include cobalt, copper, iodine, iron, zinc, manganese, molybdenum, boron and selenium, and are important for the function of enzymes and hormones in the body. Since they support immune function, an imbalance could have a negative health impact.

Vitamins: Vitamins A, D, and E are important for immune cell function and many metabolic pathways. The precursor of Vitamin A is ß-carotene, which is plentiful in the green chlorophyll of fresh pasture, and the skin synthesizes Vitamin D when exposed to sunlight. During the grazing season a reduction in vitamin supplementation may be appropriate (except for Vitamin E, which needs to be supplemented year-round), but vitamins should be increased upon returning to winter confinement feeding.

Before feeding any mineral or vitamin, check with your certifier to be sure it is approved for use under organic regulations. Do not accept the word of a salesperson, as the list of approved supplements is constantly changing and different certifying agencies accept different supplements. Salt, a required mineral, is a good example as it must be from a naturally mined source and only a few companies make an approved product.

Continued from page 139 » ingredients are not forages, when included at a low rate they add digestible fiber when needed and maintain butterfat. These may or may not be available as certified organic.

FORAGE TESTING

In order to ensure you meet the nutrient needs of all the animal groups on your farm, it is important to test your forages periodically. The samples you send in should represent what you will feed your animals. Test the primary stored forages and pasture you feed, as they are your ration's foundation. For alternative crops, request a wet chemistry analysis for a more accurate nutrient composition. For more information on forage testing see Box 6.3.

BOX 6.3 FORAGE TESTING

When to test forages

- Before you begin to feed them
- Throughout the year when you notice a change in the forage or in production
- From pastures when there are noticeable changes in seasons or weather patterns
- From purchased forages if you do not know the growing or harvesting conditions
- If growing new forage crops or unique combinations of forage crops (such as peas and triticale together)
- When there are few book values

Sampling procedures

- *Dry hay*: Identify similar bales from each cutting, sample a number of them, and mix the subsamples together. Use a hay-corer to sample multiple layers of each bale. Once you are ready to send samples to the lab, identify them by type of forage species, cutting, or another method to help you track them for future use when formulating a ration.
- *Baleage*: Same method as above. Identify similar bales and mix subsamples together. Reseal the plastic after sampling to prevent mold.
- *Haylage or corn silage*: For an upright, bunker silo, or bag, take samples as the forage is unloaded. Grab handfuls of forage, mix, and subsample. Sample more frequently if cuttings or quality change through the silo or bag.
- *Pastures*: Watch cows to determine what they eat and take samples reflecting their choices of grasses, clovers, and weeds. Avoid areas of contamination from manure and urine, or overmature plants, just as the cows will. To sample, "graze" the pasture by wrapping the forage around your hand and rip it off at the post-grazing height left by the cows. Take many samples in a paddock, mix together and take a smaller sub-sample, pack the plant material tightly in a plastic bag, and freeze to prevent the plants from changing quality due to continued photosynthesis or fermentation in the bag. After freezing for 12 to 24 hours, send the sample to the lab.
- *Alternative crops*: Request a wet chemistry analysis for a more accurate nutrient composition of alternative crops.

SOURCING CERTIFIED ORGANIC FEEDS

A challenge feeding dairy animals under organic management is finding adequate quantities of certified organic grains and forages, although many people forget about sourcing forages because grain seems to be foremost in their minds. Consider the following things before transitioning.

- Make sure you have a reliable supplier of organic grain.
- Consider growing organic grains yourself if you have difficulty finding a supplier (see Chapter 5, "Organic Crop Management" for more discussion).
- Many feed companies sell certified organic grains, but supply is still lower than the demand, resulting in high prices. Produce your own high-quality forage to reduce grain needs.
- Calculate the forage production on your farm and identify potential shortfalls that may result by increasing the total forage in the ration.
- Plan in advance if forages need to be purchased so you can purchase higher-quality forage at a lower cost before your inventory is low.
- Make sure you see what you are going to purchase before buying to make sure it is high-quality forage. Ask for a forage analysis in advance of purchasing and request both the results and crop organic certification records to verify that what was tested is what you purchase.

IDENTIFY NUTRITIONAL GOALS

To identify your nutritional goals under organic management, it is important to identify the milk production level you need to ensure a profit and meet your lifestyle goals. Plan to set these milk production and nutritional goals for age at first calving, calving interval, and cull rates. Once you have determined your costs of production and potential profit with organic management, set your nutritional goals to meet the necessary milk production levels. Table 6.1 shows three milk production levels and management considerations for each.

The best way to use Table 6.1 is to identify the level of milk production you hope to achieve, then look down the column at each management factor. These indicate the minimum management level you will need to reach that milk production goal. For example, if you have chosen a medium level of milk production, you will achieve it as long as you grow your heifers to reach the target ages and weights at breeding and calving, have high-quality forages, and breed cows to calve every 13 to 14 months. Alternatively, if any of your management practices fall below these benchmarks, that is an area you need to improve. Anything above the benchmarks indicates you may achieve a higher production level than anticipated. There are a number of different feeding systems available to help you meet your nutritional goals. The challenge is how to integrate the system you have with your goals for organic management. Sometimes, the system will change over time, based on changing resources, ideas, and goals. There is no one right way to feed cows organically, and there are many options to reach your goal.

DIFFERENT FEEDING SYSTEMS AND METHODS

In your conventional dairy system, you likely used either a total mixed ration (TMR) or you fed different feeds to your cows individually (also known as component feeding). You may already have incorporated pasture into your year-round feeding system. Regardless of your current feeding setup, you can successfully transition to organic. There are, however, advantages and disadvantages to each system. If you are comfortable with your current system and are not considering making a change, feel free to skip to the next section: Requirements and Rations on page 146.

PASTURE-BASED RATIONS

Organic dairy farms are required to give all animals more than six months of age access to pasture. The NOP pasture rules (§ 205.2 Terms defined, §205.237 Livestock feed, § 205.239 Livestock living conditions, §205.240 Pasture practice standard) were under review at the time this book was printed and are scheduled to be released in 2010. Check the current NOP rules for the organic pasture

TABLE 6.1 MANAGEMENT CONSIDERATIONS FOR ASSESSING MILK PRODUCTION GOALS UNDER ORGANIC MANAGEMENT

		LOW PRODUCTION (35–50 lb. milk/cow/day)	MEDIUM PRODUCTION (50–65 lb. milk/cow/day)	HIGH PRODUCTION (65–80 lb. milk/cow/day)
Age/weight at breeding	SMALL BREEDS	More than 17 months 600 lb. or less	15–17 months 600 lb. or less	14–16 months 600–700 lb.
	LARGE BREEDS	More than 17 months 800 lb. or less	15–17 months 800 lb. or less	14–16 months 800–900 lb.
Age/weight at calving	SMALL BREEDS	More than 25 months 950 lb. or less	23–25 months 950 lb. or less	22–24 months 950–1,100 lb.
	LARGE BREEDS	More than 25 months 1,200 lb. or less	23–25 months 1,200 lb. or less	22–24 months 1,200–1,400 lb.
CALVING INTERVAL		More than 14 months	13–14 months	12–13 months
FORAGE QUALITY		Low to average	Good to excellent	Excellent
FEEDING SYSTEMS		Low to no grain Component fed All hay-type crops Low or no management grazing	Moderate grain fed (1:5 or less) Component or TMR fed Hay crop + moderate-quality corn silage Good to excellent grazing management	Higher grain fed (1:4 or more) TMR or intensively managed component fed Hay crops + high-quality corn silage Excellent grazing management

requirements. Based on recommendations made to the NOP, it is likely you will have to document a specific percentage of dry matter intake from pasture. Chapter 4, "Pasture Management" discusses how to estimate pasture intake. Pasture will likely be a large part of your feeding program for 4 to 6 months of the year, and up to 8 months if you can extend it into spring and fall.

In all feeding systems, pasture should be the basis of your ration and everything else should supplement the pasture. Treating pasture as a supplement to your regular feeding system will result in less success and higher costs. Pasture should be managed for high quality and high intake, and is a different management system than confinement. Managed grazing is complex, and learning how to graze before you begin your transition to organic is recommended.

If pastures are well-managed and high quality, you can achieve 60–100% of the animal's intake from pasture. It will be high in protein, as high as 30% in early spring, which can reduce or eliminate all purchased protein during the grazing season. Since organic protein sources are expensive, the more pasture you use, the lower your costs will be. Recent research has shown that overfeeding protein in the barn reduces pasture intake because cows can sense nutrient overload.[1] Protein from pasture is highly rumen degradable, and rumen bacteria convert extra degradable protein into ammonia if they cannot utilize it. This costs the cows energy to first convert and then to excrete the ammonia as urea, and contributes to loss of milk production and body condition.[2] Thus, the protein balance is essential. The most important nutrient to feed is energy, which encourages cows to seek out high-protein pasture. Be sure to read Box 4.3.

EXAMPLE 6.1 TIESTALL BARN WITH COMPONENT FEEDING

Brian and Liz Bawden, organic dairy farmers in northern New York, feed grain solely to maintain body condition, not to push production. Their grazing season feed ration includes pasture, dry hay, a 16% protein grain, and a high-calcium mineral fed in the tiestall barn. They provide free-choice Redmond salt and kelp to the milking herd and baleage in the pasture. They individually feed 5–10 pounds of grain each milking. Their cows begin grazing in mid-May and remain outside until the end of October. In the winter, they feed the cows dry hay and grain in the tiestall and baleage outdoors unless the weather is inclement. During bad weather they feed baleage in the barn. Liz and Brian visually monitor the cows for production and manure consistency to determine if they need to adjust their feeding program.

In early spring, feed a higher rate of magnesium, up to 0.35–0.37% of dry matter, to avoid grass tetany or "grass staggers." During the rapid spring pasture growth, the grasses do not take up as much magnesium as they do during slower growth periods. If magnesium levels are low or barely adequate in the ration, it can cause tetany, a malfunction in the nervous system's interactions with muscles that prevents the muscles from contracting and expanding properly. If potassium is high in the diet, tetany increases because high potassium inhibits the uptake of magnesium from the diet. Signs of grass tetany include the inability to stand or walk and noticeable muscle tremors. It is sometimes confused with milk fever, especially in cows between early lactation and peak milk production.[3]

WORKING WITH YOUR CURRENT FEEDING PROGRAM

When you transition to organic farming, you will have to incorporate pasture into your feeding program, but organic farmers work with many of the same feeding systems that conventional farmers do. Table 6.2 (pages 148–149) outlines the various feeding systems organic farmers use, their advantages and disadvantages, possible improvements, and how to incorporate high pasture intake into each system.

EXAMPLE 6.2 TIESTALL BARN WITH MOLASSES COMPONENT FEEDING

Jim Gardiner of Otselic, New York advocates feeding molasses as the only energy source for milking cows on pasture and has been doing so for the last few years. From late April to early November, his cows are fed high-quality pasture and a maximum of 7 pounds of molasses per cow per day in two feedings based upon milk production and body condition. This rate is based upon a philosophy that each pound of molasses is the energy equivalent of 3–4 pounds of corn meal. The cows milk well and maintain good body condition through the grazing season with this feeding program. Jim varies the mineral and vitamin supplementation based on the weather conditions, and feeds primarily Sea-Agri 90 as his trace mineral salt source. From early November until cows go out to pasture in April, he feeds baleage and dry hay harvested from hay fields and pastures as well as from cover crops of peas with oats, barley, or millet. He also feeds molasses in the winter until mid-January, then begins to feed a mix of barley, peas, corn, and oats with less molasses.

REQUIREMENTS AND RATIONS

NUTRITIONAL REQUIREMENTS

Scientists have estimated nutritional requirements for every age, size, class, and breed of dairy animals through research done in a confinement conventional system. Although this research does not account for many of the organic or grazing systems, it provides a starting point. Nutritional requirements are set by the average herd needs. There are three different rations (formulated, fed, and consumed) because animals do not read computer printouts, people are not always consistent, and sometimes the animals know better than we do what they want and need to eat. To maintain healthy animals, consistency in management and nutrition goes a long way, and may be more critical than meeting all nutritional requirements exactly.

One challenge with organic dairying is that few nutritionists understand the ration changes that encourage good health, productivity, and profitability in an organic system. Many organic dairy farmers have decided to do their own ration balancing. They make it work by knowing their forages, the principles of nutrition, and closely monitoring their cows for changes. Of course, if you currently have or can find a nutritionist who is willing to work with you during and after your transition to organic, that may be helpful. Computer programs exist for ration balancing, but you may need to take a class or learn how to use them from a nutritionist, another farmer, or your local cooperative extension.[4]

It is important to offer a consistent diet to dairy animals and meet the needs of the group average. To do this, you must know the nutritive values of your forages and supplement the nutrients lacking. The amount of feed your cows receive on a daily basis should be as consistent as possible, especially for lactating cows. If dry matter intake changes frequently, the nutrient supply also changes. The result of an inconsistent diet is inconsistent milk production, inconsistent growth, and stress resulting in possible health issues.

DEVELOPING RATIONS

Forage sample results are a starting point for developing rations for animals on the farm. They help prioritize which forages to feed to different groups of animals, feeding the best quality to those with the highest nutritional needs such as calves and milking cows. The reports list important nutrients such as protein, energy, fiber, and minerals, and can be used to formulate a basic ration either by hand or with computer software.

Table 6.3 (page 150) identifies the recommended minimum forage quality you will need to produce on your farm to keep purchased feed costs low. Compare your current forage nutrients to the recommendations. Will you need to supplement nutrients?

REQUIREMENTS OF MILKING COWS

Lactating dairy cows have the most complex nutritional requirements and their requirements change through early, mid, and late lactation. Thus, it is difficult to balance one ration for a whole herd or group of cows without over or underfeeding some of them at all times unless you are a seasonal dairy. The greatest demand for nutrients comes 50 to 90 days after the cow calves, when she approaches and is at the peak of her milk production.

Nutritional considerations during peak production

- Energy and protein needs are greater than what she can consume in her feed and body condition goes down.
- Early-lactation cows mobilize body energy stores like fat and other tissue to meet those needs.

Continued on page 150 »

EXAMPLE 6.3 TIESTALL BARN WITH TMR

In central New York, Kathie, Rick, and Bob Arnold milk 130 cows split into two housing and milking groups in a tiestall and a small freestall barn. They intensively graze 200 acres of pasture. They move the herd into fresh pasture every 12 hours and the cows consume approximately 40–50% of their DMI this way. In the summer they feed a TMR of haylage, corn meal, barley or triticale, wheat midds, salt, and minerals to supplement pasture intake. The haylage is stored in a bunker silo and the dry grains are stored in open bay commodity sheds. They use alternative grains such as ground peas or sunflower meal when wheat midds are not available. The Arnolds use a mobile mixer wagon to make the TMR and add ingredients with a bucket tractor. They drive the wagon into the barn's hayloft and empty it into a chute that drops the TMR into an electric feed cart. In the freestall, they feed the TMR in a bunk. During the winter, the TMR includes haylage, high-moisture corn, wheat midds, salt, and minerals. The high-moisture corn is augured from an upright silo into the mixer wagon.

TABLE 6.2 FEEDING SYSTEMS AND CONSIDERATIONS			
BARN	**FEEDING SYSTEM**	**DAIRY COMPATIBILITY**	**ADVANTAGES**
TIESTALL	Component feeding	Smaller dairies	Can feed cows individually for milk production or body condition. Promotes preventive healthcare: can assess daily individual cow health concerns, changes in behavior or appetite, and signs of stress
	TMR	Smaller dairies	Reduces labor Daily cow observation With mobile mixer wagon, can feed more in a bunk area outside barn Option of partial TMR if feeding TMR outside barn and extra forages or grain to the cows in the barn
FREESTALL BARN	TMR	Midsize to large farms with milking parlor	Can group and feed cows according to production level Less labor
	Grain fed in parlor	Older barns without drive-through feed alley or access for tractors or mixer wagons	Better control over amount of grain fed: •Some feeders adjust feed amounts when cows come into the parlor •Electronic transponders give preset amount of grain based on cow's production.
	Computer feeders (programmed transponder on cow releases set amount of grain)	Older freestall barns not conducive to TMR feeding	Control over individualized grain feeding without slug feeding too much at once
ANY	Forage only, no grain rations	Any	Farms with well-managed pasture and high-quality forages, good cow genetics, and exceptional cow management have had good success with this model Avoid high purchased-feed costs

DISADVANTAGES	POSSIBLE IMPROVEMENTS	PASTURE FEEDING CONSIDERATIONS
High daily labor requirements: • Many different feeds fed more than once or twice daily • Cows milked in their stalls—less efficient than a parlor	Reducing number of different feeds Establishing group feeding outside barn Include minerals in grain or free choice	Grazing season: Reduce variety and amount of stored forages Limit corn silage to 15–25 lb./day Supplement pasture with only molasses and minerals in barn
Requires more dry hay to increase amount of forage in ration Certain models of TMR mixers do not handle high levels of hay, so it must be fed separately or chopped Requires additional equipment: stationary mixer, mobile mixer, and/or automated cart	If you plan to make changes with your TMR, first discuss with other organic dairy farmers who have made the same change.	Make partial TMR by reducing the amounts and types of forage Reformulate partial TMR mix based on pasture quality and protein and use as supplement to pasture
Less attention paid to individual animals to detect health problems or stress—easier to miss a sick cow Farmer may look at the group as a whole	Increase forage by adding haylage, corn silage, baleage, or chopped hay into mix Make partial TMR by feeding dry hay or baleage separately and free choice	Same as Tiestall; TMR (above)
If grain amount cannot be preset, allows more room for operator error Tendency to overfeed grain—when one cow does not eat her full allocation, the next cow eats the leftovers plus her portion	After and during transition, many farmers change to feeding all cows same amount of grain, or eliminate parlor feeding all together Convert to conveyor system	Few changes to grain delivery method Change forage types and amounts fed Reduce grain protein levels
Limited by number of computer stalls—some cows may wait to eat, especially during the grazing season when in the barn for short periods of time. Dominant cows may block submissive cows from entering and/or eat their grain If there is a computer problem, no one eats.	Many organic farmers have made the same adaptations as above (grain fed in parlor). Some have reduced amount of grain fed to reduce cow waiting time, others include grain in forage	Same as Grain Fed in Parlor (above)
Difficult to get right Risk of loss of body condition if not getting enough nutrients	Forage sampling Willingness to be flexible and supplement when needed.	

Continued from page 147 »

- It is critical to feed enough nutrients at peak production to minimize stress.
- Ensure that intake increases as quickly as possible after calving—offer high-quality forage and supplement with grain to boost energy and protein intake.
- Total dry matter intake of forage and grain at peak production should be between 3.5% and 4.0% of her body weight.

Nutritional considerations during peak production for less grain
Organic dairy farmers often feed less grain than the average conventional farm, which requires extra attention during peak production.

- Providing a grain to milk ratio at peak production of 1:4 or less (i.e., 1:5 or 1:6) provides extra nutrients less expensively.
- Feeding some grain prior to calving so the rumen adjusts to it.
- Increase grain feeding rates by 1 pound every 3 to 4 days. Otherwise, metabolic problems such as acidosis and additional stress may compromise the immune system.

Once cows pass peak production, they eat what they need for production, plus maintenance and breeding. Whereas before peak they were in "negative balance," they return to a positive balance after peak (see Figure 6.1), allowing them to start ovulating and return-

TABLE 6.3 MINIMUM RECOMMENDED FORAGE QUALITY FOR ORGANIC MILK PRODUCTION

Forage	% NDF (Fiber)		% Protein		NEL, Mcal/lb. (Energy)		% NFC (Carbs)		% ADF (Fiber)	
	MINE	GOAL	MINE	GOAL	MINE	GOAL	MINE	GOAL	MINE	GOAL
Dry hay (grass) / mixed legume		50–60		13–18		0.52–0.61		15–20		35–42
Dry hay (alfalfa)		42–47		20–22		0.55–0.58		22–25		33–35
Grass/legume haylage or baleage		49–55		16–18		0.49–0.58		18–22		32–38
Alfalfa haylage/baleage		37–45		21–23		0.55–0.61		22–28		30–35
Managed pasture: spring & fall		45		>25		0.68		17		25
Managed pasture: summer		50		>20		0.64		15		30
Corn silage		45		8.8		0.65		38		28
Other:										
Other:										

EXAMPLE 6.4 FREESTALL BARN WITH TMR

Susan and Vaughn Sherman of Jerry Dell Farm in Dryden, New York, along with their sons and nephews, milk 340 cows year-round. The Shermans use a nutritionist to balance the herd's rations and feed a TMR in the freestall barn to supplement summer pastures. About 25% of the ration is TMR at peak pasture and they feed complete TMR during the winter months. Their TMR includes salt and minerals, haylage, and a grain combination of high-moisture corn, shell corn, ear corn, soy, and small grains such as spelt and triticale. The summer ration includes 15 pounds of grain and the winter ration contains 25 pounds. Their milking herd grazes 250 acres and the family is developing more pasture for the dry season. Vaughn believes that providing high-quality pasture is the most cost-effective way to maintain high milk production while reducing grain expenses. They operate an intensive grazing system and move their milkers every 12 hours. They have divided their milking herd into three groups—high, medium, and low production—and keep them in individual pastures.

EXAMPLE 6.5 FORAGE ONLY, NO GRAIN RATION

Rob and Pam Moore, in the New York Southern Tier, have not fed any grain to their cows since starting their organic dairy in 1996. They began with a herd of heifers not bred for extreme high production and who had never eaten grain. The first few years after those heifers calved and began milking were challenging, as body condition and milk production both dropped. Over the years, the Moores have learned the combination of forage quality, genetics, and management to make a no-grain feeding program work. They are the first to admit that it is not as easy as it sounds. In recent seasons with extreme weather such as drought or heavy rain, they have supplemented the cow's energy intake with 1–4 lb. of molasses/cow/day. However, they do not feed molasses on a routine basis. They also feed kelp, Redmond trace mineral salt, and Sea-Agri 90.

Currently they manage a spring seasonal herd and feed dry hay and baleage in the winter while the cows are dry. They begin to graze in the spring when the grass is 4 to 6 inches tall and gradually reduce hay and baleage until May 1, when it is eliminated. They usually graze until late December, providing there is forage still available, slowly adding in dry hay and baleage in the beginning to mid-October. From January to April they feed the cows a 100% hay and baleage ration.

ing body reserves. In Figure 6.1, the shaded area between the milk energy and feed energy lines represents the amount of energy that the diet is lacking compared to the need for energy for milk production—the negative energy balance.[5]

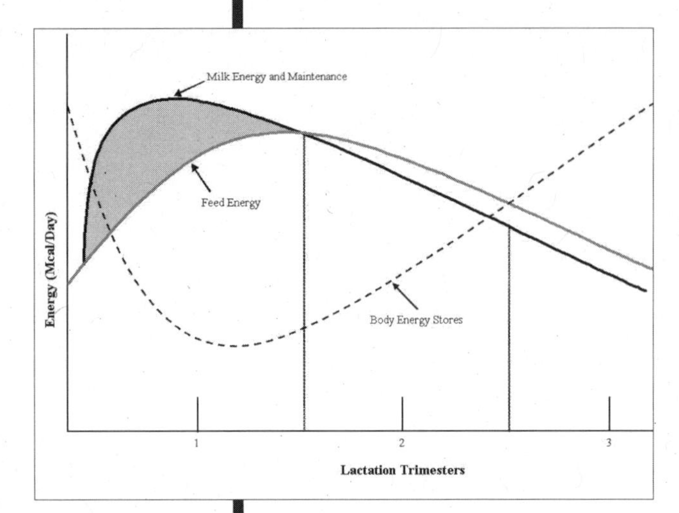

FIGURE 6.1 *Generalized milk energy to feed energy relationship*

At positive balance, a ration balanced for total dry matter intakes of approximately 3.25%–4% of body weight that meets the guidelines in Table 6.4 will allow the cows to breed back, gain condition, and produce milk. Table 6.4 includes nutrient ranges that account for differences in breeds and milk production. Generally, the lower end of the ranges are for smaller breeds or those producing less milk.

Learning to balance rations: Examine Table 6.4. If you have the ability to feed cows individually or group them by production, you can use this data to identify the changes you should make in different lactation stages. Example 6.6 calculates a ration for cows at peak production, since they have the greatest need for nutrients.

Double-check your rations to ensure you meet the protein and energy requirements. Check different subsets of cows, such as those in later lactation producing less milk. If you change forages, reanalyze them and check that requirements are still being met. Lastly, be aware of other changes such as significant shifts in production or body condition and determine if the ration is still on target.

TABLE 6.4. LACTATING COW NUTRITIONAL REQUIREMENTS

STAGE OF LACTATION	DRY MATTER INTAKE		PROTEIN %	NEL	
	LB/DAY	% BODY WEIGHT		MCAL/DAY	MCAL/LB
Early (0–50 days)	20–35	2.1–2.6	16–18	19–30	0.85–0.95
Peak (50–90 days)	35–55	3.5–4.5	14.5–17	24–40	0.70–0.75
Midlactation (90–180 days)	30–50	3.0–3.5	13.5–17	22–37	0.70–0.75
Late lactation (180–300 days)	25–45	2.5–3.0	12–16	20–25	0.60–0.70

Source: National Research Council. 2001. Nutrient Requirements of Dairy Cattle, 7th rev. ed. National Academy Press, Washington, DC.

EXAMPLE 6.6
BALANCING A RATION FOR PEAK PRODUCTION USING TABLE 6.4

Holstein-Jersey crossbred 1,200-pound cows peaking at 80 pounds of milk on pasture represent the average performance of cows on many Northeast farms. This example calculates for dry matter and then converts it to "as fed."

STEP 1. DETERMINE TOTAL DRY MATTER INTAKE (DMI)

Weight of cow × % body weight (using Table 6.4) = DMI
1,200 lb. × 3.50% = 42 lb. DMI

During pasture season, high-quality forage is on hand, so we can target 70% forage in the diet, which is 29.4 pounds of forage (42 × 70%) and 12.6 pounds of grain (42 – 29.4), a grain to milk ratio of 1:5.

STEP 2. DETERMINE HOW MANY POUNDS OF PROTEIN THEY NEED:

Lb. DM × % protein (using Table 6.4) = lb. of protein
42 lb. DM × 17% protein = 7.14 lb. of protein.

STEP 3. CALCULATE HOW MUCH OF THAT PROTEIN IS MET BY YOUR FORAGES:
The forage analysis for this example is 21.4 lb. of 22% crude protein (CP) pasture, 5 lb. of 8% CP corn silage, and 3 lb. of 12% CP hay.
The cows will eat 21.4 lb. of pasture dry matter (29.4 lb. DMI – 8 lb. other forage DMI).
The calculations of protein are:

DM WEIGHT × % CP =	LB. PROTEIN
21.4 lb. pasture × 22% CP =	4.71 lb. protein
5 lb. corn silage × 8% CP =	0.40 lb. protein
3 lb. hay × 12% CP =	+ 0.36 lb. protein
Total protein =	5.47 lb. protein

The ration is short of the protein requirement (7.14 lb.) calculated in Step 2 by 1.67 lb., so more protein will need to be provided by the grain mix:

1.67 lb. protein ÷ 12.6 lb. grain = 13% CP needed in the grain

A 13% CP grain mix would be affordable. If the pastures were higher quality in protein, at 27% CP, the ration would only be short by 0.60 lb. of protein, and a little corn meal would help meet the requirements.

STEP 4. HOW MUCH ENERGY DO THEY NEED FROM GRAIN?
They need between 35 and 38 Mcal/day based on Table 6.4. The pasture will provide 0.68 Mcal/lb., the corn silage 0.75 Mcal/lb., and the hay 0.53 Mcal/lb. (Table 6.3).

4A. TO CALCULATE THE TOTAL ENERGY IN THE FORAGE:

FEED WEIGHT X ENERGY FROM FEED MCAL/LB =	ENERGY IN RATION FROM FEED
21.4 lbs. pasture × .68 Mcal/lb =	14.55 Mcal
5 lbs. corn silage × .75 Mcal/lb =	3.75 Mcal
3 lbs. hay × .53 Mcal/lb =	+ 1.59 Mcal
Total Energy =	19.89 Mcal/day

The grain must provide at least 15.11 Mcal for the ration (35 Mcal/day – 19.89 Mcal/day from forage).

Continued on next page ≫

Continued »

4B. TO CALCULATE THE TOTAL ENERGY NEEDED IN THE GRAIN:

Needed energy ÷ Total lbs. of grain in ration = Energy needed in grain

15.11 Mcal ÷ 12.6 lb. grain = 1.20 Mcal/lb. of grain fed

It is important to note that the number above (1.20 Mcal/lb.) is energy per pound of grain fed, not the total amount of energy in the total grain fed. Unfortunately, not many certified organic grain products meet this energy requirement, as energy density of grains is usually less than 1.0 Mcal/lb. Thus there are only a few options for this ration—feed more grain and less forage, improve forage quality to meet the guidelines in Table 6.4, or include molasses as an energy source. It may make sense to feed a higher grain rate, such as 1:4, until cows are past peak production and in positive energy balance. If no changes are made to the ration, it is likely that these cows would lose milk production, body condition, and/or have trouble breeding back.

Formulating your own grain mix

If you decide to formulate your own grain mix, a simple and old-fashioned method is Pearson's Square. Draw a box as below, with the amount of protein you want to achieve in the center. In our first example above, we needed a 13% CP mix. On the left side of the square, list one ingredient and its protein content next to the top corner, and a second ingredient and protein content next to the bottom corner. In this case, corn meal (9% CP) and sunflower meal (28% CP) form the base of the grain mix. Follow the arrows diagonally across the box, subtracting the number in the middle of the box from those on the left side. Then add the numbers that result on the right.

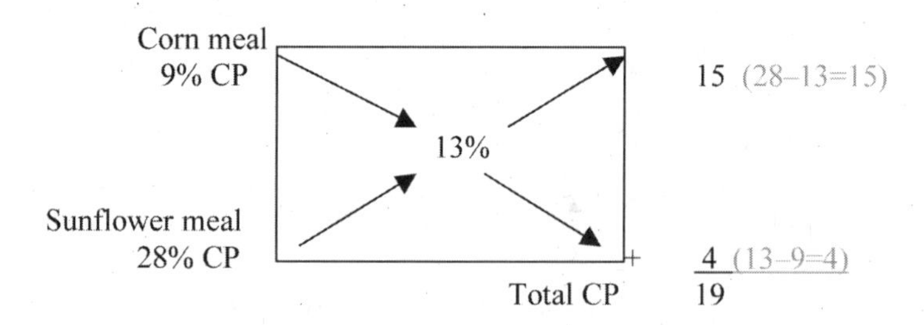

To determine the percentage of the grain mix that each ingredient contributes toward the desired 13% protein, use each number on the right side of the box for the ingredient on the left side. For example:

Corn meal = 15 ÷ 19 = .79 × 100 = 79% of grain mix
Sunflower meal = 4 ÷ 19 = .21 × 100 = 21% of grain mix

If you want to use more than two ingredients, the method is the same as above but repeat the Pearson's Square after you mix the first two ingredients. The corn meal/sunflower meal mix (19%) will be at the top left side and your third ingredient will be in the bottom left corner. If you want more than three ingredients, you will first have to calculate the average protein content of the two new feeds together, for each of the top and bottom places on the left side of the square, and then run the calculations as above. Table 6.5 shows protein and energy values for common organic feed grains.

To convert the ration for peak-lactation cows to an "as-fed" basis, you need to know the dry matter content of each feed. If we performed a forage test, we would likely find our pasture is 20% DM, corn silage is 35% DM, hay is 84% DM, and grain is 89% DM. The calculations for an "as-fed" ration are as follows.

21.4 lb. pasture ÷ 0.20 = 107 lb. as fed
5 lb. corn silage ÷ 0.35 = 14.3 lb. as fed
3 lb. hay ÷ 0.84 = 3.6 lb. as fed
12.6 lb. grain ÷ 0.89 = 14.2 lb. as fed

This process works for all major nutrients to check if you meet the nutritional requirements. The general rules of thumb for other nutrients include the following.

Acid Detergent Fiber (ADF)	17–21% of DM
Neutral Detergent Fiber (NDF)	25–33% of DM
Non–Fiber Carbohydrate (NFC)	35–40% of DM
Rumen Degradable Protein (RDP)	60% of CP
Rumen Undegradable Protein (RUP)	40% of CP
Soluble	30% of CP

Minerals and vitamins for milking cows are not usually formulated to meet their needs specifically, even by nutritionists. In most cases, nutritionists or feed sales representatives include a premixed mineral and/or vitamin mix. Make sure the premix is approved by your certifier because the NOP does not allow certain anti-caking agents

TABLE 6.5 PROTEIN AND ENERGY VALUES FOR COMMON ORGANIC FEED GRAINS

ORGANIC FEED	PROTEIN	NET ENERGY LACTATION	COST/TON (April 2009 prices)
48% SOYBEAN MEAL	48%	0.95 Mcal/lb.	$1,100–$1,200/ton
ROASTED SOYBEANS	43%	1.23 Mcal/lb.	$850–$925/ton
SUNFLOWER MEAL	28% nondehulled seeds 42% dehulled seeds	0.62 Mcal/lb.	$600–$675/ton
FIELD PEAS	25%	0.91 Mcal/lb.*	$550–$650/ton
FLAX MEAL	32%	0.71 Mcal/lb.	$1,000–$1,100/ton
CORN MEAL	9%	0.91 Mcal/lb.	$375–$450/ton
OATS	13%	0.80 Mcal/lb.	$375–$450/ton
WHEAT	14%	0.90 Mcal/lb.	$375–$450/ton
BARLEY	12%	0.84 Mcal/lb.	$400–$475/ton

*Value is for conventional field peas; data on organic field peas was not available.

Source: National Research Council. 1989. Nutrient Requirements of Dairy Cattle, 6th rev. ed. National Academy Press, Washington, DC.

BOX 6.4 MINERALS

A few organic herds manage their minerals more exactly. For example, Jim Gardiner of Otselic, NY changes the amounts of vitamins A and D that he feeds based on weather conditions. If the weather is cool and wet during the grazing season, he feeds one ounce of each vitamin per cow to compensate for less vitamin A in the plants and less vitamin D from sunlight. He increases the amount in the winter for the same reasons, and may use as much as 2 ounces of vitamins depending on forage quality. He feeds 5 ml/cow/day of zinc and copper (feed-grade) during drought or wet conditions. In drought, it helps boost the immune system so if cows graze too closely and pick up parasite larvae, they are better able to defend themselves. During wet conditions, cows need to maintain good foot health, so he adds zinc and copper to help maintain healthy hoof tissue.

such as yellow prussiate of soda. Kelp is often fed free choice because it provides a broad spectrum of minerals, vitamins, and antioxidants that provide both nutrition and immune system benefits. It also is frequently mixed into a grain or mineral mix.

LATE LACTATION MANAGEMENT

As cows approach the end of their lactation, they should have good body condition, but not overly fat. If they still milk fairly well close to 60 to 45 days before their due date, you can change the ration to help them reduce their production before drying off.

Important nutritional considerations during end of lactation

- Body condition should score around 3.5 to 3.75 (discussed in more detail later in this chapter).
- *Thinner cows*: Feed more energy to increase body condition before drying off. (Milking cows gain weight more easily because their metabolism is still high and they convert energy to fat more easily. Metabolism slows down in the dry period and trying to increase condition at this stage will require more feed.)
- *Fatter cows*: Reduce the amount of energy in their diet to drop body condition. (Fat cows in the dry period may develop health and/or metabolic problems, including fatty liver and ketosis after calving and into the first month of lactation.)
- *High producers*: Restrict intake and feed lower-quality forage. (This will minimize udder pain when dried off and minimize the chance of mastitis in the next lactation due to related stress or immune system effects.)
- *To dry-up a cow quickly*: Restrict water intake, but no more than a 30% reduction in water intake, as other health problems due to dehydration may result if water is more limited.

REQUIREMENTS FOR DRY COWS

There are two stages of the dry period to consider from a nutritional standpoint. The first of these is the early dry period, the first

30 days after cows dry off, and the second is the transition or close-up period, the final 2 to 4 weeks before the cow calves when her nutritional requirements increase.

Nutritional considerations during early dry period (first 30 days)

- Cows can be maintained on forages and minerals alone.

Nutritional considerations during transition or close-up period

The more challenging part of the dry period is the transition or close-up period. To manage cows during this time, it helps to understand the metabolic changes a cow undergoes during the transition.

- Through the dry period the cow primarily uses nutrients for fetal growth.
- The unborn calf grows rapidly during the last trimester (90 days) of gestation.
- Protein and energy requirements significantly increase as calving date approaches.
- Dry matter intake decreases as calving approaches, so cows require a higher nutrient density ration.
- Table 6.6 lists the recommendations for ration formulation with transition dry cows.

Metabolic changes

The dry period is important because many metabolic changes occur. If cows are managed poorly, these changes can get out of hand and lead to metabolic disorders. Immediately before calving, the cow dramatically drops her dry matter intake as much as 70% compared to just two to three days before. This drop in intake causes the cow to begin mobilizing body fat for energy, which breaks down into nonesterified fatty acids (NEFA) that release into the blood. Body fat mobilization also releases glycerol for energy. NEFA in the blood increases with less intake, and the longer intake is reduced, the longer NEFA is elevated. The liver absorbs NEFA and converts it either to usable energy, ketone bodies, or triglycerides. When NEFA levels are too high, the liver makes more triglycerides, which are stored in the liver leading to fatty liver syndrome. High levels of NEFA have also been associated with other metabolic disorders such as dystocia, retained placenta, ketosis, displaced abomasum, and mastitis. The cost of these disorders is high and thus worth managing transition cows properly.[6] Under organic management, it becomes even more critical to manage the transition period well to avoid these disorders.

Energy

Another important strategy in controlling NEFA levels is to ensure that cows receive the dietary precursors of glucose before and after calving. Glucose is the primary energy source for both fetal growth and milk production and is primarily made in the liver from products of rumen digestion. If dry matter, carbohydrate, and protein intake levels are all sufficient, the cow's liver will make glucose and minimize the mobilization of body fat (and thus NEFA levels).[7]

Diet changes
Major diet changes around calving time can have a negative influence on dry matter intake. Changing forages, TMR recipes, or concentrates can put a cow off feed for a few days. This is especially important in grazing herds because fresh pasture forage is significantly different from stored forages. If you feed close-up dry cows stored forages before calving, the switch to fresh grass after calving may cause them to decrease their intake significantly. Likewise, close-up cows on grass that are confined after calving also drop their total intake. Thus, it is best not to change the diet for transition cows. Dry cows on pasture should stay on pasture after calving, and dry cows on all stored feeds should continue with stored feeds for the first 21 days of lactation.

From a practical standpoint, producers have had good results letting close-up dry cows graze in paddocks the milking cows have already grazed. They put the dry cows on the paddock during one grazing period per day (i.e. days or nights only) and feed additional forages and grain in the barn the other part of the day. This ensures that the dry cows have some fresh forage in their rumens, but also receive a diet balanced for protein and carbohydrates. This works especially well for farms that do not have extra acreage to designate for dry cow pasture or where the traditional dry cow pasture is poor quality. Producers with additional pasture may keep dry cows out all the time and deliver the remainder of the ration to an outdoor feedbunk or feed wagon.

Mineral content of close-up ration
A concern for close-up dry cows is the mineral content of the diet. Potassium (K) levels should be as low as possible—less than 2% K—to minimize the risk of milk fever. Forages from fields with high rates of manure application should be tested for K before feeding to dry cows. Potassium's effect is to tie up magnesium (Mg) and make it unavailable to the cow. When K levels are high, Mg can be increased to compensate for the effect. See Table 6.6 for the K and Mg requirements of close-up dry cows.

Selenium and vitamin E are important supplements during the dry period to boost the immune system of the cow and help the calf after it is born. Selenium and vitamin E prevent the calf from developing white muscle disease and the cow from developing metritis and cystic ovaries. Northeast soils are deficient in selenium, making forages grown there also deficient, so it needs to be supplemented in the diet. Most vendors comply with the legal limit of supplemental selenium (0.3 mg per kg [2.2 lbs.] of DM) in purchased feeds or minerals. Check with your certifier for approved sources of selenium and vitamin E.

Managing transition cows is different for each farm based on its resources. Table 6.6 outlines the nutrient guidelines for transition cows. The key factors that determine a successful transition cow program are:

TABLE 6.6 NUTRIENT GUIDELINES FOR EARLY DRY, CLOSE-UP DRY, AND FRESH COWS

ITEM	UNITS	FAR-OFF COWS	CLOSE-UP COWS[a]	FRESH COWS[b]
GENERAL				
DMI	Lb.	28–30	27–29	42+
DMI	kg	13–14	12–13	19+
NEL	Mcal/d	15–17	16–18	
NEL	Mcal/lb	0.59–0.63	0.62–0.66	0.78–0.80
NEL	Mcal/kg	1.30–1.39	1.40–1.45	1.72–1.76
ME allowable milk	% of req.	N/A	N/A	80–90%
PROTEIN				
Crude protein	% DM	< 16	<15	<18.5
Rumen N balance	%	100–120%	100–120%	105–130%
Rumen peptide balance	%	95–130%	95–130%	95–130%
Metabolizable protein (MP)	g	1,000–1,100	1,200	N/A
MP from bacteria	% of MP	no guideline	> 60%	> 50%
MP allowable milk	% of req.	N/A	N/A	90–95%
Urea cost	Mcal/d	no guideline	< 0.5 Mcal	< 0.5 Mcal
CARBOHYDRATE				
Forage NDF, optimum	% BW	0.7–0.8	0.6–0.7	0.85–1.0
peNDF	% DM	30–40	28–35	24–26
Starch, maximum	% DM	16	19	27
Sugar	% DM	<3	<3	5
NFC	% DM	26–30	30–34	38–42
FAT				
Total fat, optimum	% DM	3.5	3.5	4
Total fat, maximum	% DM	4.5	4.5	5
MACROMINERALS				
Calcium (Ca)[c]	% DM	0.5–0.7	0.9–1.1	0.8–1.0
Phosphorus (P)	% DM	0.3–0.35	0.3–0.35	0.35–0.4
Magnesium (Mg)	% DM	0.2–0.25	0.4–0.45	0.3–0.4
Potassium (K)	% DM	< 2.0	< 1.3	1.5–2.0
Sodium (Na)	% DM	< 0.1	< 0.1	0.4–0.6
Chloride (Cl)[c]	% DM	0.4–0.8	0.4–0.8	0.3–0.4
Sulfur (S)	% DM	0.3	0.3	0.25–0.3
TRACE ELEMENTS				
Selenium (Se)[d]	PPM	0.3	0.3	0.3
Cobalt (Co)	PPM	0.3	0.4	0.4
Iodine (I)	PPM	0.8	0.8	0.8
Copper (Cu)	PPM	15–20	15–20	15–20
Manganese (Mn)	PPM	60–80	60–80	60–80
Zinc (Zn)	PPM	60–80	60–80	60–80
Iron (Fe)[e]	PPM	60	60	60
VITAMINS				
A	IU	100,000	125,000	150,000
D	IU	25,000	30,000	45,000
E	IU	1,000	1,800	1,000

Note: Your nutrition advisor may adjust these amounts for your herd situation.

a If on an anionic ration, ration calcium should be increased to 1.1–1.4%. Magnesium and sulfur would be increased to 0.4–0.45% and chloride up to 0.85%.

b First 30 days in milk or until cows are coming on feed rapidly.

c If DCAD formulation in close-up diet, increase Ca to 1.2–1.3% and use Cl sources to titrate urine pH.

d FDA limit for added selenium (recommend organic source if available for dry and fresh cows).

e Iron will generally not be added to the diet.

Sources: T. R. Overton and L. E. Chase, Cornell University. Updated 1/2009.

National Research Council. 2001. Nutrient Requirements for Dairy Cattle. National Academy Press, Washington, DC.

- maintaining intake
- minimizing body fat mobilization, and
- feeding appropriate levels of protein and carbohydrates during the 42-day period.

Rations: Rations for dry cows in the early part of the dry period, approximately 45 to 60 days prefresh, are relatively easy to formulate because the cows' metabolism has slowed down to maintain themselves and grow the unborn calf primarily. They can eat lower-quality forages than other groups of animals on the farm and can reduce or eliminate their grain intake. However they still need their mineral supplements. The process of developing the ration is similar to the process used for milking cows (see Example 6.7).

EXAMPLE 6.7 DEVELOPING A DRY COW RATION FOR 45–60 DAYS PREFRESH

To develop a ration for dry cows, use the same 1,200-pound Holstein-Jersey crossbred cows as in Example 6.1, but assume their body weight has increased slightly due to a gain of body condition and the increased weight of the calf. They are most likely closer to 1,300 pounds at the end of lactation, so the estimation of DMI is 26 pounds per day based upon this weight (1,300 lbs × 0.02).

A typical feed for this group might simply be average quality baleage, or a mix of hay and a small amount of corn silage. Most baleage meets the requirement of 10–12% CP and 0.44–0.48 Mcal/lb. of NE_L, as do many good-quality grass-legume or legume hays. During the grazing season, this group of cows can follow milking cows and graze the lower-quality forage left behind. Even though it is lower quality, it is still higher-quality than early dry cows need so you can feed lower-quality hay as well if time and labor allow. Free-choice minerals and salt may be offered to this group whether they are grazing or eating stored forages.

Physiological changes during the dry period

When dry cows used to eating grain shift to a diet of almost all forage, the rumen bacteria shift from carbohydrate digesters to fiber digesters. This causes the rumen pH to become more neutral and the rumen papillae (which absorb nutrients) lining the inside of the rumen to become smaller, reducing the surface area in the rumen for nutrient absorption.

To manage this physiological change, reintroduce grain 2 to 4 weeks before calving so the rumen bacteria have time to increase the populations of carbohydrate digesters and the papillae have time to lengthen *before* grain is increased after calving. This is also important with low-level grain feeding after calving, because if the papillae do not have time to regrow, acid builds up in the rumen and cannot be absorbed. This can lead to metabolic health issues such as subclinical acidosis, laminitis, and ketosis. To keep the cows healthy, the rumen needs to be healthy as well.

HEIFERS

Requirements and rations

All certified organic animals over the age of 6 months are required to have access to pasture during the grazing season. It is highly recommended that the majority of their intake is pasture forage, with a minimum of 30% dry matter from pasture. The NOP may soon require this pasture intake (check the new pasture rule scheduled to be released in 2010).

Nutritional considerations for heifers

- Target growth rates of up to two pounds per day are possible on well-managed pasture with little to no supplemental feeding.
- A little corn meal or corn silage can help the heifers utilize the pasture's protein more efficiently.
- Offer free-choice minerals and kelp.

Table 6.7 includes the post-weaning requirements for both small and large breeds of various ages and their nutrient needs. Most of the values are a range because of breed differences and varying growth rates—faster growing animals have higher requirements. You can meet these requirements with high-quality pasture or forages and minimal grain feeding.

TABLE 6.7 POST-WEANING REQUIREMENTS FOR VARIOUS AGES AND BREEDS OF HEIFERS

AGE	DRY MATTER INTAKE		% PROTEIN	NEG	
	LB./DAY	% BW		MCAL/DAY	MCAL/LB.
Weaning to 6 months	7–10	2.3–2.5	15–16	1.3–2.0	0.20–0.22
6 months to 1 year	10–16	2.1–2.4	12–13	1.35–2.74	0.13–0.17
1 year to breeding	17–20	2.1–2.3	12	2.05–3.15	0.12–0.16
Breeding to last third of gestation	18–26	2.0–2.4	12–14	2.35–3.53	0.13–0.14
Last third gestation	26–32	2.2–2.4	12–14	2.77–3.9	0.10–0.12

Source: National Research Council. 2001. Nutrient Requirements of Dairy Cattle. National Academy Press, Washington, DC.

Heifers between the ages of 6 months and breeding age (approximately 15 months) have similar nutrient requirements, but their intake needs to increase as they grow. Thus, you can give a group of heifers on well-managed pasture the same ration and as long as there is adequate quantity, it will meet their needs. This is illustrated in Example 6.8.

EXAMPLE 6.8 RATION COMPARISON OF 6-MONTH HOLSTEIN HEIFER TO BREEDING-AGE HOLSTEIN HEIFER

Dry matter intake:
6 months: 400 lb. × 2.4% = 9.6 lb. DMI
15 months: 850 lb. × 2.3% = 19.55 lb. DMI

Protein requirements:
6 months: 12.5% CP × 9.6 lb. DMI = 1.2 lb. CP
15 months: 12% CP × 19.55 lb. DMI = 2.35 lb. CP

Energy requirements:
6 months: 2.5 Mcal/day NE_G
15 months: 3.0 Mcal/day NE_G

As long as you manage the heifers' pasture similarly to how you manage the milking cows' pasture, the two groups of heifers in Example 6.8 will easily meet their protein requirements. In fact, they will probably eat a bit more than their requirements, but are able to handle the excess protein more easily than milking cows since their metabolism works more slowly and feed stays in their rumen longer.

The NE_G level of pasture is usually in the 0.40–0.50 Mcal/lb range, which also meets the nutritional needs of the heifers:

6 months: .40 Mcal/lb. x 9.6 lb. = 3.84 Mcal/day
15 months: .40 Mcal/lb. × 19.55 lb. = 7.82 Mcal/day

When you feed heifers stored feeds, feed as much forage as possible to minimize costs. Generally, a good-quality hay, haylage, or baleage at least 12% CP and free-choice feed will meet their protein requirements. It may also meet their energy requirements as long as it tests over 0.25 Mcal/lb. of NE_G. Feed salt and minerals either free choice or top-dressed on the forages. If you do not have high-quality forages, you may need to feed a small amount of grain. To calculate how much protein and energy you would need from the grain, refer to the calculations explained in Example 6.6.

From breeding to calving, pasture does a good job meeting nutritional requirements without supplemental feed other than salt and minerals. Some farmers graze this group with the milking cows or use them to clean up a pasture after the milking cows. However, if you have a history of Johne's Disease in your herd, or buy animals from herds that are not guaranteed Johne's-free, you may be better off establishing clean, separate grazing systems for heifers.

Short-bred Holstein heifers at 16 months old would be 900 to 1,000 pounds body weight, and heifers in the last third of gestation at 22 to 24 months old would weigh approximately 1,200 to 1,300 pounds. They can graze together and eat a similar ration, although when they are within a month of calving they should begin a transition diet similar to a dry cow. Example 6.9 illustrates how to calculate this transition diet.

Confinement rations for 16- to 24-month-old heifers must also meet the same parameters as for the younger group and provide enough forage to achieve their intake requirements. If the forage quality is low, you may need to calculate a grain mix to meet their needs. One precaution for growing heifers is not to overfeed high-energy feeds, especially corn silage. Heifers easily gain weight if provided with too much energy, which can cause metabolic problems such as ketosis after calving. If you plan to feed corn silage or another high-energy feed, be sure to balance it with protein feeds such as pasture, hay, or haylage.[8] This is critical in an organic system to minimize health problems.

EXAMPLE 6.9 RATION COMPARISON OF 16-MONTH HOLSTEIN HEIFER TO HEIFERS IN THE LAST THIRD OF GESTATION

Dry matter intake	16 months:	900 lb. × 2.4% = 21.6 lb. DMI
	22 months:	1,200 lb. × 2.4% = 28.8 lb. DMI
Protein requirements	16 months:	13% CP × 21.6 lb. DMI = 2.8 lb. CP
	22 months:	13% CP × 28.8 lb. DMI = 3.7 lb. CP
Energy requirements	16 months:	3.4 Mcal/day NE_G
	22 months:	3.8 Mcal/day NE_G

As long as pasture is high quality, it will provide over 13% CP and 0.40 Mcal/lb. of NE_G and meet the needs of these bred heifers as it did with the younger group.

CALVES

Requirements

Calves require organic whole milk from birth until weaning. Milk replacer is not allowed by the NOP organic regulations. Young calves have very high nutritional requirements. As they wean and move toward breeding age, their requirements gradually decline. As discussed, once heifers are bred their requirements reduce and they can meet their requirements on quality forages alone.

Organic farms wean their calves at different ages and feed whole milk at a rate of 8–10% of body weight or more until they begin weaning. Additionally, you can offer a calf starter grain with at least 19 to 20% CP and 0.60 Mcal/lb. of NE_G after the first few days of life. Starter grains usually include molasses to assist with palatability and are sold bagged or in bulk. Fresh water should also be offered free choice to calves, as this will encourage them to eat the grain. Starter grains assist early rumen development. After calves have adjusted to eating it, you can add forages at any time. High-quality hay is the best forage to introduce first, followed by fermented feeds as consumption of hay increases.[9]

BOX 6.5 NURSING

A unique aspect of Rob and Pam Moore's farm is that they allow their heifer calves to remain with their mothers from the time they are born until they are 3 to 5 months old, when they are put on nurse cows until they are weaned at 8–9 months of age. They do this for two reasons—they feel the calves are healthier and they learn how to be good grazers from their mothers. Under these management conditions, they do not have sick calves and few, if any, parasite or pinkeye problems. Although they have lost significant milk income, they feel it is a long-term investment, creating a more efficient grazer and thus more profitable milking cow.

If you cannot find a premixed calf starter grain, you can formulate and mix your own if you have the ingredients on hand. To formulate it, use Pearson's Square as discussed earlier in this chapter. Typical ingredients in calf starters include corn, oats, wheat midds, flax meal, soybean meal, molasses, and minerals. The minerals usually include limestone, trace mineral salt, and dicalcium phosphate. If you formulate your own, be sure your certifier approves the minerals. You may want to reformulate it during the grazing season due to the higher levels of protein in the forage.

Some farmers do not feed calf starter to their calves and keep them on whole milk until weaning. This strategy does work, but with a few challenges. Calf starter helps the rumen develop rumen papillae, which are not there when the calf is born but are important to help the animal absorb nutrients. The rumen papillae begin to develop when bacteria are introduced to the rumen from the calf's environment and when sugars and starches from calf starter ferment, producing acids that stimulate the rumen wall to begin growing papillae. Calves are not able to digest forages well, even if they begin to graze at a young age, until the rumen bacteria multiply and the papillae develop.

When calves drink milk, the milk bypasses the rumen through the esophageal groove into the abomasum, the "true stomach" where enzymes and acids help digest nutrients. Thus, it takes much longer for the calf's rumen papillae to develop without calf starter, as there are few sugars and starches available to ferment. Whole organic milk is valuable, so keeping calves on milk longer to avoid feeding calf starter has financial implications. Before you try feeding only whole milk, weigh the economics against the possible long-term benefits that may not have an immediate financial return, such as less disease and animals that do not have to be weaned off grain.

It is important to offer organic calves forages at an early age, so they begin to learn what to eat and establish a population of fiber-digesting rumen bacteria. Pasture, dry hay, or baleage are good choices, although you can feed a small amount of corn silage as they approach weaning. After weaning and before 6 months, you can feed calves high-quality forage and possibly a small amount of grain (if they were fed calf starter) because their rumen is not yet big enough for them to meet their requirements with only forage.[10] Example 6.10 describes how to calculate the nutritional requirements of a newly weaned 4-month-old calf.

MANAGING SEASONAL DIET SHIFTS

One aspect of organic dairying some farmers struggle with is the change from a stored-forage feeding program to a pasture-based feeding program and back again in the fall or early winter. Even experienced graziers can find this seasonal shift to be a challenge, due to differences in crops from year to year or the varying end date of the grazing season. If you do not have much experience with grazing, this feeding change will be a significant challenge during your transition to organic dairy.

SPRING SHIFT

Changing from lower-quality stored feeds to high-quality pasture is much like changing silos. If changed too quickly, milk production

EXAMPLE 6.10 RATION CALCULATION FOR A 250-LB., NEWLY WEANED 4-MONTH-OLD CALF

Use Table 6.7 to follow the recommended requirements used in the example.

Dry matter intake	250 lb. × 2.4% = 6 lb. DMI
Protein requirements	16% CP × 6 lb. DM = .96 lb. CP
Energy requirements	1.8 Mcal/day

Pasture would meet both the protein and energy requirements, but stored forages such as dry hay or baleage need at least 16% CP and 0.30 Mcal/lb NE_G. While many farms' hay crop forages meet this level, farmers usually feed high-quality forages to milking cows. Thus, if these forages are unavailable for the younger calves, their diet should be supplemented with grain.

If you fed about 5 lb. of dry hay with 12% CP and 0.25 Mcal/lb. of NE_G, this would provide:

5 lb. × 12% CP = 0.6 lb. CP
5 lb. × 0.25 Mcal/lb. = 1.25 Mcal

This ration is short by 0.36 lb. of CP (0.96 – 0.6) and .55 Mcal of energy (1.8–1.25). If you supplement the calf's diet with 1 lb. of grain, it would need:

0.36 lb. CP ÷ 1 lb. grain DM = 36% CP
0.55 Mcal ÷ 1 lb. grain DM = 0.55 Mcal/lb. NE_G

A pound of corn meal would provide .67 Mcal, but not enough protein. Pearson's Square indicates that a mix of 31% corn meal (9% CP) and 69% soybean meal (48% CP) would provide the 36% CP needed in the grain. This mixture also provides .73 Mcal/lb. of energy, exceeding slightly what the calf requires. This should not be a problem as it may help her gain a little more weight. Use a scrap piece of paper and practice Pearson's Square to see if you calculate the same answer as above.

drops until the cows and rumen microbes become accustomed to the new feed. The rumen microbes are especially sensitive to sudden changes, because they need time to shift their numbers and types to those that are more adapted to higher-quality forage.

Spring shift management considerations

- Start grazing when the grass is 3–4 inches tall. Cows should be out 1–2 hours.
  - Left on grass longer, they eat too much and refuse much of ration in barn.
- Over next few days, gradually increase time on pasture until out full time.
- 3- to 4-inch forage height quickly becomes 6 to 8 inches, due to rapid spring growth.

  - Recommended minimum grazing height is 6 to 8 inches after spring flush.
  - Grazing at 3 to 4 inches past spring flush results in over-grazing and unproductive pastures.
- Cows will gradually refuse more feed in the barn.
- First reduce protein forages such as haylage, baleage, and dry hay.
- Next cut back the amount of protein from grain or concentrate as cows increase intake of protein from pasture.
- If feeding a TMR, mix for five to ten fewer cows (depending on herd size) each day as refusals increase.
  - When TMR is less than 70% of full ration, reduce protein supplement 1 pound every 3 days.
  - When TMR is less than 50% of normal, check protein and NFC levels to ensure balance, and evaluate if reformulation is needed.
- After 7 to 10 days, ration should have less than 10 pounds dry matter from stored forage, pasture dry matter intake greater than 15 pounds.
- Grain mixes should be below 12% protein.
  - Protein concentrates fed at less than 1 pound per cow—**if** other forages are fed.
- Do not feed protein while the cows are on pasture if it is the main forage in diet.

Through the spring, grasses and clovers grow rapidly and can easily become overmature and unpalatable to your animals. Strive to turn cows onto pasture in the vegetative stage of growth when it is highest in quality, approximately 6 to 8 inches in height. If pasture height gets ahead of you, look for where the grass is the right height and quality and put the cows there. Harvest the overmature pasture for hay, haylage, or baleage—there are no rules that say you have to graze all the paddocks before going back to the first one, that you cannot skip paddocks, or that you have to follow the same sequence of paddocks every rotation. If you go where the grass is best, cows will maintain their intake, rumen bugs will not have to adapt, and consistent milk production will occur.[11] For more detailed information, refer to Chapter 4, "Pasture Management."

FALL SHIFT

In the fall, concerns about changing to new feeds are essentially the same. There are, however, a few new challenges. The stored forages are most likely from the recent growing season. Since no two growing seasons are the same, the quality of the forages will be different from what was fed in the spring. It is hard to predict how the cows will respond to the new forages in terms of both intake and performance. Predicting the end of the grazing season is difficult. If you shift to stored forages too early, you may not maximize all remaining high-quality pasture. Likewise, if the feed change begins too late, the pasture could run out before introducing the stored forages.

Predicting the end of the grazing season is different every year due to weather and management of the pastures. It is important, however, to try to predict the last day of grazing using simple planning techniques.

Techniques to predict final grazing day

- Walk the paddocks at least once a week beginning in September.
- Measure the total amount of pasture DM available.
- Once total pasture cover is known, divide it by the total amount of pasture DM needed per day.
- The total pasture DM needed/day approximates the remaining number of grazing days if the feeding program stays the same.
- When significant differences in total available grass vary week to week, begin the shift to stored forage.

Strategies for transitioning feeding in the fall are similar to in the spring except that the steps happen in reverse. You will introduce or increase stored forages in the barn. Cows will stay in the barn at night once the temperatures begin to fall below 35°F (unless the plan is to outwinter). Eventually the amount of time the cows spend on pasture gradually diminishes, especially after a frost kills the grass and there is little to no new growth. At this point, the winter ration begins because you will provide the majority of their intake in the barn.

WINTER SHIFTS

Changing from stored forage to pasture and back again may not be the only ration changes you will need to work through. Rumen bugs cannot adapt quickly to abrupt diet changes. Thus, manage diet changes slowly if you have two silos of haylage, major differences between cuttings of hay or baleage, or purchase significant quantities of forage.

With a little planning, you can predict when you will need to change forages and can test them for quality beforehand. If there are significant quality differences, begin introducing the new forage before the old one runs out. The easiest way to do this is to substitute two pounds of new feed for old per cow every few days. If there are large differences in protein or energy between the two forages, gradually shift the supplements to account for that difference.

Hopefully diet transitions will be minimal during the winter confinement and stored-forage feeding time. Maintaining consistent high-quality forages will reduce the number of changes you need to make, help to keep your costs down, and maintain milk production.

EXTENDING THE GRAZING SEASON

Many forage-type crops can extend the grazing season, both in the spring and in the fall. These include cereal grains such as winter rye or winter wheat for early spring grazing and a variety of brassicas for late fall grazing. Stockpiling some species of grasses also works well in the fall, although this strategy is better suited for animals that are not milking heavily.

From a nutritional standpoint, cereal grains planted in the fall and grazed in the spring are very high in quality. Similar to cool season pasture, they have over 20% CP and fiber digestibility over 75%, as long as the cows graze them in the vegetative growth stage before the boot stage.[12] The ration on these grains can be the same as your spring pasture ration. These crops begin growing in late March or early April, allowing you to graze two to three weeks earlier than traditional pasture and three to four times in the spring, depending on weather conditions. This may allow you to harvest more pasture acres for hay in the spring. Furthermore, cereal grains work well as a cover crop on land you are renovating for pasture or on land where you plan to plant annual crops.

Brassicas such as turnips, rape (swedes), and kale are an option for either midsummer grazing or extending the grazing season in the fall. Nutritionally, brassicas range from 12–18% protein and at more than 85% digestibility, are some of the most digestible forages. They also have high water content. Cows should graze brassicas for only a short time each day, obtaining the majority of their intake from other forages. Health problems can result from too much wet and highly digestible forage, and they can cause an off-flavor in the milk, especially if grazed within a few hours of milking. If you try brassicas, transition animals to them slowly to prevent rumen upset and continue to feed the same ration, but increase the amount of fiber with additional dry hay.

Stockpiling grasses, by allowing pastures to grow without harvesting from mid-August onward, is another method to extend the grazing season. However, the nutritional quality of the grass declines over time, so it may not be appropriate for milking cows. If you are a seasonal dairy and the cows are in late lactation through the fall or if you are going to graze youngstock or dry cows on stockpiled forage, this may be a good option for your farm.

Research from the University of Wisconsin showed that stockpiled tall fescue and orchardgrass maintained the best quality after frost, but reduced the CP levels by 2% from October to December.[13, 14] If you use stockpiling on your farm, take a pasture sample to determine the protein levels in case you need to increase supplemental protein in the ration. Refer to Chapter 5, "Organic Crop Management" for more information on growing small grains and Chapter 4, "Pasture Management" for more information on grazing brassicas.

EVALUATING YOUR FEED COMPONENTS, RATION BALANCE, MILK PRODUCTION, AND HERD HEALTH

Once you have decided how to feed your cows under organic management, it is important to assess your feeding program. Observation is essential to evaluating your success, and you will need to learn how to look closely at your cows for signs of problems or changes. It is easy to become engrossed in the day-today activities on the farm and to look at cows without really seeing them. This section outlines some tools you can use to evaluate and troubleshoot, as well as corrective actions you can take to solve problems.

MILK PRODUCTION TRENDS

One of the easiest ways to determine if the feeding program is on target is to watch the bulk tank. If milk production is consistent from day to day and week to week, that is a good sign the cows are eating a consistent ration that meets their intake and nutrient needs. However, if milk production varies more than 5 pounds per cow per day, or more than 7 pounds per cow per week, there may be a problem with the feeding program. Variability can result from differences in total intake, differences in amount fed over the course of the day, or differences in pasture yield from paddock to paddock.

Two tools will help you identify the source of milk production variability. The first one is a chart to record daily milk weights so you can see the trends in production. Essentially, it is a graph plotted with the daily or bi-daily production per cow. If your milk is picked up every other day, you can use the weights listed on the slip from the milk truck driver. An example of a milk weight chart is shown in Appendix G.

If your milk weights show a significant daily difference, examine your feed delivery in reference to the following factors.

- If more than one person feeds the cows, are there differences in how people mix a TMR or weigh a scoop of grain?
- Do these differences result in different rations?
- Which person is making the biggest feeding difference influencing production?

Meet with all members of the farm who feed the cows to create standard operating procedures (SOP) for feeding and identify where the differences are. This may be as simple as doing one feeding together to determine feeding differences. Include all members of the farm in this discussion and post the SOPs in an easy-to-find location.

The second tool is used during the grazing season to record information on the paddocks cows graze every day.

- Record when cows are in each paddock.
- Identify which paddocks result in a drop in milk production.
- Formulate a plan to improve those paddocks.
- Increase the amount of forage fed in the barn on days cows graze those paddocks.
- Monitor regrowth periods to be sure pasture pregrazing height is 6–8 inches. If the pregrazing height is tall enough, recalculate the paddock size to make sure it is large enough for the current herd.
- Alternate grazing between paddocks with different quality or yields after each milking.

Taking this tool a step further includes measuring pasture yields and growth rates to determine a paddock sequence for grazing. It also indicates when to reserve a paddock for mechanical harvest due to overmaturity. Refer to Chapter 4, "Pasture Management" for more detailed information on monitoring paddock size and quality.

MILK COMPONENTS
Milk components (fat, protein, and other solids) can provide insight into your feeding program, provide sensitive indicators of health, and manage a positive bottom line. Genetics, lactation stage, and the feeding program all influence a cow's ability to produce components. Under organic pricing, maintaining higher components can be profitable due to higher component prices compared to conventional.

Butterfat
Of all the components, the feeding program most easily affects butterfat. Butterfat is partially the result of rumen bacteria digesting fiber and producing acetic acid. Acetic acid is used by the mammary cells to produce fat. The more acetic acid, the higher the milk fat content. Likewise, when acetic acid in the rumen is low due to a lack of fiber digestion, butterfat is lower. Rumen pH is also a factor of fat content. On high-fiber diets, rumen pH increases because there is more acetic and less propionic and lactic acid production. Rumen pH is important relative to polyunsaturated fats (PUFA) in the diet because PUFA is used to form other fatty acids that depress milk fat synthesis. More acid in the rumen uses more PUFA, and butterfat drops.

To maintain high butterfat, note the following.
- Feed a high forage diet to maintain or increase fat production.
- If fat is low, too much grain and not enough forage may be the cause.
- Milk-fat depression may be caused by feeding too much fat—minimize fat sources such as whole soybeans.

Milk protein
Milk protein is more difficult to influence with changes in the diet and is mostly a result of genetics.

To increase milk protein, note the following.
- Try feeding a small amount of extra protein, but be careful not to overfeed protein as cows use energy to excrete it.
- If the milk protein percent is greater than the milk fat percent, the diet includes too much grain and not enough forage.
- Immediately examine the ration and increase forage intake.

Ratio of first-test fat percent to protein percent
The ratio of fat percent to protein percent at the first DHIA test after calving can indicate if your fresh cows are at risk for early lactation metabolic problems. If more than 40% of the cows have

a first test ratio greater than 1:4, the herd is at risk for ketosis and displaced abomasums. You can calculate the ratio for individual cows with this equation:

% fat at first test ÷ % protein at first test

If you note abnormalities (greater than 1:4), you should review the dry cow ration and management to be sure that you are not over-conditioning your cows.[15]

Fat-to-protein inversions
Fat-to-protein inversion occurs when the milk protein level is equal to or greater than the level of milk fat. Dietary fiber directly relates to the fat percentage in milk. One cause of low milk fat is low dietary fiber/high concentrate rations, which puts the herd at risk for subclinical acidosis and laminitis. Other diet-related problems may result from:

- forage chopped too finely,
- cows able to sort feed while eating, thus not getting the full ration, and
- poor hay quality.

Temporary solutions to resolving acidosis include feeding a buffer (sodium bicarbonate) or feeding a yeast additive that has been approved by your certifier.

MILK UREA NITROGEN (MUN) LEVELS
If cows consume too much protein and not enough carbohydrates, rumen bacteria convert the protein to ammonia. The ammonia is transported out of the rumen and into the bloodstream, where it eventually passes through the liver. The liver converts the ammonia into urea, which ends up back in the bloodstream before going to the kidneys for excretion in the urine. On its way to the kidneys, some of the urea also passes through the udder, where it is absorbed by the mammary cells, and excreted in the milk as milk urea nitrogen (MUN).

MUN levels above 14 mg/dl indicate

- cows consumed too much protein,
- cows did not eat enough carbohydrates, or
- both of the above.

Most milk handling companies run MUN tests from the bulk tank sample on a regular basis, and testing services such as DHI also do individual cow samples. This is a good troubleshooting tool, especially during the grazing season when the potential for overfeeding protein is highest. If the bulk tank MUN level is within the range of 10 to 14 mg/dl, the ration is on target. If a group or some individuals are outside the desired range, determine if their feeding program is out of balance.

To adjust MUN levels

- If levels are high, greater than 20, assess the feeding program

and reduce the protein or increase NFC levels to no higher than 40%.
- If levels are low, less than 4, feed a little more protein.
  - Increasing protein may also increase milk production.
  - Make changes slowly until MUN increases to the target range.

RUMINATION (CUD CHEWING)

Cows on a high forage diet with adequate fiber levels should spend about a third of their time chewing their cuds, or ruminating. This indicates good rumen health as cud chewing helps to buffer the rumen through saliva production. If cows are not chewing their cud and making saliva, they may have acidosis or subclinical acidosis, which can lead to laminitis in their hooves.

Periodically assess how many cows are chewing their cuds while at rest to help you troubleshoot problems associated with the feeding program. Since cows spend approximately 8 hours a day eating, 8 hours resting, and 8 hours ruminating, at any one time you should count 60 to 80% chewing their cuds.

If very few cows are chewing their cuds
- The diet may contain too much grain and not enough forage.
- There may be inadequate fiber or NDF levels in the ration.
- Cows may be off-feed.
- There may be low forage available in the paddock, changing the forage-to-grain ratios. Measure pasture yields and growth rates, to ensure correct pasture intake.

BODY CONDITION SCORING (BCS)

Routinely taking the time to look at body condition on your cows, bred heifers, and dry cows is important. Body condition scoring (BCS) indicates your cows' energy status as they mobilize fat in the early stages of lactation. They should not lose more than one condition score from calving until just past their peak milk production. If they do, it will be difficult for them to gain it back before they dry off. You should also score your bred heifers and dry cows to ensure they are neither too fat nor too thin at calving time.

The system of body condition scoring for dairy cattle uses a scale of 1 to 5, with 1 being extremely thin and 5 being extremely fat. The average of the milking cows in herds that calve year-round should be close to 3. In seasonal herds the average is 3 by 150 to 175 days in milk. There should not be any 1s or 5s in your herd. If there are, determine why and develop a plan to improve condition. Dry cows and heifers should have a body condition of at least 3 and no more than 4.

The BCS system uses fractions of whole numbers or a plus/minus system to describe cows that are in between scores. A cow with a little more condition than a 3 might be described as a 3.25 or 3.5, and one with less condition might be a 2.75 or 2.5. The most important places to look for condition are on the short ribs, spine, and around the tail head. The more bone definition in these areas, the lower in condition the cow is, and vice versa. Grazing cows have better muscle tone due to walking, so do not confuse an athletic cow with a thin cow. Review the condition scores in Appendix H.

One condition score is equivalent to 200 Mcal of NE_L, so if the cows are too thin by 100 days in milk, they will need to be fed 1 Mcal/day extra for the last 200 days to put the condition back on before drying off. If you wait longer than 100 days to assess them, the ration will need to have even more energy.

Assessing and recording BCS every month or two will help you discover problems before it is too late to fix them. Have someone who does not see the cows every day do the scoring for you, as their assessment may be more objective. Also, recognize there are some cows that are simply thin all the time, no matter how well you feed them. You may want to exclude those cows from the average, as well as any cows that have struggled with other health problems and become thin.

LOCOMOTION SCORING

Locomotion scoring can help you assess rumen health and the overall well-being of your cows. This tool helps you identify lame cows and the degree of their lameness. In many cases, lameness results from laminitis, caused by an acidic rumen. Heel warts, bruises, abscesses, or foot rot can also cause lameness. Identifying even mildly lame cows before the problem progresses and checking their hooves can help solve problems early.

Locomotion scoring uses a scale of 1 to 5, with 1 being normal and 5 being severely lame. A cow that is not lame stands and walks normally with a flat back. Once cows begin to feel any pain in their feet, they begin to walk with a slightly arched back and a slight change in their gait. From there, as the lameness progresses, their back arches most of the time, their gait becomes progressively awkward, and they may favor the legs that are not painful to walk or stand on. A severely lame cow will have difficulty walking or standing at all. As a benchmark, fewer than 2% of your cows should have lameness. See Appendix I for photos detailing the stages of locomotion scoring.

Once you have identified cows with a locomotion score greater than 1, you or your hoof trimmer need to lift the affected feet to assess the problem. If it is a nutritional problem resulting from acidosis, you will see sole ulcers, usually in all four feet. Unfortunately, laminitis caused by acidosis does not appear until a few months after the cows were acidodic, so the ration may have changed since the actual problem occurred. If you have not changed the ration in awhile, acidosis may still be a problem and you should check your feeding program for adequate forage and fiber.

Fortunately, if you feed a high forage diet and minimal grain, the feeding program will likely not cause lameness problems. When cows are on pasture, they are less likely to have acidosis unless pasture intake is limited.

MANURE SCORING

Manure scoring can help evaluate how well the ration is balanced for protein, fiber, and carbohydrates. It can also assess if the cows are digesting the feed correctly and if water intake is appropriate. This tool is helpful in evaluating pasture-based diets, as the manure gives a good indication of whether too much protein is being supplemented, too much forage is being fed in the barn, or whether the rumen bugs are utilizing grain for energy.

Once again, this system uses a score of 1 to 5, with 1 being very liquid and 5 being very stiff. In practical use, a score of 1 would slide off your boot, a score of 3 would stick to your boot, and a score of 5 would bounce off your boot.[16]

Photo by Robert DeClue

Score 1: Manure is runny with a consistency of pea soup. The manure may arc from the rump of the cow. Excess protein or lack of fiber can lead to this consistency, typical of early spring pasture-based diets. If Score 1 continues past early spring, too much protein may be supplemented in the barn. Cows with diarrhea or Johne's disease will also score in this category.

Photo by Robert DeClue

Score 2: Manure is runny and does not form a distinct pile, rather it splatters when it hits the ground or concrete. This is typical of cows on pasture after the spring flush and if little supplemental forage is fed in the barn. If seen in full stored-forage confinement feeding, the cows may lack adequate fiber in their diet and have acidosis.

NRCS Photo Gallery

Score 3: Manure has the appearance of pudding, and will form a pile with concentric rings around a small depression in the middle. It makes a plopping sound when it hits the floor or ground and sticks to shoes, pasture sticks, and the flanks of cows that lay in it. Cows fed pasture, some stored forage, and adequate energy without extra protein will attain this manure score. Cows on stored forages in the winter with properly balanced grain mix, will also produce this kind of manure. Score 3 is optimal.

Photo by Robert DeClue

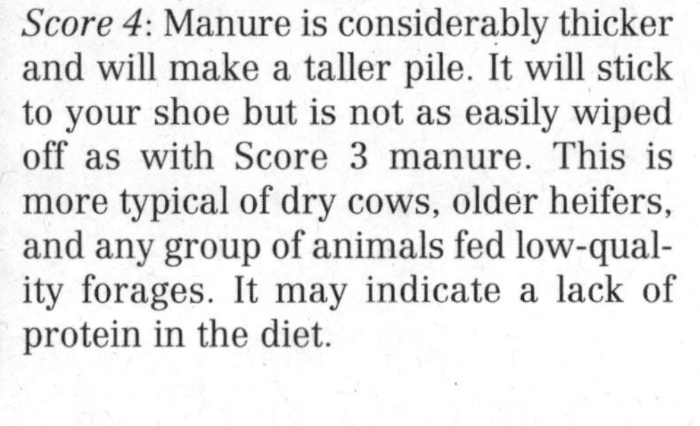

Score 4: Manure is considerably thicker and will make a taller pile. It will stick to your shoe but is not as easily wiped off as with Score 3 manure. This is more typical of dry cows, older heifers, and any group of animals fed low-quality forages. It may indicate a lack of protein in the diet.

Photo by Robert DeClue

Score 5: Manure is stiff, firm, and possibly in the shape of a ball. This indicates extremely low quality forage, a digestive blockage, or dehydration. If you feed relatively good-quality forage and see manure of this consistency, you need to call a veterinarian because death of the animal is possible due to digestive blockage or dehydration.

If you are willing to feel manure for texture and consistency, you can score it by hand. The following guidelines provide a reference for this examination.

Score 1: Manure feels like a creamy homogenous mixture. There are no visible undigested food particles.

Score 2: Manure feels like a creamy homogenous mixture. There are a few visible undigested food particles.

Score 3: Manure does not feel homogeneous. Some undigested particles are visible. When you squeeze the manure in your hand, some undigested fibers will stick to your fingers.

Score 4: Undigested food particles are clearly visible and are larger than in Score 3. When you squeeze the manure in your hand, a ball of undigested food will remain.

Score 5: Larger food particles are clearly visible in the manure and it is easy to distinguish the undigested components of the feed ration.[17]

To troubleshoot your feeding program with manure scoring, walk through the barn or pastures and look at the consistency of the manure.

Cows on pasture

Cows on pasture tend to have looser manure due to the combination of high protein, low fiber, low dry matter content (80% water, 20% dry matter) and highly digestible feed. This is acceptable as long as it is close to 2 on the score. If cows on pasture have scores of 4 or 5 or if there is a lot of grain in the manure, you need to re-

FIGURE 6.2
Happy lines. Photo courtesy of Hubert J. Karreman, VMD

assess how much you feed in the barn. If the manure is very stiff, you may not be meeting the 30% pasture requirement and feeding too much forage in the barn. If there is a lot of grain in the manure on pasture, you may need to grind it more finely for the rumen bacteria to be able to utilize it.

Cows in the barn
When cows are in the barn on stored feeds, the manure should average a score of 3. If the manure is closer to a score of 2 and you can smell ammonia in the barn, you may be feeding too much protein. If the manure is very stiff and scores above a 4, you either need to feed higher-quality forage or increase the protein level in the grain.

OVERALL NUTRITION ASSESSMENT
The cows themselves provide you the best assessment of whether or not the feeding program meets their needs and if they feel healthy. Although the tools discussed above help you pinpoint areas that need changes based on physical signs, there are also behavioral indicators.

Well-fed, healthy cows are bright-eyed and approach you with their ears up and forward, walk with energy in their step, and are curious about their environment. If they keep to themselves, put their ears back, and are slow to move, they may have a nutritional or health problem.

Another sign of good nutrition and health are health lines; some farmers and veterinarians refer to them as "happy lines."[18] These lines appear on the side of the cow in the rib area, usually running horizontally. They are usually faint, and you have to stand in the right light and angle to the cow to see them. Scientists do not know what causes them, but they indicate good health and possibly good rumen function. Figure 6.2 shows health lines.

Successful organic dairy farming is dependent upon your ability to be observant of all aspects of animal health and nutrition and to make changes quickly once a problem is identified. High-quality forages, high forage intake, and a minimum of grain are key to profitability. Use Appendix R, Nutrition Troubleshooting Guide to assess the cause of problems and how to address them.

CONCLUSION

Good organic dairy nutrition will respond by supporting high-producing, healthy dairy cows. It is a critical component to a successful organic dairy business. If you are not comfortable balancing and monitoring your own feeding rations, make sure that you find a nutritionist who is knowledgeable about pasture-based rations and the organic rules. Balancing and managing dairy cow rations is a valuable skill and worth spending the time to learn because your cows will be healthier, produce better quality milk, and you will see a better bottom line.

RESOURCES

Crowley, J. et al. 1991. *Raising Dairy Replacements*, North Central Regional Extension Publication #205.

Hedtcke, J. L., et al. 2002. "Quality of Forage Stockpiled in Wisconsin," *Journal of Range Management.* 55:33–42.

Hoffman, Sullivan, K., R. J. DeClue, and D. L. Emmick. 2000. *Prescribed Grazing and Feeding Management of Lactating Dairy Cows*, NYS Grazing Lands Conservation Initiative/USDA-NRCS.

Karreman, Hubert J. 2007. *Treating Dairy Cows Naturally*. Acres USA, Austin, TX.

National Research Council. 2001. *Nutrient Requirements of Dairy Cattle*, 7th revised ed. National Academy Press, Washington, DC.

Overton, T. R. and M. R. Waldron. 2004. "Nutritional Management of Transition Dairy Cows: Strategies to Optimize Metabolic Health," *Journal of Dairy Science.* 87:(E. Suppl.):E105–E119.

Paine, L., and K. Barnett. 2007. "Stockpiling pasture." *Grass Clippings.* volume 2, number 3. (Retrieved September 13, 2007 from http://www.cias.wisc.edu/pdf/pasturenews807.pdf).

Website Resources

Dairy One Cooperative, Inc.: http://www.dairyone.com/

DeLaval: http://www.delaval.com/Dairy_Knowledge/EfficientCowComfort/Feeding.htm

National Organic Program: http://www.ams.usda.gov/nop

Northeast Grazing Guide, Northeast Pasture Consortium: http://www.umaine.edu/grazingguide

Penn Dutch Cow Care, Dr. Hubert J. Karreman, DVM: http://penndutchcowcare.org/

Premier Molasses Company: http://www.premiermolasses.ie/silage.htm

Thorvin: http://www.thorvin.com/

NOTES

1 D. L. Emmick, D.L. 2007. "Foraging Behavior of Dairy Cattle on Pastures." Ph.D. dissertation, Utah State University, Logan, UT.

2 Hoffman Sullivan, K., R. J. DeClue, and D. L. Emmick. 2000. Prescribed Grazing and Feeding Management of Lactating Dairy Cows, NYS Grazing Lands Conservation Initiative/USDA-NRCS.

3 Ibid.

4 There are many computer programs available to help balance nutritional rations. Contact your local extension office for information about helpful programs.

5 Hoffman Sullivan, K., et al. 2000.

6 T. R. Overton and Waldron, M. R. . 2004. "Nutritional Management of Transition Dairy Cows: Strategies to Optimize Metabolic Health," Journal of Dairy Science. 87:(E.Suppl.):E105–E119.

7 Ibid.

8 J. Crowley, et al. 1991. Raising Dairy Replacements, North Central Regional Extension Publication #205.

9 Ibid.

10 Ibid.

11 Hoffman Sullivan, K., et al. 2000.

12 P. E. Cerosaletti and L. A. Fields. 2008. "Final Technical Report: Alternative Continuous Cover Forages 2." Northeast SARE Project LN05-215.

13 J. L. Hedtcke et al. 2002. "Quality of forage stockpiled in Wisconsin," Journal of Range Management. 55:33-42.

14 L. Paine and K. Barnett. 2007. "Stockpiling Pasture," Grass Clippings. volume 2, number 3. (Retrieved September 13, 2007 from http://www.cias.wisc.edu/pdf/pasturenews807.pdf).

15 Todd Duffield et al. Sept. 2002."Herd Level Indicators for the Prediction of High-Risk Dairy Herds for Subclinical Ketosis" Dept. of Population Medicine, Ontario Veterinary College, Univ of Guelph, Ontario. The AABP Proceedings, vol. 35.

16 Delaval website (http://www.delaval.com/Dairy_Knowledge/EfficientCowComfort/Feeding.htm), 3/10/2009.

17 Hubert J. Karreman. 2007. Treating Dairy Cows Naturally, Acres USA.

18 Ibid.

Rob and Pam Moore.

ROB AND PAM MOORE

KAREN HOFFMAN AND BETHANY WALLIS

Rob and Pam Moore are the eighth generation to farm at the Moore family's 150-year-old farm in the Southern Tier Region of New York. Rob began farming with his father and uncle under a conventional operation, milking registered Holsteins and feeding grain. Rob's father and uncle were forced to retire early when they both experienced health problems within the same year, leaving Rob to manage the farm alone. Rob had always enjoyed working with the cows and quickly became frustrated doing crop work. As a result, he sold his milking herd and worked off the farm installing high tensile pasture fences and raising heifers. This experience allowed Rob to visit many farms and see how they managed their herds, giving him time to rethink how he wanted to farm. He also spent time reading about the New Zealand system of grass dairying and considering how to set up a similar system here. Rob returned to farming intending to graze intensively, and he purchased a herd of heifers with what he calls "hearty scrub genetics." He certified in 1996 and credits his neighbor, Kevin Engelbert, with influencing his decision to do so.

Currently Rob and Pam manage 50 cows on about 250 acres of pasture. They certify additional hay ground, using rented land as an insurance policy to provide enough feed in poor growing seasons and sell excess feed during good seasons. They custom-harvest all of their forages. The Moores milk their cows seasonally and winter animals outdoors, providing woods for shelter. Rob has found that keeping the animals outside year-round has greatly reduced disease and labor expenses. They usually graze until late December, providing there is forage still available, slowly adding in dry hay and baleage in the beginning to mid-October. From January to April they feed the cows a 100% hay and baleage ration. The Moores switch from twice-a-day to once-a-day milking when milk production drops

Calves nursing at Moores' farm.

below 30 pounds in the late fall or early winter. Then Rob and Pam dry the cows off according to body condition and the weather. While the cows are dry they feed dry hay and baleage. Since the Moores do not feed any grain, they keep a close eye on body condition throughout the year to maintain good herd health. When weather conditions have been extremely dry, they have fed 1–4 lb. of molasses per cow per day to supplement energy and maintain good body condition, but they do not feed molasses on a routine basis. They also feed free-choice kelp and Sea-Agri 90 sea minerals.

Cows freshen from April to June and calves nurse with their mothers until mid- to late August. In late August they separate the calves from their mothers during night milking until the morning milking, giving the calves about 9 hours with their mothers and 14 hours without them. In September they completely remove the calves from their mothers and place them with nurse cows until December or January, weaning them at the end of the year. They do not feed the calves any grain, only milk and pasture. Rob and Pam acknowledge this is a more expensive way to raise calves, but feel it is an investment in the future of the farm. Allowing the calves to nurse virtually eliminates pink eye and parasite problems. Rob explains, "We don't miss the other calf problems associated with bottle feeding; we do not have sick calves under our current management system." Nursing their calves has ended the herd health repercussions many grazing calf systems face—they do not worry about putting their calves on "clean pasture," pasture that has not yet been grazed by adult animals. Their calves have stronger immune systems and are better acclimated to their grazing environment.

The Moores enhance their soil health by aging the bedded manure from their winter calf barn. This practice in concert with grazing has positively changed the soil health. They also evenly distribute baleage feedings throughout the pastures because cows drop ma-

nure where you feed them. Aerial farm maps illustrate fertile dots across each paddock where they fed round bales. This helps evenly build soil health across the whole farm. To monitor soil conditions, Rob and Pam observe the clover stand percentage in their pasture and hay crops. A good grazing rotation and proper forage height is critical to maintain and improve soil health. Most of the nutrients in their soils are well balanced, but they do add lime and frost seed some pastures to increase productivity.

Rob and Pam follow holistic management planned grazing. Although they winter their herd outside, they do not begin their grazing season until April when pasture is 4 to 6 inches tall and dense enough for the cows to receive a new 1- to-2 acre paddock after each milking. During the remainder of the grazing season, they graze the pastures at 10 to 16 inches tall and limit the amount of time the herd remains in a paddock to two days. They begin with a 20-day rotation and increase to a 45-day rotation in dry conditions. The furthest rotation is two miles away and requires two hours to move the herd back and forth. Under extreme drought conditions, however, it has saved their pastures. Under drought conditions, they supplement grazing with baleage. They manage weeds by clipping and grazing. If they begin grazing their cows when they are young, they learn to eat many of the weeds. The Moores clip or mow all of their pastures at least once during the grazing season to maintain high forage quality. They provide water in every pasture with above ground water lines and portable water tanks. During the winter, the cows travel to frost-free water hydrants.

While managing a high-production conventional herd, Rob saw many herd health issues that resulted in monthly vet checks and ever-increasing hormone expenses to breed cows. Since transitioning to organic, Rob's cows do not have retained placentas, experience few metabolic problems, and have great breeding results, which Rob attributes to being in harmony with nature. His cows still have an occasional bout with milk fever (about one per year), but he credits his overall good herd health to grazing. Pam agrees, "We focus on the health and vitality of our cows first, not production. We monitor body condition and try not to take too much from our cows." The Moores believe herd health starts with the newborn calf. "Allowing calves to remain with their mother provides the moms an opportunity to teach them how to be a cow."

They milk their cows with a swing-milking parlor housed in a coverall building and work to monitor milk quality. Their milking procedure includes pre- and post-dipping and using a clean, dry towel on each cow. They regularly test for mastitis using the California Mastitis Testing system. If they identify a cow infected with *Staph aureus* bacteria, they sanitize the milking machine after milking her. They cull cows with high somatic cell counts or use them as nurse cows as quickly as possible. They have been able to reduce somatic cell counts (SCC) through these practices and currently average around 200,000 SCC.

The Moores have not vaccinated their cattle in more than 10 years. They use homeopathy, herbs, garlic, and Crystal Creek products to control any herd health problems that arise. Their cows eat a diverse diet that keeps them healthy. Rob explains that their cows supplement their own diets by "grazing the hedgerows and grazing the leaves off of the trees. They eat all of those herbs and forbs, not just your clovers and grasses, but all of those other weeds that bring in an array of vitamins and minerals typical forages do not." Pam and Rob rely on many sources for information on alternative therapies including books and seminars featuring veterinarians Paul Dettloff, Hubert Karreman, and Linda Tikofsky. They read *Graze* magazine, are members of NODPA's electronic mailing list, Odairy, and seek advice from fellow organic dairy farmers and graziers.

When the Moores first transitioned, the organic milk market was new and presented challenges. Rob reflects "After getting through all of the hard times, things are finally falling into place." He suggests that farmers interested in transitioning find an organic dairy mentor. "You are going to need to go to someone weekly or even daily in the beginning, so find someone close to home." Rob and Pam work closely with a farmer grazing network and have found them to be the most influential people in their farm management and the cornerstone of their success.

Pam does an excellent job maintaining the recordkeeping for organic certification. Keeping these records has improved their management by providing a reference of past successes and challenges. Pam is upgrading their recordkeeping with custom Excel charts for grazing management and is searching for a compatible grazing financial tracking system. Monitoring their average production per cow has not, however, given them a holistic view of the farm, and they would like to develop standards for monitoring overall farm productivity.

JIM GARDINER, HIDDEN OPPORTUNITIES FARM

KAREN HOFFMAN AND BETHANY WALLIS

Hidden Opportunities Farm is an organic dairy operation with 32 milking cows. The farm is managed by Nancy and Jim Gardiner with the help of their children Jake, Clinton, Katie-Lynn, and Joyful. They have managed their farm organically since 1989 and certified in 2000 when the organic milk truck came to their region and made it possible to market their milk organically.

Jim grew up on a dairy farm. When his father passed away at the early age of 45 from cancer, Jim was only 16 years old. The premature death of his father caused Jim to question their farming practices, the chemicals used, and the long-term effects on the farmers' health. He began to study human anatomy, agronomy, natural nutrition, and herbology. He also asked retired farmers about their practices for treating animals prior to the chemicals of modern-day farming. Jim began to apply this knowledge of natural health to his dairy animals and farm. Through years of research, he has developed sound natural herd health practices including herbal, homeopathic, essential oil, and nutritional therapies. Jim says, "Using materials that are in harmony with the earth's natural flow and function, we can accomplish our goal to leave the farm better than we found it."

Jim Gardiner.

The Gardiners graze 32 milkers with 26 heifers and dry cows on 55 acres of permanent pasture. They have enough land to expand their pasture system to 130 acres if they need to lengthen their grazing season or to accommodate dry weather conditions. The herd begins their grazing season around April 20 and remains on pasture until approximately November 10, or as long as the weather permits. Their rotational pasture system is more intensive in the spring when they move 48 animals every 12 hours on quarter- to half-acre paddocks. When pasture growth slows in the summer, they give the herd 4- to 5-acre paddocks and move them every 24 hours. This system helps reduce labor during cropping season and provides the herd with unlimited pasture intake. During this time they also supplement the herd's ration with baleage.

In the spring, the cows often graze a pasture once and then, if needed, the Gardiners mow a cutting from the permanent pasture system. Their fencing is high tensile perimeter with polywire movable interior paddock dividers. The farm has springs in each field and seven streams that provide drinking areas on the outskirts of the fields.

Jim advocates feeding molasses as the only energy source to milking cows on pasture. From late April to early November, he feeds his cows high-quality pasture and a maximum of 7 pounds of molasses per cow per day in two feedings based on milk production and body condition. His on-farm research has shown that each pound of molasses is the energy equivalent of 3–4 pounds of corn meal. The cows milk well and maintain good body condition through the grazing season with this feeding program. Jim varies their mineral and vitamin supplementation based on the weather conditions and feeds primarily Sea-Agri 90 for trace mineral salts. From early November until the cows go out to pasture in April, he feeds baleage and dry hay harvested from hay fields, pastures, and cover crops of peas with oats, barley, or millet. He also feeds molasses in the winter.

Katie-Lynn Gardiner feeding molasses.

Jim and Nancy manage 200 acres of cropland including pasture and do not currently grow corn. They rotate between 20 and 30 crop acres yearly and leave an eighth of the total acreage fallow each year. Jim says, "By letting the land rest, which includes no spreading or harvesting, you'll see almost double the production in the following years." Once a portion has rested for one year, they seed a small grain cover crop with field peas and a new seeding. Jim leaves well-drained soils longer between crop rotations, rotates poorly drained soils every four to five years, and leaves clay soils up to eight years. Avoiding plowing heavier soils allows them to build more organic matter. A longer rotation also reduces his fuel expenses. "I am just not into recreational tillage," says Jim.

The Gardiners manage soil health by visually observing weeds, performing soil tests, reducing tillage, planting cover crops, applying aged manure, and leaving land fallow. They test soils every seven years using a lab that tests for micronutrients. Jim suggests waiting until the ground has reached 50°F before taking soil samples. "Otherwise you'll be told that you need phosphorus when it's actually just that the microorganisms that process phosphorus aren't active yet so phosphorus is not being released." Jim avoids excessive tillage and plows only so oxygen and water exchange is optimal. He maintains soil balance through proper mineralization and claims it takes three years to detoxify the soils. He recommends that transitioning farmers feed the soils with apple cider vinegar, raw milk, and molasses to enhance microbial activity in early spring "when you can smell the dirt coming to life." Jim applies raw milk because it contains the same nutrients available in fish emulsion but in greater quantities. Jim's application rates are: 3 gallons of raw cows' milk, 1 gallon of apple cider vinegar, and 2 gallons of organic molasses added to 14 gallons of water per acre.

The Gardiners age their manure with lots of bedding (chopped hay) to optimize soil nutrient levels. Jim explains, "The bedding has a high viscosity so the carbon [from the bedding] binds to the nitrogen in the manure, slowing the release of the nitrogen. You should avoid wood chips because they add acids to your manure and avoid sand because it promotes the growth of *E. coli* bacteria by removing air from the manure." He spreads aged manure 2–4 tons to the acre.

Jim reduces crop yield loss and maintains high forage quality by testing his soil and feeding it with seeds and cover crops that return nutrients to the soils. He suggests planting dual-purpose nurse crops such as buckwheat, oats, rye, and wheat that you can either plow down to add nutrients or harvest to provide additional grains. Jim recommends, "If you have available organic land near you where you can expand your farm for a minimal cost, it is a good idea to incorporate this into your crop plan. You can sell additional forages in surplus years, and it provides risk management during more difficult seasons and during your transition when you adjust to organic management." He also delays his planting until later in the season to obtain better stands, which helps to reduce feeling pushed in the spring to get crops in the ground and allows him to wait for better weather stability.

Jim's wife Nancy influenced his decision to manage their farm organically. They both felt it was important to know that every time they handled something it would be safe and not toxic. This basic view matured into realizing the need for nurturing what they are responsible for. Jim read William Albrecht's papers during their transition and since then has continued to read piles of books and learn through applied research on his farm. He says, "Don't be afraid to take time to consider what nature is doing and how it is trying to help you." He attributes his success to his willingness to learn new things that are compatible with the natural system. His biggest success is raising a family that is knowledgeable about organics and how to benefit from them. Jim believes, "Cows are ruminants, which is why they require grass. They are not birds who require seeds. You can only fool with Mother Nature for so long and then she will have you pay her back for it."

CHAPTER SEVEN

ORGANIC DAIRY HERD HEALTH

LINDA TIKOFSKY

This chapter provides an overview of organic dairy health, areas and concepts that may differ from health management on conventional farms, basic descriptions of common ailments and best management practices, and allowed treatments appropriate for organic farms. Resources are provided throughout the chapter for readers that wish additional information. This chapter is not meant to be the single source for the organic farm, but a quick reference guide for transitioning organic farmers.

KEY ORGANIC TRANSITION POINTS

- Follow the NOP rules regarding animal management and treatments (§205.236–§205.239, §205.290, and §205.603–§205.604). **Always** check with your certifier before using a new treatment or substance! Many animal health treatments commonly used on conventional farms are NOT permitted in organic management (e.g., antibiotics, hormones, propylene glycol, calcium proprionate gel tubs, many udder salves, etc.).
- Dairy health on organic farms is more than knowing the allowed and prohibited substances. Your focus needs to be on preventive practices, management, and sanitation.
- Observe! Watch your animals carefully and often and understand their behavior. Most illnesses evolve over time, so careful observation will allow you to detect symptoms early and act before you hit a crisis.
- Know how to perform a basic physical exam and record your findings (see Appendix J, “Cow Assessment”).
- She is what she eats. Good nutrition is essential to good animal health and a properly functioning immune system.
- Cleanliness! Disease is related to the number of bacteria, viruses, etc. presented to the animal. Keep stalls and yards free of manure, eliminate access to damp and muddy areas, and be careful about contaminating youngstock areas with adult manure.
- Pasture is your partner. The nutrition provided by “Dr. Green” will help prevent many health problems and is your ally in treatment. Natural ventilation, breezes, and UV radiation from the sun will act as natural antimicrobial agents.
- Reduce stress! Prevent overcrowding, provide superior nutrition, and practice good animal husbandry. During the winter housing season, ensure superior ventilation to prevent respiratory diseases. Always provide a fresh supply of good-quality water. Would you drink from your cattle’s water troughs?
- Practice good recordkeeping. This will keep you in good

stead with your certifier and allow you to identify problem trends early.

- Vaccinations are allowed and should be used where indicated.
- Foster a good relationship with a local veterinarian. Many established organic farmers find that the number of veterinary treatment visits decrease in a successful organic system. However, your veterinarian can be a good consultant and will help you establish your best management practices to prevent disease and will be available to you for critical situations.
- Familiarize yourself with complementary and alternative therapies through reading, attending meetings, and speaking with established organic farmers. Know what works well in your hands and on your farm (keep records of treatments and responses).
- When alternative treatments fail and animal welfare is in jeopardy, you are obliged by the NOP rule to provide necessary medical treatment, even if it uses prohibited substances and you must remove the animal from your herd.

CONCEPTS OF ORGANIC DAIRY HEALTH

Organic dairy herd health is based on a holistic philosophy where soil, the environment, nutrition, and animal health are integrated. In conventional farming, a reductionist approach to health is taken. When disease occurs in an organ or system, that diseased area is the target for examination and treatment, and, often, the rest of the body is irrelevant or ignored. Organic farmers and veterinarians use a different approach and look at the animal as an integrated unit that is a part of the whole ecosystem in which it lives. When disease becomes a problem on organic farms, the farmer cannot just look at the symptoms of the sick animal but must consider the symptoms of the farm as well. What is lacking in soil health, nutrition, housing, and management that is predisposing the cows to disease?

The *prevention* of disease through best management practices is essential. Organic farmers do not merely substitute alternative medicine or treatment for one they may have used conventionally. Preventive practices, such as excellent nutrition, vaccination as necessary, stress reduction, and attention to sanitation greatly enhance the health of the herd and reduce disease. When disease does occur, early diagnosis and intervention is essential. To be most effective, alternative treatments need to be introduced earlier and more intensively than conventional treatments. The advantages to animal health under organic management include higher forage diets from predominantly good-quality pasture, more exercise and fitness from actively grazing, reduced stress from lower production, crossbreeding for hybrid vigor, and the ability to exhibit natural behaviors. Farmers new to organic may have concerns about not being able to reach for a bottle of penicillin or prostaglandins each time a problem occurs, but in a well-tuned organic system, the instances when these products are needed are often reduced.

"With herd health, the hardest thing to learn is the point at which you are not going to be able to pull an animal through with alternative remedies and knowing the point when you need to resort to antibiotics. The problem in some conditions is if you wait until they look like they are going to die, [the cows] probably still will die anyway. [The NOP] requires that you do not let animals suffer, although a treated animal must then be removed from the herd. The longer you are in organic management, the healthier your cows are."
—Liz Bawden, organic farmer in northern New York

RISK ASSESSMENTS AND BEST MANAGEMENT PRACTICES

Risk is the possibility of an event happening (like disease) that will have a major effect on the health and financial profitability of the farm. A risk analysis helps you identify potential problems and determine how to manage them most effectively. Your herd veterinarian can help with this process. Many states have cattle health assurance programs that assist the development of farm plans and sometimes provide financial support for testing. One of the most comprehensive sites is provided by the New York Cattle Health Assurance Program (www.nyschap.vet.cornell.edu).

Risk analysis consist of three basic steps

- Identify the problems, contributing factors, and extent of the problem.
- Design the best management practices to reduce and prevent disease on the farm.
- Communicate the issues and management strategies to all members of the farm team (workers, consultants, veterinarians, etc.) so that control is a community effort.

Using best management practices on your farm will help increase your herd's health and will reduce disease, so it is worthwhile to take the time to include these in the herd health section of your Organic Systems Plan. Best Management Practices (BMP) are methods that reduce the risk of disease in the herd and may be as simple as keeping stalls clean and comfortable or may be more complex, such as designing and testing a segregation program to manage Johne's on the farm.

Key steps in instituting BMPs are

- identifying risks for diseases,
- designing management changes to reduce risk,
- implementing those changes,
- monitoring the effects of those changes on animal health, and
- re-evaluating regularly and adjusting management as needed.

ANIMAL WELFARE ON ORGANIC FARMS

Most consumers of organic dairy products purchase them with the idea that the cows producing the products are housed comfortably, are allowed to express natural behaviors, and are outside on pasture. They also assume that animal stress, illness, and suffering are reduced.

The National Organic Program (NOP) rule specifically states that "organic livestock producers must establish and maintain livestock living conditions which accommodate the health and natural behaviors of animals" and that "no producer shall withhold medical treatments from animals in an effort to preserve its organic status." If properly administered alternative treatments are ineffective, farmers must use prohibited conventional substances as a last resort if it might save the animal. That animal must be identified and its food products (meat and milk) must be eliminated from the organic food chain *forever.*

ANIMAL HEALTH RECORDKEEPING

The backbone of every organic farm is good recordkeeping so you can verify your organic management practices. Although individual animal health records are not mandated by the NOP, they are strongly recommended as a good business practice and required by many certifiers. Good records help you make well-informed decisions about preventative healthcare, develop appropriate culling strategies, and fine-tune your reproductive program.

Keeping detailed information on estrous (heat) cycles, feeding, and production is also recommended and makes good business sense. Appendix K provides an example of an individual animal health record. Your certifier or milk processor may have other recordkeeping templates or suggestions. Finally, the most important piece of recordkeeping is actually *using* the information you collect to make sound financial and management decisions by being able to cull a repeat breeder or a cow with chronic mastitis.

Minimum data to track in your health recordkeeping

- Animal identification
- Birthdate or date of purchase
- Sire and dam
- Lactation number
- Calving dates
- Milk production and components
- Dates and outcome of testing (Johne's, BVD, etc.)
- Treatments (product, date administered and dose, site administered and by whom, withholding time, and outcome)
- Date of culling, sale, or death

VALUE OF A VALID VETERINARY-CLIENT-PATIENT RELATIONSHIP (VCPR)

Even though most conventional pharmaceuticals are not allowed in organic management, veterinarians can still play an important role on the organic farm in preventing, diagnosing, and managing disease. As the number of organic farms increase, we will likely see an increase in the number of veterinarians skilled in the use of alternative treatments, or at least familiar with managing diseases on organic farms.

A VCPR exists when the veterinarian has sufficient knowledge of the animal and the farm management through examination and farm visits. The veterinarian can work with you to develop a preliminary organic treatment plan and should be available for follow-up in case of adverse reactions or worsening of disease conditions.

Most alternative treatments are not approved by the Food and Drug Administration (FDA) and are therefore considered "extra label drug use" (ELDU) under the Pasteurized Milk Ordinance (PMO).[1] Alternative treatments (e.g., aloe vera, homeopathic drugs, and botanical tinctures) must comply with the drug labeling and storage requirements of Section 15r of the PMO. All treatments must be labeled with the proper ELDU documentation. Without proper labeling and storage, you may face debits on farm inspections and risk losing your permit for shipping milk.

Labelling requirements for alternative and allowed conventional treatments

- The name and address of the authorizing veterinarian (one who is personally familiar with the farm)
- The name of the active ingredients (this is met by displaying the drug's common, generic, or scientific name, *not* the trade or brand name)
- Adequate directions for use
- Withholding times for meat and milk, even if zero
- Any necessary cautionary statements

PERFORMING A PHYSICAL EXAM

Performing physical examinations on your animals will help you identify illnesses earlier, communicate with your veterinarian, and monitor response to treatments. It is worthwhile to develop a consistent routine for your physical examinations and to record your findings as you go. The information in Appendix J was developed by Cornell University's PRO-DAIRY and is a useful guide to physical exams for all farms.

ALTERNATIVE AND COMPLEMENTARY TREATMENT AND MEDICINES

Complementary and alternative veterinary medicines (CAVM) are treatments not currently considered part of conventional medicine. The term complementary defines treatments that are used in conjunction with conventional therapies, and the term alternative indicates those that are used instead of conventional medicine. CAVM includes most common botanical (herbal) medicines, homeopathy, acupuncture, and chiropractic medicine.

The goal of this chapter is to give the reader a general overview of therapies commonly used on organic dairies, but it will not go into specific details. There are several excellent texts by Drs. Karreman, Sheaffer, and Dettloff, among others, that should be in every transitioning farmer's library (see the list of Resources at the end of this chapter). Additionally, many organic groups (NOFAs, NODPA, MOFGA, etc.) organize workshops and meetings with experts in these fields. These meetings should be a priority for you if you wish to become skilled in alternative medicine.

BOTANICALS

Botanical medicine uses plants and plant substances as medicines. These are some of our most ancient treatments and are still in use as primary medical therapies in many parts of the world. Many modern conventional medicines have their origins in botanical therapies.

Botanicals can be administered in a number of ways. The most common forms are tinctures (alcohol extractions), tisanes (hot water extractions), essential oils topically applied, or the consumption of whole or parts of dried herbs. Dried herbs may also be administered in gelatin boluses or capsules. Table 7.1 lists commonly used botanicals.

Potential hazards of botanical medicine

- Not inherently safe and dose is important. May be a narrow line between beneficial effects and toxicity.
- Difficult to standardize dose. Growing conditions vary by location and year and affect the concentrations of medically active substances in plants.
- Production methods may include impurities.
- Milk and meat withdrawals have not been established.
- Few botanicals have been critically evaluated in ruminants. Since we produce food (milk and meat) we need to be cautious of potential residues from plant medicines just as conventional farmers are careful about residues from conventional medicines.

HOMEOPATHY

Homeopathy, developed in Western Europe, bases treatments on the Law of Similars or "like treating like." The concept is similar to conventional medicine vaccinations where small amounts of dead or live bacteria or viruses are introduced into the body to stimulate the body's immune response, helping the *body* do the real disease fighting.

In homeopathy, substances that would create adverse symptoms in a human or animal are extremely diluted and shaken, or succussed (a homeopathy term), at each step of dilution. These dilute remedies (10X = diluted 1:10 ten times; 30C = diluted 1:100 thirty times; 10M = diluted 1:1,000 ten times) are used to treat symptoms that the original product would have created. The more dilute a remedy, the greater its potency. These dilutions are administered as a liquid or by lactose spheres onto which homeopathic companies have sprayed the remedy. The liquid or tablet homeopathic remedies are placed in contact with mucous membranes (mouth, nasal passages, or vulva). Typically, remedies are administered more frequently (up to six times a day) early in the disease and then tapered down as symptoms regress.

A basic philosophy of homeopathy is that each animal is unique and so there are no "one size fits all" recommendations. Successful homeopathic treatment relies on fully assessing or repertorizing (a homeopathy term) the animal, taking into account not only the symptoms present (fever, diarrhea, cystic ovary, etc.), but the location (e.g., cystic *right* ovary) and the animal's *Continued on page193 »*

TABLE 7.1 COMMONLY USED BOTANICALS ON ORGANIC FARMS

COMMON AND SCIENTIFIC NAME	PARTS USED	USES	TOXICITIES
Aloe vera *Aloe* spp.	Leaves	Topically: rashes, cuts, burns Internally: Digestive aid, anti-spasmodic, immune stimulant	Digestive upset, colic, diarrhea
Black cohosh *Cimicifuga* spp.	Rhizomes	Reproduction	Liver disease
Black walnut *Juglans nigra*	Hulls	Intestinal parasites, diarrhea	Laminitis, seizures, respiratory failure
Boneset *Eupatorium perfoliatum*	Leaves, flowers	Immune stimulant, bone pain, fever	Diarrhea, skin irritation
Burdock *Arctium* spp.	Root, leaves	Blood cleanser, diuretic, skin disease	None known
Calendula *Calendula officinalis*	Flowers	Incorporated into salves or ointments for abrasions, eyewash, mouth ulcers, skin irritations	None known
Cayenne *Capsicum* spp.	Ground fruit	Stimulate local circulation, pain, antimicrobial	Irritation
Cedar *Thuja* spp.	Leaves, twigs	Topically for warts, chronic skin problems, upper respiratory problems	Abortion, digestive upset
Chamomile *Matricaria recutita*	Flower heads	Digestive disorders, mild sedative	None known
Comfrey *Symphytum officinale*	Leaves, root	Bone healing, commonly used as a poultice for wounds, acute mastitis, teat injuries	Internal uses can be extremely toxic: liver toxicity and failure; liver cancer
Damiana *Turnera diffusa*	Leaves	Used to stimulate estrus	None known
Dandelion *Taraxacum officinale*	Leaves, root	Cleansing tonic for liver, udder edema	None known
Fenugreek *Trigonella foenum-graecum*	Seeds, leaves	Stimulate milk production	Muscle disease, anemia
Foxglove *Digitalis lanata*	Leaves, roots, seeds	Heart disease	Extremely toxic: 6–7 oz. of fresh leaves can kill a cow; nausea, vomiting, diarrhea, slowed/abnormal heartbeat, hallucinations
Garlic *Allium sativum*	Bulb, cloves	Antibiotic, antifungal, dewormer	Bleeding, anemia
Ginger *Zingiber officinale*	Roots	Digestive upset	Bleeding
Ginseng *Panax ginseng*	Root	Immune stimulant, increase fertility, mastitis	None known
Goldenseal *Hydrastis canadensis*	Whole plant	Anti-inflammatory, antimicrobial, laxative, digestive disorders, increase bile flow	Diarrhea, seizures at high doses

COMMON AND SCIENTIFIC NAME	PARTS USED	USES	TOXICITIES
Marshmallow *Althaea officinalis*	Root	Digestive and urinary disorders, diarrhea, chronic coughs	None known
Milk thistle *Silybum eburneum*	Seed, leaves	Liver damage, giardia, ketosis prevention	None known
Pau d'arco *Tabebuia impetiginosa*	Inner bark	Antibacterial, antifungal, immune stimulant	Skin irritation
Pokeweed *Phytolacca* spp.	Berries, root	Most commonly used homeopathically: tumors, mastitis	Extremely toxic: colic, diarrhea, respiratory failure, weakness, death
Purple coneflower *Echinacea purpurea*	Root	Immune stimulant, antiviral	None known
Sage *Salvia* spp.	Leaves	Indigestion, decrease lactation	Large doses: dizziness, increased heart rate
St. John's wort *Hypericum perforatum*	Flowers, plants, stems	Antidepressant, nerve pain	Restlessness, confusion, depression, circling; skin irritation; abortions, reduced milk production
Tea tree oil *Melaleuca alternifolia*	Oil from leaves	Antimicrobial, skin infections: bacteria, fungal, yeast	NOT FOR ORAL USE
Valerian *Valeriana officinalis*	Root	Sedative	Very safe
White willow *Salix* spp.	Bark	Anti-inflammatory: fever, injuries, pain	Digestive upsets
Wolfbane or leopard's bane *Arnica montana*	Leaves	Bruising, inflammation	NOT FOR ORAL USE, most commonly used homeopathically
Wormwood *Artemisia absinthium; Artemisia annua*	Leaves	Dewormer, insect repellant	Digestive disorders, hallucinations/delirium, paralysis, death

Continued from page 191 » behavior. Once all of these conditions are noted, you can look up the appropriate therapies in the *Homeopathic Materia Medica*.[2]

IMMUNE STIMULANTS

Organic dairy production and the integration of soil and crop health with animal health enhances the animal's immune system. As a standard, try to increase the cows' immunity through holistic methods: reducing stress, proper nutrition, considering comfort and welfare, vaccinating, and good husbandry. On occasion, organic farmers use other products to stimulate the immune system.

Colostral-whey injections

These products are produced from the colostrum of hypervaccinated cows. These pasteurized products contain antibodies (immunoglobulins) and other immunologically active proteins (lymphokines, cytokines, lactoferrin, and enzymes) that attempt to generally stimulate the immune system. These products are available from a variety of companies and some are more specific than others (e.g., *Strep* specific, *Staph aureus* specific). You must ensure that the colostrum-whey product you use comes from cows managed organically.

Immunoboost™
This product is a biological stimulant produced from the cell walls of dead mycobacteria. The USDA licenses it to treat calf scours due to *E. coli*. Some veterinarians use this product off-label as a general immunostimulant.

Nosodes
Nosodes are the "vaccinations" of homeopathy created from a diseased organ or discharge (e.g., mastitis culture) in the same dilution and succussion process described earlier. Commonly, they are administered before anticipated exposure to a disease or as part of a treatment regimen. Currently there is no scientific evidence in conventional veterinary literature proving that nosodes confer the same amount of immune protection as conventional vaccinations.

Hyperimmunized serum
These products provide passive antibodies and are most often used to combat diseases (pneumonia and scours) caused by gram-negative bacteria (*E. coli*, *Salmonella*, and *Pasteurella*) or by clostridial diseases. Companies produce these products by hypervaccinating adult cattle for the target diseases, harvesting their blood, and removing red and white blood cells to create a serum. They are administered to treat a disease when symptoms are apparent.

TABLE 7.2
CHECKLIST FOR EVALUATION OF NEW TREATMENTS

TABLE 7.2 CHECKLIST FOR EVALUATION OF NEW TREATMENTS
Does this product fit a need or disease you currently have?
If you have a disease on your farm, have you evaluated management deficiencies so that you can make changes to prevent the disease?
Can the company provide you with the published results of independent research (not done or funded by the company) on the product?
Does the company have safety data on the product and information on milk and meat withholding?
Will the company share information with or take the time to consult with your veterinarian?
Does the company have sufficient contact information so that you can contact it for support in the event of a bad reaction to the product?
Have you contacted other farmers who are familiar with the product?

EVALUATING COMPLEMENTARY AND ALTERNATIVE TREATMENT REGIMENS
Most alternative treatments presented to organic dairy producers have not been subjected to extensive testing under controlled circumstances and may only be accompanied by testimonials. What works on one farm may not work on every farm because of differences in climate, management, genetics, and nutrition. It is necessary that you critically evaluate whether a new product's potential benefits on your farm outweigh the cost of the product. Table 7.2 provides questions to ask yourself and the company representatives about a new product. If most of your answers are "no," think twice before writing that check!

THE NATIONAL LIST
The NOP prohibits most synthetic substances from use in organic

livestock production, so always check with your certifier when evaluating a product for use with your animals. The basic rule of thumb is that the NOP prohibits all synthetics unless specifically allowed and allows all natural substances unless specifically prohibited (Table 7.3). Some "natural" products, however, contain prohibited substances as carriers or additives. Send a product label to your certifier to be sure! It is necessary to review products with your certifier *before* you need them and definitely before you *use* them. All products you use on your farm must be listed in your organic systems plan.

NOP Rule §205.603 contains a list of all substances allowed for use in organic livestock production, provided in Table 7.3 for your convenience. Over time, there may be additions or deletions to the National List so check the most current version at the National Organic Program website (http://www.ams.usda.gov/nop/NOP/). Your certifier may also provide you with an allowed products list.

Ensuring adequate colostrum intake is the single most important step in maintaining neonatal health.

YOUNGSTOCK

Heifer calves are the future of the herd. Since the greatest risks for sickness and death are during the first eight weeks of life, spending time and attention developing good youngstock health during the first months of life will be a profitable investment for years to come.

Good calf management begins during the dry period of the dam by providing proper nutrition for calf development (particularly supplementing her diet with selenium and vitamin E to enhance immune function) and a clean and comfortable environment for calving. Maternity pens or stalls should be separated from other adult housing areas and cleaned out between calvings in the winter months to reduce the risk of infecting newborn calves with Johne's disease. In warmer weather, maternity pastures should be large enough so that calving cows have adequate space.

Youngstock best management practices

- Maximize calf's immunity through nutrition.
- Maintain excellent sanitation to reduce the number of germs the calf encounters.
- Avoid overcrowding pens. Allow enough space for feeders and waterers so that even the smallest calf has access.
- Maintain adequate ventilation and moving air.
- Isolate sick calves to better observe and to prevent disease spread.
- Delay weaning if calves are unthrifty or ill.
- Use dedicated youngstock pastures. Do not graze calves on adult pastures.

COLOSTRUM

Ensuring adequate colostrum intake is the single most important step in maintaining neonatal health. Colostrum is the milk produced during the few days prior to and after calving. It is rich in

TABLE 7.3 NATIONAL LIST OF ALLOWED SYNTHETICS FOR LIVESTOCK HEALTH	
SUBSTANCE	**ALLOWED USE**
Ethanol	As disinfectant or sanitizer; NOT as a feed additive
Isopropanol	As disinfectant only
Aspirin	Allowed to reduce inflammation
Butorphanol	Veterinary use only. 8-day milk withdrawal, 42-day meat withdrawal
Vaccines	All allowed
Chlorhexidine	As a teat dip when other germicidal agents or barrier have lost effectiveness; allowed for surgical procedures performed by a veterinarian.
Chlorine materials: Calcium hypochlorite Chlorine dioxide Sodium hypochlorite	Allowed to disinfect and sanitize facilities and equipment. Residual levels in water must meet limit under the Safe Drinking Water Act.
Electrolytes	Allowed as long as they do not contain antibiotics.
Flunixin	Allowed with twice the withdrawal period.
Furosemide	Allowed with twice the withdrawal period.
Glucose/Dextrose	Allowed
Glycerine	Allowed as a teat dip ingredient, but must be produced from hydrolysis of fats and oils
Hydrogen peroxide	Allowed
Iodine	Allowed as disinfectant and topical treatment
Magnesium hydroxide	Veterinary use only
Magnesium sulfate	Allowed; used for treatment of grass tetany
Oxytocin	Allowed for postcalving emergencies; NOT allowed for milk letdown. Some processors will not allow it because it is a hormone.
Ivermectin	Allowed for emergency treatment of parasites in dairy and breeder stock if your organic system plan is not preventing infestation. Is prohibited in stock to be slaughtered as organic stock. Milk from treated animals must be withheld for 90 days.
Phosporic acid	Allowed as an equipment cleaner as long as it does not contact livestock or land
Poloxalene	Allowed for emergency treatment of bloat
Copper sulfate	Allowed as topical treatment, hoof treatment
Lidocaine/Procaine	Allowed as local anesthetic. Milk withdrawal is 7 days, meat withdrawal is 90 days
Hydrated lime	Allowed as external pest control; NOT allowed as a bedding agent
Mineral oil	Allowed for topical use and as a lubricant; NOT FOR INTERNAL USE
Tolazoline	Used to reverse effects of xylazine; veterinary use only; 8-day meat withdrawal, 4-day milk discard.
Trace Minerals	Allowed as feed additive as long as they are FDA approved
Vitamins	Allowed as feed additive as long as they are FDA approved
Xylazine	Tranquilizer; veterinary use only; 8-day meat withdrawal, 4-day milk discard

protein, fat, antibodies, vitamins, and minerals, and contains many bio-active substances (lactoferrin and enzymes) that help prevent disease in the newborn. Calves receive very few antibodies while they are in the womb. They rely on antibodies absorbed across the gut from the colostrum during the first day of life. After 24 hours, the intestinal wall no longer permits the absorption of antibodies so it is essential that colostrum intake happen immediately. Vaccinating cows with appropriate vaccines (especially for scours) during the dry period will increase antibodies passed on to the calf.

Calves should receive two to four quarts (depending on breed and size) of good quality colostrum within the first hour of life and again 12 hours later. Nursing the dam is acceptable in a Johne's-free herd; otherwise, colostrum should be fed by bottle or carefully with an esophageal feeder.

There may be times when banking colostrum is appropriate for farms. Second-lactation and older animals produce richer colostrum higher in antibodies than first-calf heifers. If a herd is Johne's-positive, it is a good idea to freeze colostrum from test-negative cows to feed to the calves from Johne's-positive dams.

Directions for banking colostrum

1. Freeze colostrum from Johne's-negative second lactation and older cows in gallon-size resealable plastic bags. They can then be placed flat with wax paper between bags in an ordinary freezer and stored for up to six months.
2. Thaw colostrum slowly in warm water (do NOT microwave).
3. Stir before feeding and feed in a sanitized bottle or esophageal feeder.

Nutrition programs for calves vary among organic farms, but the common denominator among all of them is fresh, whole milk.

HOUSING

Calf housing varies with farmer preference and the level of disease on an individual farm. Regardless of the option chosen, all calves older than six months must have access to the outdoors. The advantages and disadvantages of various housing options are listed in Table 7.4.

NUTRITION

Nutrition programs for calves vary among organic farms, but the common denominator among all of them is fresh, whole milk. Although some farms feed low-quality milk (high somatic cell count or abnormal milk) to calves, whole milk fit for human consumption is preferred. The first eight weeks of a calf's life are important and it should be fed high-quality feed to boost its immune system.

Calf nutrition best management practices

- Calves should be fed milk equivalent to 8–10% of body weight per day. (e.g., an 80-lb. calf should receive 6–8 lb. of milk or about a gallon a day, divided into 2–3 feedings.
- There are many delivery systems for milk (bottles, buckets, mob feeders). All equipment should be washed, sanitized, and allowed to dry thoroughly after each use.

TABLE 7.4 ADVANTAGES AND DISADVANTAGES OF CALF HOUSING OPTIONS

TYPE OF HOUSING	ADVANTAGES	DISADVANTAGES
Individual hutches	Completely separates calves and spread of disease Can be disinfected after use Easily moved and modified	Worker discomfort feeding calves during winter months or severe weather No socialization among calves
Greenhouse with individual pens	Worker comfort Can alter ventilation in greenhouse according to weather	May increase risk for spread of disease if pens are close together Expensive
Existing barn with individual pens	Worker comfort Less expensive	Often reduced ventilation Difficult to clean or maintain
Tied in lactating barn	Low cost Ease of care	Risk of disease spread Difficult to clean and sanitize Restricted movement for calves
Group pens	Reduced cost Inter-calf socialization	Risk for disease spread Difficult to observe individual animals (manure, appetite, urine)
Pastured with dam	Natural nutrition Access to pasture, ventilation, and sunlight Natural socialization	Potential spread of Johne's and other diseases Spread of contagious mastitis Difficult to observe individual animals (manure, appetite, urine)

- Calves should have access to fresh, clean water at all times. Amounts will vary with grain and hay intake, recommended amounts are 2–5 gallons per day.
- Good quality calf starter, dry hay, and pasture can be introduced free choice after the first week.
- The most common weaning age is 8 weeks, but if parasites or scours are an issue, milk can be fed longer.
- Weaning is stressful, so do it gradually when calves are eating enough starter, quality hay, or pasture. Do not wean when calves are adjusting to other stresses (housing changes or dehorning).
- Some farms raise calves in groups of 2–3 on nurse cows, which is quite efficient both for growth and labor. Nurse cow should be test-negative for Johne's disease.
- Avoid feeding milk with contagious mastitis pathogens (*Staphylococcus aureus*, *Streptococcus agalactiae*, and *Mycoplasma*) if calves are group housed. Group housed calves may cross-suckle, spreading these organisms among each other, which can result in first-calf heifers freshening with contagious mastitis.

PREVENTIVE HEALTHCARE FOR CALVES

An ounce of prevention is worth a pound of cure, especially when raising youngstock. Having simple best management practices in place and protocols for workers to follow will help ensure that all calves get optimal treatment every time. Table 7.5 lists basic calf health preventive practices appropriate for all farms.

An excellent, though conventional, source of calf management information, written by Dr. Sam Leadley, is available at: www.calfnotes.com. Information on management, hygiene, and calfhood issues are of benefit to organic farmers.

SCOURS (NEONATAL DIARRHEA)
Scours are one of the most common health problems affecting calves on any dairy (roughly 50% of calf deaths are due to scours). However, good management can go a long way in reducing the number of cases per year and the severity of the disease. As always, if a farm is experiencing an outbreak of scours, the question to ask is "What is wrong with management?" not "What should I treat it with?" Table 7.6 lists basic causes of scours in calves, methods of diagnosis, and specific management recommendations.

Scours management and treatment practices

1. *Oral electrolytes*: Dehydration is the number one reason calves with scours die. Electrolyte solutions are available com-

TABLE 7.5 ROUTINE PREVENTIVE HEALTHCARE PRACTICES FOR CALF MANAGEMENT

PREVENTIVE HEALTHCARE FOR CALVES			
PRACTICE	AGE	COMMENTS	
Navel disinfection	Birth	Navel should be dipped in tincture of iodine (2–7%) to disinfect and help seal umbilical cord.	
Animal identification	Birth	Especially important for herds with multiple calvings in a week Allows you to begin your animal health recordkeeping easily and early	
Dehorning	2–3 months	Multiple methods available (mechanical gougers or electric/butane iron) Local anesthesia with lidocaine or procaine is strongly recommended. During warmer months, an approved fly repellant should be used on area to prevent fly strike.	
Extra teats	2–3 months	Remove at time of dehorning	
Vaccination	4–6 months and at breeding	*Need and type of vaccinations will vary by farm.*	
		Vaccines to consider: IBR (infectious bovine rhinotracheitis) BVD (bovine viral diarrhea, Type I and Type II) BRSV (bovine respiratory synctitial virus)	Pinkeye (Moraxella bovis) Leptospirosis Brucella vaccination (4–8 months)
Ventilation	Always	To check the barn ventilation, sit on the floor where the calves' heads are and check the air quality. It may seem fine six feet above the ground, but barn walls and pen dividers can make it quite stale at calf level.	

mercially or you can make them at home. Provide them at the first signs of scours. If the calf will drink, electrolytes can be fed in buckets or bottles. If the calf will not drink, administer the electrolytes with an esophageal feeder. Feed 1 pint per 10 lb. 3 to 4 times a day between milk feedings.

BOX 7.1 HOMEMADE ELECTROLYTE SOLUTION

- 1 tsp. low-sodium salt
- 2 tsp. baking soda
- 1¾ oz. (1 packet) fruit pectin
- 1 can vegetable consommé

Add water to make 2 quarts. Feed at the rate of 1 pint per 10 pounds of bodyweight 3 to 4 times a day. Feed milk 2 to 3 hours before or after due to bicarbonate content

2. *Probiotics*: These products contain "good" bacteria (*Lactobacillus*, etc.) that compete with the scours bacteria in the gut. They can be fed preventively or as a part of a treatment regimen for scouring calves. Review these products with your certifier first to ensure there are no GMOs and that all ingredients are allowed.
3. *Nutrition*: Calves need nutrients to help them fight disease and to recover. Withholding milk is not recommended.
4. *Mannan-oligosaccharides*: These are complex sugars from yeast cell walls. "Bad" bacteria, such as *E. coli*, bind to these yeast sugars, which travel through the intestinal tract and out into the manure. Because the "bad" bacteria are bound to the mannan-oligosaccharides, they cannot attach to the gut wall and cause disease. These products can be fed preventively or as part of treatment regimen.
5. *Immunoboost™*: See the section, Immune Stimulants on page 193.
6. *Passive antibodies:* See the section, Immune Stimulants on page 193.
7. *Dewormers*: Document best management practices and natural dewormers used. The only allowed conventional dewormer for organic cattle is ivermectin, but this product can be used only *on an emergency basis* after documented alternatives have failed and after a veterinarian has checked the calves' manure samples and determined that worms (nematodes) are the problem for the farm. There is a 90-day mandatory milk withhold for organic animals treated with ivermectin, and ivermectin can *never* be used in organic slaughter stock. Contact your certifier *before* treating.
8. *Alternative therapies*: These may not be scientifically evaluated or appropriate for all farms. Make sure you consult a reference recommended in the Resources list at the end of this chapter for specific instructions.
    - putting the calf back on the dam to nurse. Best for mild scours in a Johne's-free herd.
    - ferrous iron and tannin supplements
    - black walnut hulls and/or wormwood: both have been used as dewormers in folk medicine
    - slippery elm bark powder
    - aloe vera/garlic tincture
    - crushed garlic
    - homeopathy, determined by calf's presentation and symptoms
    - diatomaceous earth

- chamomile or fennel teas
- organic yogurt

PNEUMONIA AND RESPIRATORY DISEASES IN CALVES
Pneumonia is any infection of the lungs. Signs are fever, coughing, and labored breathing. After scours, it is the #2 cause of death in heifers and calves. All the best management practices for scours also apply for the prevention and management of respiratory disease.

Pneumonia may be caused by virus, bacteria, or mechanical damage. Weaning, dehorning, transporting, and mixing calves from dif-

TABLE 7.6 SCOURS MANAGEMENT

AGENT AND SOURCE	AGE AFFECTED	SYMPTOMS	METHOD OF DIAGNOSIS	COMMENTS
Escherichia coli "white scours" (environment)	First 1–2 weeks	Severe diarrhea, fever, dehydration, death	Fecal culture	Dry cow vaccination; calve in clean dry area
Salmonella (environment: adult cattle, rodents, birds)	Any age, but usually more than 10 days	Watery diarrhea ± blood, loss of appetite, high fever, membrane-like substances in feces, death	Fecal culture	*Salmonella* DT–104 is a major human health hazard
Clostridium perfringens A, C, & D (soil and intestinal flora, overfeeding, stress, hot weather)	Usually less than 10 days	Painful abdomen, bloating, acute death without symptoms	Culture gut loops of dead calves	Vaccination
Corona and Rota virus (adult cattle)	10–14 days	Watery diarrhea, usually no fever, depression, drooling, dehydration	Fresh colon from necropsied calf	Can be complicated by *E. coli* infection
Bovine viral diarrhea (BVD) (infected cattle)	Any age	Long-term diarrhea, fever, rapid breathing, may have sores in mouth and nose	Blood test, fresh colon, intestinal lymph node, ear notch	Vaccination
Cryptosporidium (environment)	7–21 days	Watery diarrhea, abdominal pain, no fever	Fecal flotation	Can spread to humans
Coccidiosis (environment)	>21 days; commonly at weaning	Watery diarrhea ± blood, dehydration	Fecal flotation	Associated with unclean environment, overcrowding, and stress
Intestinal parasites (environment)	2–3 months or older	Diarrhea, poor growth, pot-bellies, poor hair coat, pale gums (from anemia)	Fecal flotation	Associated with unclean environment, overcrowding

ferent farms are stressful events that may trigger episodes of respiratory disease. Pneumonia is also associated with changes in weather, particularly in poorly ventilated calf barns. Good ventilation is essential. Table 7.7 lists the common causes of respiratory disease in calves, symptoms, and good management practices.

Respiratory disease management and treatment practices

- Improve ventilation.
- Increase hydration.
- Vaccinate with one of the nasal vaccines early in the outbreak.
- Administer passive antibodies for *Pasteurella*, Immunoboost™, vitamin B and C injections.
- Administer anti-inflammatories (aspirin) to reduce fever and prevent lung damage.
- Consider antibiotics. In cases where the calf is not responding to the above treatments, antibiotics **must** be given to prevent suffering. Animals treated with antibiotics must be removed from organic production.
- Alternative therapies: These may not be scientifically evaluated or appropriate for all farms. Make sure you consult a reference recommended in the Resources list at the end of this chapter for specific instructions.
  - herbal antibiotic tinctures
  - garlic
  - homeopathy, determined by calf's presentation and symptoms
  - essential oils (eucalyptus)

TABLE 7.7 RESPIRATORY DISEASE MANAGEMENT: COMMON CAUSES (AGENTS), SYMPTOMS, AND MANAGEMENT COMMENTS

AGENT AND SOURCE	SYMPTOMS	COMMENTS
Bovine respiratory syncitial virus (BRSV) Infectious bovine rhinotracheitis (IBR) Parainfluenza Type 3 (Diseased cattle or healthy carriers)	Coughing, trouble breathing, fever, eye and nose discharge, death	Best prevented through good management, nutrition, and vaccination
Pasteurella species (respiratory tract of normal animals)	Depression, nasal discharge, high fevers (107°F), coughing	Often follows a viral respiratory disease
Hemophilus species (respiratory and reproductive tracts)	Difficulty swallowing, bawling, labored breathing	Prevent with good management, vaccination
Mycoplasma species (respiratory tract, infected milk)	Mild to severe pneumonia, head tilt or ear infections, eye infections, swollen joints	Do not feed milk from *Mycoplasma*-positive cows, or pasteurize milk
Aspiration pneumonia (poor technique with esophageal feeder, careless drenching)	Cough, fever, sudden death	Care during feeding and drenching

COW COMFORT

HOUSING

Housing for adult cattle varies with climate, finances, and farmer preference. Housing on organic farms varies from year-round housing on pasture with run-in sheds or woodlots for shelter to freestalls and greenhouse constructions. Certain cow comfort basics must be met in all housing designs. Discomfort, dirty conditions, and poor ventilation will stress the immune system, and cows will be more susceptible to disease. All adult cattle must have daily outdoor access and exercise unless severe weather conditions exist.

All adult cattle must have daily outdoor access and exercise unless severe weather conditions exist.

Ventilation

Good air exchange is essential. Fresh air should enter the barn and move warmer, contaminated air out. The temperature difference between the outside and inside during winter months should be no more than 10°F.

Flooring

Dry pasture that is free of small stones is the ideal surface for cow health and lameness prevention. Standing on hard concrete all day can cause hoof damage and can strain the cows' feet and legs. Cows sometimes do spend the majority of the day indoors, for example during harsh Northeast winters. Avoid slick concrete flooring so cows do not slip and can feel secure enough to express natural heats. Concrete floors can be grooved to provide better traction (grooves should be 0.5 inches deep, 0.5 inches wide and 2–3 inches apart) or rubber mats may be used for comfort and to provide good footing.

Access to feed and water

Cows housed in tiestall and stanchions have their own assigned bunk space for feeding. In freestall situations, at least 30 inches of bunk space per cow is recommended. If there is only 16–24 inches of bunk space per cow, the risk for fresh cow disease increases.

Daily water requirements for dairy cows vary with weather, type and quality of feed, and production. A 1,500-lb. cow producing 40 lb. of milk on a 40°F day requires 18 gallons of water/day; on a 80°F day, she will need 25.5 gallons of water/day. The recommended water trough size (if individual waterers are not available) is 3 inches per cow.

Stalls

Cows prefer to spend much of their day lying down (12–14 hours). Blood flow through the udder increases by 30% when cows are resting, thereby increasing milk production. Time spent lying in stalls also increases rumination and rests the cows' feet and legs.

Properly designed stalls should allow cows to have the freedom to lunge and move forward and side to side while rising and lying down, but they should not be so wide that cows lie improperly in stalls and cleanliness is an issue. Stalls should be properly sized for both breed size and age of cattle (heifers require smaller stalls than adult cattle). Guidelines for sizing stalls (both tie- and freestalls) are included in Appendix L.

Bedded packs

Bedded packs are increasing in popularity. Loose housing barns provide cows with one large resting area rather than individual stalls. These types of facilities are also quite economical. Bedded packs can provide great cow comfort and reduced lameness when managed properly. Ideally, a 100-square-foot area (or more) is allotted per cow. Sawdust, straw, or shavings are the most common bedding materials for bedded packs.

Good management practices are essential to maintain cleanliness and to reduce bacterial growth in packs. The top 10 to 12 inches of bedding should be stirred twice daily while the cows are being milked. Fresh bedding should be added daily so the top surface (which is in contact with the cow) remains clean and dry.

Bedding

Bedding is essential for cow comfort and cleanliness and falls into two categories (organic and inorganic). The choice of bedding for your farm depends on your manure handling system, availability, price, and personal preference. Remember, if the bedding is a material the cow might possibly consume, it must be certified organic. It is also important to remember that with inorganic bedding, bacterial numbers can soar into the millions if it is contaminated with manure, milk, feed, or urine. Any bedding can be acceptable if the stalls are groomed regularly and fresh clean bedding is added on a regular basis. Table 7.8 reviews the advantages and disadvantages of commonly available bedding materials.

Pasture

Well-maintained pastures are nature's perfect housing, complete with good footing, ventilation, and sunshine to enhance animal health. It is no wonder then that graziers refer to it as "Dr. Green." On the other hand, poorly maintained pastures can have a severe negative impact on animal health and increase the risk of disease. Care should be taken to fence off wet areas (creeks, ponds, and swampy sections) so that cows do not stand in water or lie in mud. Shaded pasture areas can become contaminated and harbor millions of mastitis-causing bacteria if pastures are not rotated.

Evaluating cow comfort

You can assess cow comfort on your farm by using the NYSCHAP hock scoring guide (see Appendix M). More than 95% of cows should have normal hocks, without swellings or abrasions. If more than 5% of the herd has abnormal hocks, you should reassess your housing, bedding, and pasture conditions and identify areas of improvement.

TABLE 7.8 ADVANTAGES AND DISADVANTAGES OF COMMONLY USED BEDDING MATERIALS			
BEDDING TYPE	**CLASS**	**ADVANTAGES**	**DISADVANTAGES**
Straw or hay*	Organic	Easily accessible, absorbent Works with most manure systems Less expensive than others Very comfortable if sufficient amounts used Can be grown on the farm	Good nutrient source for bacteria once wet Promotes Strep mastitis bacteria and flies
Sawdust/ shavings (nonpressure-treated wood)	Organic	Easily sourced in Northeast Works with most manure systems Low to moderate expense	Supports bacterial growth when wet Associated with Klebsiella mastitis
Newsprint	Organic	Easily accessible Inexpensive Can work with most manure systems	Supports bacterial growth Glossy/colored inserts must be removed
Rice hulls*	Organic	Inexpensive Very dry Work well with manure systems	Supports growth of mastitis bacteria when wet
Manure solids	Organic	Readily available Inexpensive Reduces waste to be spread Requires investment in drying or composting system Comfortable Can be made on the farm	Easily supports bacterial growth when wet
Old feed* (TMR, silage, refusals)	Organic	Readily available Inexpensive	Supports bacterial growth Contaminated with yeast/mold,leading to increased yeast mastitis
Sand	In-organic	Available Does not support bacterial growth unless contaminated Excellent cow comfort	All manure systems may not handle well
Gypsum	In-organic	Does not support bacterial growth	Decreased cow comfort Must not come from recycled wallboard

*Must be certified organic.

REPRODUCTION

THE FUNDAMENTALS

A sound reproductive program is essential to the financial health of the organic dairy. A fundamental goal should be to breed cows back in a timely manner so that daily milk production remains high and a steady supply of new heifers is available for replacements or sale. Organic dairies cannot use artificial hormones for breeding or for treating reproductive problems. Instead, they must rely on an understanding of the heat cycle, good heat detection, and natural alternative treatments to keep the pregnancy rate high.

Reproduction, however, can be negatively impacted by a number of factors including (but certainly not limited to) poor nutrition, inadequate heat detection, mishandling of semen and insemination technique, diseases, weather stresses, and housing (e.g., slippery floors). Having a basic understanding of the anatomy and physiology of your dairy cows will help you to better troubleshoot their reproductive problems. More information on the female bovine reproductive tract and the estrous cycle is contained in Appendix N.

BREEDING
Organic dairies cannot use artificial hormones for breeding or to treat reproductive problems. Instead they must rely on an understanding of the heat cycle, good heat detection, and natural alternative treatments to keep the pregnancy rate high.

Observing a standing heat is the best way to judge the time to inseminate or breed a cow. Some farms may use paint heat detection devices (check with your certifier to see if these are allowed) or other physical methods but nothing beats watching your cows for heats twice daily. This will occur most naturally on pasture where cows have good footing, but can also be accomplished inside if floors are not slippery and if the cows have room to move.

Cows ovulate about 12–16 hours after a standing heat so most producers use the "Morning-Evening" rule for breeding. If a farmer observes a cow in standing heat in the morning, he/she will breed her that evening; if a cow is observed in standing heat in the evening, she will be bred in the morning.

Although frozen semen contains small amounts of antibiotics, artificial insemination is allowed on organic dairies for safety reasons and to improve genetics by breeding for selected traits. Some farms still use natural service (breeding by a bull) if good heat detection is difficult, for "clean-up" (breeding cows that have not been successfully bred by AI), or on virgin heifers. Additionally, some farmers feel that available semen does not possess the genetic traits they desire. All bulls are potentially dangerous animals, and you should manage them with caution and respect and follow some basic considerations.

Guidelines for working safely with bulls
- Raise bull calves in a group or on a nurse cow for proper socialization.
- Know aggressive behavior in bulls.
- Never run from bulls.
- Remove bulls from herd at the first sign of aggressive behavior.

REPRODUCTIVE CHALLENGES
After mastitis, reproductive problems are usually the second biggest

headache for dairy producers (both conventional and organic), but a systematic approach can help you solve many of these problems. A brief summary of the major bovine reproductive problems is summarized in Table 7.9.

Alternative reproductive therapies
These may not be scientifically evaluated or appropriate for all farms. Make sure you consult a reference recommended in the *Resources* list at the end of this chapter for specific instructions.

- homeopathy, dependent on cow's presentation and symptoms
- botanicals, wild yam/cramp bark/black cohosh, flax seed, evening primrose, dong quai, and vitamin B6 daily until heat
- herbal antibiotic tincture infusion
- metritis: garlic/aloe infusion
- pyometra: pulsatilla infusion
- failure to cycle: damiana

CARE AND MANAGEMENT OF THE FRESH COW

Management of the fresh cow actually begins during the dry period. Nutrition (energy balance, minerals, and vitamin levels) is one area of greatest impact (see Chapter 6, "Managing Dairy Nutrition"). The length of the dry period is also a consideration for healthy fresh cows. It should be long enough (more than 45 days so that udder dry off is complete and we can take advantage of natural healing factors) but not too long (more than 60 days is too long and cows risk becoming fat). Rather than sending dry cows to a back-forty pasture and forgetting about them, you should observe them regularly.

Fresh cows are fragile cows and should be monitored closely so that problems can be detected promptly, allowing for timely interventions. What happens to a cow during the first month after calving can set the stage for the rest of her lactation. Designing a fresh cow protocol (Table 7.10) to ensure that all cows get consistent evaluation and management is recommended.

Disease and problems happen on every dairy farm, but if fresh cow problems increase, the total system must be reviewed (soil balance, crops, management of dry cows, housing) so changes can be made to prevent problems in the future. Table 7.11 lists common fresh cow problems, contributing factors, signs, and treatments.

Alternative therapies for the fresh cow
These may not be scientifically evaluated or appropriate for all farms. Make sure you consult a reference recommended in the *Resources* list at the end of this chapter for specific instructions.

- *Milk fever*: 2 oz. apple cider vinegar twice daily for 2 weeks prefresh; homeopathy (Calc phos)
- *Ketosis*: molasses orally
- *LDA*: ginger, gentian, cayenne, sodium bicarbonate, or caffeine twice daily.

TABLE 7.9 SUMMARY OF REPRODUCTIVE CHALLENGES, POTENTIAL CAUSES, AND MANAGEMENT CONSIDERATIONS

PROBLEM	DEFINITION	FACTORS INVOLVED	MANAGEMENT CONSIDERATIONS
Anestrus-True	Failure to come into heat; ovaries are not cycling	Lack of energy in diet Low hormone levels because of feeding excessive stored feeds Cystic ovaries Uterine infections Anemia	Evaluate ration for energy balance Feed adequate amounts of trace minerals (selenium, iron, copper, phosphorus) and vitamins Feed fresh forage or stored forages that contain estrogens (red clover) Have veterinarian perform physical to detect anemia, cystic ovaries, or infections
Anestrus	Failure to detect heats	Poor animal identification Poor recordkeeping Cows in an environment where they cannot express estrus (crowded) Not enough time to watch cows	Keep good estrous cycle records Increase time spent watching cows Provide good footing Consider tail paint
Cystic ovaries	Large cysts on ovaries that last more than 10 days; cows may be constantly in heat or not cycling	High estrogens in feeds (legumes or molds) Genetics Calcium:phosphorus ratio greater than 2:1 Older cows	Forage analysis (calcium, phosphorus, mycotoxins) Avoid cows and bulls that produce cystic daughters
Persistent CLs	CLs that are present on the ovary beyond 20 days in a nonpregnant cow	High milk production Uterine infection	Evaluate diet for energy imbalance (too little energy)
Retained placenta (RP)	Failure to drop placenta within 24 hours of calving	Twin births or difficult calvings Selenium/vitamin E/vitamin A deficiencies Fat dry cows Infection (bacterial or viral) Low-grade milk fever	Review selenium and vitamin A and E levels in dry cow diet Avoid weight gain in dry period Select bulls for easy calving Calcium supplements
Metritis	Infected discharge that lasts more than 2 weeks postcalving	Secondary to retained placenta Difficult calving Poor hygiene during an assisted calving	Calve in clean areas and use sanitized equipment when assisting calving Address factors that result in RPs Feed maximum amounts of vitamin E/selenium
Pyometra	Severe infection of uterus with yellow, foul smelling discharge	See metritis (above) Poor infusion technique when treating metritis	See metritis (above) Uterine infusion by veterinarian (iodine, chlorhexidine-check with certifier) Antibiotics, prostaglandins, and removal from herd in severe cases

TABLE 7.9 SUMMARY OF REPRODUCTIVE CHALLENGES, POTENTIAL CAUSES, AND MANAGEMENT CONSIDERATIONS			
PROBLEM	**DEFINITION**	**FACTORS INVOLVED**	**MANAGEMENT CONSIDERATIONS**
Repeat breeders	Cows requiring 3 or more inseminations	Poor insemination technique Improper handling of semen Early death of embryo (rough rectals, bacterial, or viral infection) Bull infertility Sexually spread diseases Heat stress	Culture/testing for infectious diseases Good quality forage Careful timing of breeding New bull tested for high fertility Review AI technique and semen handling Veterinary examination Vaccination
Abortions	Loss of calf between 42 and 260 days	Genetic problems Fungal toxins Bacterial and viral infections Multiple calves Injury (rough palpation) Nitrate/nitrite poisoning Neospora	Forage/feed analysis Vaccination Testing of aborted fetus Water analysis Careful palpation

TABLE 7.10 RECOMMENDED FRESH COW PROCEDURES	
TIMING	**MANAGEMENT INTERVENTION**
Immediately after calving	Offer large amounts of fresh warm water. Offer excellent-quality dry hay. Note if placenta drops by 24 hours.
First 10 days (perform daily)	Take rectal temperature daily (should be less than 103°F). Note appetite, water consumption, and milk production. Examine udder. Test milk with California Mastitis Test. Observe vulva for abnormal discharges. Listen to abdomen for rumen contractions.

UDDER HEALTH AND MILK QUALITY

DEFINING MILK QUALITY

Milk quality can mean many things to many people. Consumers expect that their milk will have a consistent appearance, pleasant taste, and last a week or more in their refrigerator. According to the Pasteurized Milk Ordinance, Grade A milk has a somatic cell count of less than 750,000 cells/ml and a standard plate count (SPC) of less than 100,000 colony-forming-units (cfu)/ml and is free of added water and harmful residues above a designated tolerance limit. These are the minimum standards. Organic processors are willing to pay for high-quality, so there is opportunity for dairy producers to increase their milk check substantially by focusing on udder health

TABLE 7.11 COMMON FRESH COW PROBLEMS AND MANAGEMENT CONSIDERATIONS

FRESH COW PROBLEM	SIGNS	CONTRIBUTING FACTORS	MANAGEMENT AND TREATMENT
Milk fever	Down, trembling or wobbly cow Cold ears Normal or below normal temperature	Jersey breed Older cow Calcium:phosphorus ratio in diet Magnesium-deficient dry cow diet High-potassium forage	500 ml calcium (± magnesium and phosphorus) in the vein and under the skin Calcium tubes orally (no calcium proprionate) Forage analysis (reduce potassium, increase magnesium in dry cow diet)
Grass tetany	Nervousness Wobbly cow Muscle spasms	Low magnesium in forages, more common in rapid-growth pastures	Improve soils to increase magnesium in forages Add 1–2 oz. magnesium/cow/day to diet Limit pasture during times of rapid growth Administer 500 ml of magnesium-containing electrolyte solution in the vein
Ketosis	Decreased appetite Decrease milk production Ketone smell on breath Nervous behavior	Other disease in the cow (twisted stomach, mastitis, uterine infection) Cobalt deficiency Fat cows at calving	Prevent over-conditioning in the dry period Introduce concentrates slowly in prefresh period IV or oral dextrose Oral glycerin (from vegetable fats) Niacin boluses and other B vitamins Molasses in diet High-quality forage and feed Avoid major diet changes in early lactation
Displaced abomasum Left side = LDA Right side = RDA	Off-feed "Ping" on left or right side of abdomen	Inadequate roughage in diet Milk fever Lack of exercise Other condition that decreases appetite	Call veterinarian for surgical intervention Fluid therapy if dehydrated Manage ketosis and milk fever Increase fiber in diet Provide adequate exercise
Metabolic acidosis	Low milk fat test Lameness Diarrhea Liver abscesses Poor cud chewing	Too much carbohydrate in feed Finely chopped forage Low fiber content of diet Mycotoxins Fast changes from high fiber to high concentrate diet Foamy manure	Add sodium bicarbonate to diet or feed free choice Avoid slug feeding Evaluate fiber levels and particle size in diet Add probiotics to ration Activated charcoal in acute cases

and limiting mastitis on their farms. Many processors are reluctant to receive milk that is consistently greater than 400,000 cells/ml.

Somatic cell counts (SCC)

In a healthy udder, somatic cells are epithelial cells from the lining of the mammary gland and some types of white blood cells. A healthy gland should have cell counts less than 100,000 cells/ml and typically, counts are much lower than that (less than 25,000 cells/ml).

When bacteria invade the gland, the few white blood cells normally present send out an alarm to the rest of the body and recruit thousands of additional white blood cells from the blood stream to come into the mammary gland to fight infection. Infected glands have SCCs greater than 250,000 cells/ml and often the counts reach into the millions. If all goes well, these white blood cells kill the invading bacteria and the SCC of the gland decrease to less than 100,000 cells/ml within 30 days. We can estimate the number of infected quarters in a herd and the impact on production from the bulk milk SCC (see Table 7.12).

TABLE 7.12 ESTIMATING PRODUCTION LOST AND INFECTED QUARTERS BY BULK MILK SCC

BULK TANK SCC	PERCENT INFECTED QUARTERS IN HERD	PERCENT PRODUCTION LOSS
200,000	6	0
500,000	16	6
750,000	25	12
1,000,000	32	18
1,500,000	48	29

Source: J. Britt 1987. "Herd Linear Scores versus Bulk Tank Sampling." Proceedings of National Mastitis Council Annual Meeting. Orlando, FL.

Regularly monitoring SCC at both the herd and individual animal levels provides you with valuable information to help manage milk quality and improve profitability. Monitoring will help you detect new infections so that appropriate therapies can be administered early in the course of the disease. Monitoring chronically infected cows will help you identify those for culling, early dry off, or candidates for potential nurse cows. You can also follow SCCs after administering therapies to determine whether the treatment was successful. There are many ways to monitor SCC on farms, and they vary in terms of cost, ease of use, and reliability of information (see Table 7.13).

Bacteria counts

Several factors affect bacteria counts in raw milk: cleanliness of the cow's teats and udder, udder infections (mastitis), cleanliness of milking machines, and milk storage and cooling. Excessive bacteria counts impact shelf life, flavor, and increase the risk of food-borne diseases (e.g., bacteria that cause *Salmonella*, *Listeria*) in the milk supply. Pasteurization effectively kills a majority of the bacteria in milk, but it does *not* sterilize milk. Thus, when milk has extremely high bacteria counts, a portion of bacteria can survive the pasteurization process and be present in the final product.

Mastitis-causing bacteria may increase bacteria counts in raw milk. These are usually bacteria in the streptococcus family (*Streptococ-*

TABLE 7.13 OPTIONS FOR INDIVIDUAL ANIMAL SOMATIC CELL COUNT TESTING

CELL COUNTING OPTION	COST	ACCURACY	LIMITATIONS
Monthly DHIA testing	$0.80 to $1.50/cow/month	Excellent	Only gives SCC for the cow for that single day SCC can fluctuate day to day
Cow-side direct cell counting	$300–$3,000 for initial unit $1.50 per cow per test	Very good to adequate	More expensive devices have better accuracy
California or Wisconsin Mastitis Test (CMT/WMT)	A few cents per test	Fair	More accurate for very high cell counts Can be difficult to read for SCC< 1,000,000 cells/ml
Electrical conductivity	Hand-held unit: $300–$400 In-line monitor: thousands	Highly variable	Best used for early detection of clinical mastitis Reliability for chronic or subclinical mastitis is limited

cus agalactiae, Strep. uberis, and *Strep. dysgalactiae*) and infrequently, *E. coli.* The majority of bacteria issues in raw milk are related to poor cleaning or improper cooling of milk.

If contagious mastitis is present in a herd planning to transition to organic dairy production, the farmer should make all efforts to identify, treat, and/or cull the infected animals prior to transition.

Dairy producers can ensure that their milk has low bacteria counts by adhering to the following practices.

- Always milk clean, dry teats.
- Keep milk from cows infected with mastitis out of the bulk tank.
- Ensure that the equipment wash chemicals are appropriate for the farm's water hardness and that correct amounts are used.
- Check that rinse- and wash-water temperatures are within the recommended ranges.
- Sanitize the equipment immediately before milking.
- Cool milk to less than 45°F and preferably to 40°F.

The standard plate count (SPC), also know as the plate loop count (PLC), is the most common bacterial count done on raw milk and is the only bacteria count that has a legal limit in the pasteurized milk ordinance (PMO). Other tests include the lab pasteurized count (LPC), preliminary incubation count (PIC), and coliform count (CC). Farmers use these tests to diagnose causes of high bacteria counts: cows, poor cooling, or poor cleaning (See Appendix O).

DISEASES OF THE TEATS AND EXTERNAL UDDER

Weather and housing conditions, bedding, and infectious disease may affect the outside surface of the udder and teats. These diseases can affect milking or harbor bacteria (e.g., *Staph aureus*), resulting in new mastitis cases.

MASTITIS

Mastitis is an inflammation of the mammary gland in response to injury. Usually, this injury occurs via infection by a microorganism,

TABLE 7.14 COMMON DISEASES OF THE TEATS AND EXTERNAL UDDER

DISEASE	CAUSE	CONTRIBUTING FACTORS	SIGNS	TREATMENT OR MANAGEMENT
Mammilitis	Herpes virus	Heifers, winter weather	Blisters, scabs, "wooden teats"	Chlorhexidine teat dips in winter, vaccination
Warts	Papillomavirus	Contagious spread at milking	Small, teat-colored bump	Surgical removal, vaccination, chlorhexidine dips
Udder rot	Unknown	Udder edema, mange mites, udder shape	Skin infection between halves of udder or between leg and udder	Scrub with antiseptic soap (iodine or chlorhexidine), dry the area, calendula salve, raw honey

most commonly bacterial, but physical and chemical trauma can also cause mastitis. If contagious mastitis is present in a herd planning to transition to organic dairy production, the farmer should make all efforts to identify, treat, and/or cull the infected animals *prior* to transition. In established organic herds, contagious mastitis can be more of a headache, but it can be managed with continued effort and monitoring. Alternative therapy regimens usually yield disappointing results and are often impractical for large numbers of animals. More than 100 organisms are capable of causing mastitis in cows and most of these are bacteria. These bacteria may be considered contagious (living primarily in the cow's udder and spreading at milking time) or environmental (living in bedding, manure, and mud and infecting cows anytime), See Table 7.15.

For every one case of clinical mastitis in a herd, there may be 15–40 cases of subclinical mastitis.

There are two classifications of the degree of mastitis inflammation: clinical and subclinical.

- *Clinical mastitis*: Visible changes in milk or udder (redness, swelling, pain, clots and flakes). Most commonly caused by environmental bacteria.
- *Subclinical mastitis*: No physical changes in udder or milk, but large increases in microscopic cells and bacteria. You can detect subclinical mastitis with one of the cell counting options mentioned previously or by culturing a milk sample. Contagious mastitis commonly causes subclinical mastitis.

MASTITIS MANAGEMENT

Management of mastitis in a herd should focus on preventing new infections rather than treating infections as they occur. Even in conventional herds where antibiotic treatment is allowed, identifying the risks for new infections and adopting best management to reduce those risks is the key to producing high-quality milk and increasing the profitability of the farm.

Mastitis basic management strategies

1. Good milking procedures:
   - *Udder wash/predip*: Before you attach the milking units, wash or predip teats and wipe clean with single-use towels.

- *Forestrip into cup or gutter*: Stimulates milk letdown response and helps identify clinical mastitis early.
- *Nitrile or latex gloves*: Gloved hands are more easily disinfected between cows or when contaminated with milk or manure.
- *Milking order*: Milk cows with contagious mastitis last.
- *Overmilking/machine stripping*: Machine stripping damages teat ends causing cracks that are more likely to be colonized by *S. aureus*.
- *Post-milking teat dipping*: Apply immediately after milking to kill bacteria left on the teats by the inflations and to protect the teat end until it closes.

2. Maintain milking equipment regularly:
   - *Evaluate equipment biannually*: A qualified individual should evaluate the mechanical milking system at least twice a year and perform dynamic testing while the system is operating.
   - *Replace liners*: Replace rubber milking liners every 800 cow milkings or every 60 days, whichever comes first. Rubber

TABLE 7.15 MICROORGANISMS MOST COMMONLY RESPONSIBLE FOR BOVINE MASTITIS

CONTAGIOUS MASTITIS

ORGANISM	COMMON SOURCES	TRANSMISSION	SYMPTOMS AND IMPACT ON MILK QUALITY	BEST MANAGEMENT PRACTICES
Streptococcus agalactiae "Strep ag"	Infected udders	Cow to cow at milking time: on inflations and hands	Very high SCC Decreased milk production High bacteria counts	1. Identify infected cows through whole herd milk culture. 2. Milk infected cows last. 3. Cull high SCC cows (cull ALL Myco. cows). 4. Wear gloves to milk. 5. Post dip. 6. Control flies. 7. Culture herd replacements.
Staphylococcus aureus "Staph aureus"	Infected udders Skin	Similar to Strep ag Biting flies Cross-suckling by calves	Fluctuating SCC Occasionally, clinical mastitis Scarring of udder May cause gangrene	
Mycoplasma species	Infected udders Cows and heifers with respiratory disease	Similar to Strep ag and Staph aureus Coughing and sneezing Calves may be infected by drinking infected milk	High SCC Decreased milk production in one or more quarters Swollen joints Coughing and pneumonia Calves may have head tilt or ear infection	
Corynebacterium bovis (*C. bovis*)	Teat skin	Cow to cow at milking	Occasional clinical mastitis Minimal increase of SCC	1. Post dip with iodine-based teat dip. 2. Apply teat dip with a cup.

TABLE 7.15 MICROORGANISMS MOST COMMONLY RESPONSIBLE FOR BOVINE MASTITIS

ENVIRONMENTAL MASTITIS

ORGANISM	COMMON SOURCES	TRANSMISSION	SYMPTOMS AND IMPACT ON MILK QUALITY	BEST MANAGEMENT PRACTICES
Environmental streptococci (*Strep uberis & Strep dysgalactiae*)	Bedding, manure, and soil	Between milkings from environment At dry off and during prefresh period	High SCC and sometimes high bacteria counts Clinical mastitis	1. Clean cows 2. Clean environment 3. Well-bedded stalls 4. Good premilking teat preparation 5. Well-maintained milking equipment 6. Clean dip cups regularly. 7. Good fly control 8. Fence ponds and standing water in pastures. 9. Rotate pastures to avoid mudholes from developing under trees. 10. Store bedding properly to avoid contamination and moisture. 11. Coliform vaccination for herds with E. coli and Klebsiella problems.
Environmental staphylococci (*Coagulase negative staph—CNS*)	Bedding, manure, and soil	Between milkings from environment	Mild increase in SCC	
Escherichia coli (*E. coli*)	Manure	Between milkings from environment More common in summer	Acute mastitis: watery milk; hot, hard, quarter; very sick cow; may cause death	
Klebsiella species	Manure, soil	Between milkings from environment More common in summer	Acute mastitis: watery milk; hot, hard, quarter; very sick cow; may cause death Cows that survive will usually have chronic quarters	
Pseudomonas species	Contaminated water or hoses	Between milkings At milking time from water From contaminated mastitis tubes	Severe, chronic mastitis	
Serratia species	Contaminated teat dip Contaminated water	Between milkings from environment At milking time from contaminated dip cups	Chronic mastitis	
Arcanobacterium pyogenes "summer mastitis"	Soil, dry cow, and heifer environments Biting flies	During dry period After teat injury	Thick, smelly discharge from swollen quarter	
Prototheca species (algae mastitis)	Mud, silage juice, pasture, manure	Between milkings from environment	Chronic mastitis Cows can shed in their manure	

used longer than this deteriorates and will develop microscopic cracks and ridges that hold mastitis bacteria even through the wash cycle.

3. Dry cow management:
Organic dairy production does not allow the use of dry cow antibiotics or external/internal teat sealants. However, the dry period is an important time for udder health improvement and the rejuvenation of milking tissue.

- Reduce feed intake one week before dry off to decrease production.
- Dip teat end with teat dip twice daily for a week after dry off.
- Feed adequate amount of selenium and vitamin E in the dry period.
- Keep dry cow housing and bedding clean and dry. Hygiene is essential!

4. Mastitis management
Figure 7.1 will guide your treatment of mastitis through its decision tree. Healthy cows with good immune systems will often recover with no-to-minimal treatment. Cows with severe acute mastitis require measures to support their immune systems, allowing the opportunity for natural healing.

For specific mastitis treatments, dosages, and additional information, consult with your herd veterinarian or one of the *Resources* listed at the end of this chapter. When administering medications, administer intramammary infusions with great care so as not to introduce new bacteria into the mammary gland with the medication. Medications should be sterile and packaged in single-use in-

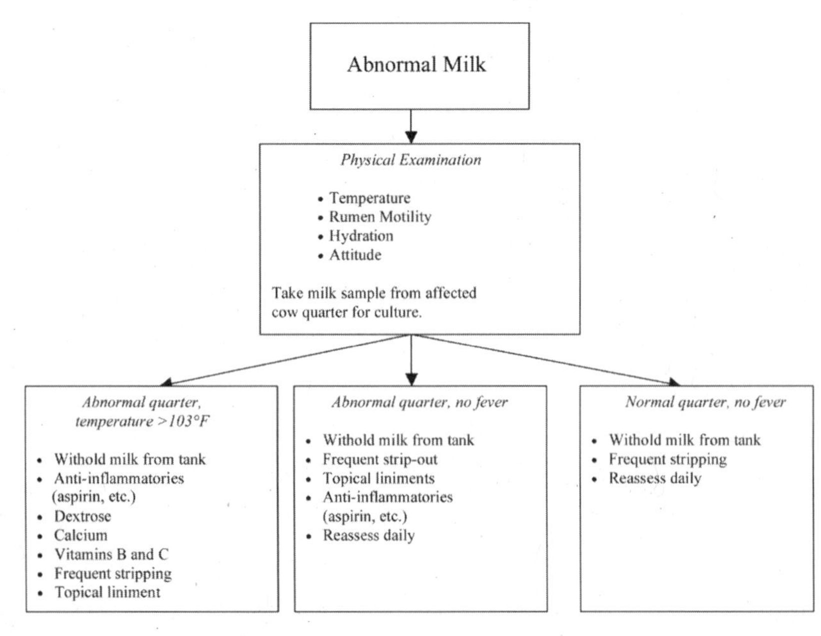

FIGURE 7.1
Mastitis treatment decision tree

fusion tubes. Proper procedures are to dip and wipe the teat, then clean the teat end with cotton moistened with rubbing alcohol, and lastly infuse the quarter. After infusion, dip the teat again.

HOOF HEALTH AND LAMENESS

Hoof health and lameness, big concerns for dairy producers, are directly related to management. The contributing factors are nutrition, walking surfaces, cow comfort (stalls), genetics, cleanliness, and biosecurity, most of which are under a farmer's direct control. On conventional farms, lameness has increased markedly over the last 20 years as intensive dairying has increased. Because organic cows are managed less intensively and production is lower, many studies have found a reduction in lameness as farms switch to grazing and organic production methods.

NUTRITION

High-starch diets and diets containing improperly processed grain can set cows up for acidosis and hoof disease. Fortunately, most organic diets limit grain intake and rely heavily on forage, so acidosis is reduced. Problems can occur, however, if slug feeding is practiced (offering grain for a short amount of time; e.g., in parlor). Chapter 6, "Managing Dairy Nutrition" includes additional information on acidosis and nutritional impact on lameness. Trace minerals (copper, zinc) and vitamins (vitamins A and E, biotin, and beta carotene) are necessary for healthy hoof growth and proper function of the immune system.

HOUSING

Confinement on concrete will increase feet and leg problems, as will uncomfortable stalls if cows are reluctant to lie down. Since organic cows spend most of their time on pasture during the grazing season, lameness problems decrease for most dairies. Cleanliness of housing can have an impact on lameness on any dairy. Moisture from wet manure, urine, mud, and standing water in pastures harbors bacteria that cause infectious foot rot and soften the hoof, making invasion by these bacteria more likely to occur.

CAUSES OF LAMENESS

As with many diseases, lameness is caused by many factors. Table 7.16 reviews the common causes of lameness (both infectious and noninfectious).

Alternative therapies for lameness

These may not be scientifically evaluated or appropriate for all farms. Make sure you consult a reference recommended in the Resources list at the end of this chapter for specific instructions.

- foot rot: homeopathic pyrogen
- abscesses: homeopathic hepar sulph, silica
- garlic tincture or crushed bulbs orally
- epsom salt/tea tree oil footwraps

TABLE 7.16 COMMON CAUSES OF LAMENESS AND MANAGEMENT CONSIDERATIONS

	NAME AND CAUSE	LOCATION AND CHARACTERISTICS	CONTRIBUTING FACTORS	MANAGEMENT AND TREATMENT
INFECTIOUS HOOF DISEASES	DIGITAL DERMATITIS (strawberry wart, hairy heel wart) Cause: bacterial	Heel; between claws (raised red area with hairs at heel or between toes; cow may walk on tip of toe) Raised patches on heel that may bleed	Concrete flooring Wet feet, mud Very contagious	Keep feet clean and dry Foot baths (copper sulfate or lime) Iodine or copper sulfate sprays Reduce pasture mud Hoof trimming Surgical removal *Must treat with antibiotics if the animal is suffering*
	INTERDIGITAL NECRO-BACILLOSIS (foot rot, foot foul) Cause: bacterial	Swelling above the hoof (may extend to hock) Occurs suddenly Very bad odor to foot Abscesses and sores Fever, loss of appetite	Mud and manure Rocks and stones Injuries from ice Walking on field stubble	Clean and trim Drain abscesses Pack and bandage with iodine/sugar paste or copper sulfate Copper sulfate/lime foot baths (wet: 5% copper sulfate) (dry: 1 part copper sulfate: 9 parts lime) Amputation of toe if severely infected *Must treat with antibiotics if the animal is suffering*
	INTERDIGITAL DERMATITIS Cause: bacterial	Mild swelling of heel +/– lameness Sores between toes	Stress Vitamin A deficiency Zinc deficiency Copper deficiency Protein deficiency	Foot trimming Topical copper sulfate Zinc and other trace minerals in feed
NONINFECTIOUS HOOF DISEASES	LAMINITIS (founder) Cause: acidosis	Bruises (hemorrhage in sole and walls) Abscesses Cow walks very stiffly on legs "Slipper feet"	Acidosis from high grain diet Uterine infection Severe mastitis	Foot trimming Diet changes
	SOLE ULCERS Cause: laminitis or poor trimming	Sores on sole Bulge of reddish tissue	Laminitis Wetness and manure Poor hoof trimming	Foot trimming Pack ulcer with topical allowed antiseptic Apply block to opposite toe to relieve pressure
	BRUISES Cause: trauma	Bluish patches on sole	Trimming hooves too short Rocky pastures Poorly constructed lanes	Confine to well-bedded area Block opposite toe
	WHITE LINE DISEASE Cause: laminitis, moisture	Swelling and hemorrhage in area of heel bulb	Laminitis Moisture Trauma	Foot trimming Establish drainage Assess diet *Must treat with antibiotics to remove toe if there is a deep infection and the animal is suffering*

EXTERNAL PARASITES AND DISEASES OF THE SKIN

FLY CONTROL

Flies are not only a nuisance, they can also impact farm profitability through reduced production and the spread of disease. Biting flies are capable of decreasing milk production by 15–30%. Because organic production prohibits synthetic insecticides, organic farmers must take a multifaceted approach to insect control. Face flies, horn flies, and stable flies are the most common types found in the Northeast.

TABLE 7.17 COMMON FLIES IN THE NORTHEAST

COMMON NAME	LOCATION ON ANIMAL	PROBLEM LEVEL	REMARKS
Face fly	Face	More than 10/ face	Problem when animals are on pasture Irritation to eyes Spread of pinkeye
Horn fly	Back and abdomen	More than 50/ side	Problem when animals are on pasture Rapid lifecycle: 10–20 days Female will lay 400–500 eggs/lifetime Eggs laid in fresh manure
Stable fly	Legs	More than 10 flies/animal	Problem in barns and on pasture Rapid lifecycle: 14–24 days Female will lay 100–400 eggs/lifetime Eggs laid in manure and dirt (around feeding areas, drainage areas)

Creating a carefully thought-out fly control program and instituting it early in fly season is essential to success. No single step will control flies; instead, it requires an integrated management plan.

Best management practices for fly control (listed in order of priority)

1. *Start your control early in the season*: Since one female fly can lay 100–400 eggs every three weeks, waiting until you notice a problem may be too late.
2. *Keep pastures and barns clean, dry, and free of manure*: Since flies breed in decaying organic material and manure, removing breeding sites will reduce the population of flies. Spread out straw, if used for bedding, as soon as possible after its removal from stalls. Improve drainage in holding areas and exercise lots. Fence wet or muddy areas in pastures. Move round bale feeder and scrape areas around feed bunks frequently.
3. *Use pheromone traps and sticky tape*: Use in barns to capture adult flies.
4. *Release parasitic wasps*: These insects feed on fly larvae and can be released regularly throughout fly season.
5. *Use essential oils and botanicals*: These are available from various organic supply companies and can be used as the

final step in fly control on your farm. Remember to check with your certifier to be sure that the product is allowed.

6. *Use hydrated lime or diatomaceous earth*: This can be used in dust bags to dry the hair coat of cows, making it less attractive to flies.

RINGWORM, LICE, AND WARTS

Ringworm, lice, and warts are diseases of the skin of cattle and are usually contagious. They affect cow comfort and production and are more commonly seen during the winter months when cows are housed inside.

TABLE 7.18 COMMON CAUSES OF COMMON SKIN DISEASES AND MANAGEMENT CONSIDERATIONS

DISEASE	CAUSE	CONTRIBUTING FACTORS	SIGNS	MANAGEMENT AND TREATMENT
Ringworm	*Trichophyton verrucosum*	Immature or compromised immune systems	Circular, grayish crusts especially on eyes, muzzle, and back	Sunlight, fresh air; trace mineral and vitamin supplementation; local treatment with tincture of iodine or tea tree oil
Lice	Sucking or biting	Crowding, winter months	Decreased production, anemia	Reduce overcrowding; improve nutrition and access to self groomers; hydrated lime or sulfur dust bags; permitted botanical and alternative products
Warts	Papillomavirus	Immature immune systems	Warts on head, neck, and teats, injury to skin	Time; vaccination

PINKEYE

Infectious bovine keratoconjunctivitis (pinkeye) is caused by the bacteria *Moraxella bovis*. Conditions that irritate the eye and surrounding structures (flies, plant material and stubble, dust, and viruses) create conditions that allow infection by the bacteria. Initial signs of the disease are tearing, redness, swelling, and sensitivity to light (squinting). Then, a cloudy area may appear on the surface of the eye that gradually spreads. The eye may bulge or eventually rupture. Conventional farms usually inject antibiotics into the conjunctiva of the eye. However, organic management does NOT allow this and must rely on good fly control and preventing irritation to the eye.

Follow good grazing practices to prevent stubble from injuring eye tissues. Consider vaccinating before fly season. Make sure nutrition is adequate. For extremely irritated eyes, place an eye patch or have your veterinarian temporarily sew the third eyelid over the globe to speed healing and lessen the pain.

Alternative therapy for pinkeye
This may not be scientifically evaluated or appropriate for all farms. Make sure you consult a reference recommended in the *Resources* list at the end of this chapter for specific instructions.

- calendula eyewash sprayed in eye twice daily

INTERNAL PARASITES OF CATTLE

Animals on pasture are at higher risk for infestation by internal parasites. In your organic systems plan, develop grazing management strategies to minimize potential infection and transfer among age groups on your farm.

NEMATODES

The most common intestinal worms affecting cattle in the Northeast are the brown stomach worm (*Ostertagia ostertagi*), the barber pole worm (*Haemonchus contortus*) and *Cooperia* spp. Larvae overwinter in cattle in the Northeast and begin to mature and produce eggs in the springtime, shedding in cattle feces. Figure 7.2 illustrates the nematode lifecycle.

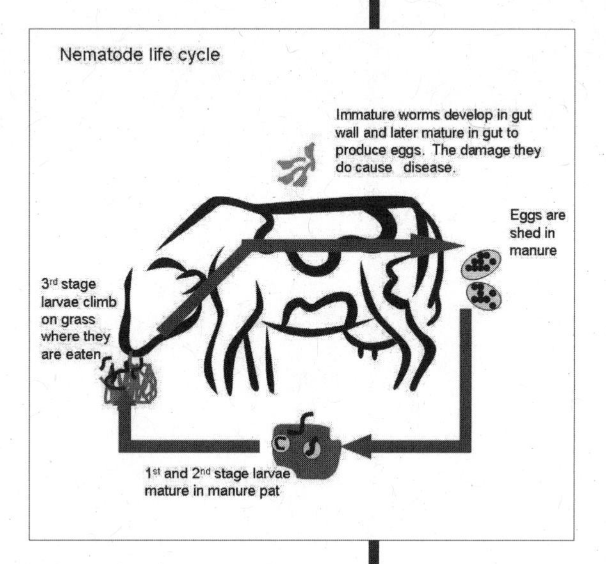

FIGURE 7.2
Nematode lifecycle

Worm infestation may be clinical (visible signs) or subclinical. Clinical infections are more common in young cattle and calves. They may have diarrhea, poor weight gain, potbellies, rough and discolored hair coats, or swelling under the jaw (bottle jaw). In subclinical infestations (most common in adult cattle), diarrhea rarely occurs but the worms negatively affect milk production and body condition. Many adult cattle develop some level of immunity to intestinal parasites.

Organic farmers manage nematodes through best management practices and alternative therapies. The NOP allows synthetic dewormers (ivermectin) *only on an emergency basis* in organic dairy cattle not designated as organic slaughter stock. Routine dewormers cannot be used. Instead, the organic systems plan must include a management plan to reduce youngstock exposure to parasites. These management steps are included in Chapter 4, "Pasture Management."

Alternative therapies for nematodes
These may not be scientifically evaluated or appropriate for all farms. Make sure you consult a reference recommended in the Resources list at the end of this chapter for specific instructions.

- black walnut hulls and/or wormwood
- garlic

COCCIDIOSIS

(see also the section, Scours [Neonatal Diarrhea] on page 199)
Bovine coccidiosis is caused by *Eimeria* spp., tiny, one-celled organisms. Most of the negative effects of coccidiosis occur in young animals (calves and heifers) and infestations are characterized by diarrhea, often with blood, weight loss, anemia, and dehydration. Death is usually due to dehydration or secondary bacterial infections.

Coccidia have a complex lifecycle. Oocysts (eggs) are shed in feces. These oocysts are quite hardy and can survive for months in moist areas, particularly where sunlight and ventilation is minimal. They are unaffected by common disinfectants. The oocysts change life stage in the environment and are then ingested by calves. Inside the calf intestine, the oocysts release sporozoites that burrow into the intestinal wall. Inside the intestinal wall they undergo a few more life stages and eventually release more oocysts into the feces and into the environment. One ingested oocyst ultimately results in 20 million additional oocysts shed into the environment.

The severity of symptoms directly relates to the number of oocysts ingested; so, the more contaminated or less hygienic an area, the greater the risk of oocyst exposure. Overcrowding, stress, diet changes, or shipping can cause an outbreak. Follow the same best management practices used to prevent scours. Ivermectins are not effective for coccidiosis.

ADULT DIGESTIVE DISORDERS

HARDWARE DISEASE

Hardware disease (traumatic reticuloperitonitis) results from a cow swallowing a foreign object (usually metal) that lodges in the reticulum. Rumen contractions may eventually cause this object to punch a hole in the reticulum, causing inflammation and infection.

Signs of hardware disease vary in severity with the level of inflammation and infection. Cows may have objects that pose no threat residing in their stomachs for years and are discovered only at slaughter. Cows with perforations may walk with an arched back and be reluctant to eat. Other symptoms include slowed or absent rumen contractions, fever, or a positive "withers pinch test." Pinch the backbone of the cow at the withers; cows with hardware disease will usually grunt.

If the object has penetrated and caused infection around the heart, you may hear abnormal heart sounds (like a washing machine) when you listen with a stethoscope. You may also be able to see a pulse in the jugular veins.

If severe signs are present (fever and pain), antibiotics and possibly the surgical removal of the foreign object is necessary. If antibiotics are used, the animal must be removed from organic production. This disease can be prevented (and mild cases treated) by administering magnets to cattle and ensuring that cows do not have access to foreign materials in the feed bunks or pasture.

WINTER DYSENTERY

This digestive disorder is characterized by an acute onset of heavy, watery diarrhea caused by a coronavirus. It frequently occurs during winter housing (November through March) and likely spreads because of close contact. Once the disease starts, it usually affects all the adult animals in a herd. Although this disease is rarely fatal, it may cause a drop in milk production.

Treatment of winter dysentery is supportive: access to plenty of water so animals do not become dehydrated. Most animals will continue to eat and drink throughout the course of the disease. After a bout of winter dysentery, the herd may be immune for a few years.

BLOAT

Bloat is caused by an accumulation of gas in the cow's rumen. It usually falls within one of two categories: frothy or gassy. In mild cases, the cow belches and releases gas, but if abnormal fermentation occurs, gas may fill the rumen. The enlarged rumen puts pressure on blood vessels and restricts breathing; if the condition is not treated, death will occur.

Bloat occurs most commonly on young, lush legume pastures (particularly clovers). Moisture on the pasture (rain or dew) increases the occurrence of bloat, particularly in springtime or early summer because of the rapid fermentation of these highly digestible, early growth forages producing excess gas in the rumen. When this happens, soluble proteins are rapidly released and attacked by slime-producing bacteria. The slime forms a stable protein foam, and fermentation gases build up under this layer and the cow cannot expel them.

Prevention

- Allow only gradual access to legume pasture.
- Feed dry hay in barn before turning out onto pasture.
- Avoid turning cattle out on wet or dewy legume pasture. Allow sunlight to dry it.

Treatment of bloat

- Poloxalene is allowed for the emergency treatment of bloat but is not allowed as a preventive.
- For mild bloat, remove cow from pasture and feed dry hay.
- Have the cow stand uphill so gas can more easily escape the rumen.
- Place a stick in cow's mouth and tie it behind the ears to encourage salivation.
- For severe cases, pass a stomach tube.
- For life-threatening emergencies, punch a hole in rumen with a trocar (have your veterinarian provide you with proper training and tools).

Alternative therapies for bloat and indigestion
These may not be scientifically evaluated or appropriate for all farms. Make sure you consult a reference recommended in the Resources list at the end of this chapter for specific instructions.

- indigestion: Aloe vera juice
- probiotics
- garlic
- bloat: vegetable oil

ADULT RESPIRATORY DISEASE

Bovine respiratory disease can be a major economic drain on organic farms, causing loss of production, increased labor costs, premature culling, and death. Upper respiratory disease (URD) affects the nostrils, throat, and windpipe; pneumonia (or lower respiratory tract disease) affects the bronchial tree and lungs.

A variety of causes (see Table 7.7) is responsible and is often triggered by stressful environmental factors. Stressors include humidity, dust, dehydration, irritating gases from manure build-up, and nutritional deficiencies. Abrupt weather changes, cattle transport, and poorly ventilated barns may also lead to respiratory disease outbreaks.

The signs of respiratory disease indicate whether the upper or lower respiratory tract is affected and the severity of the disease. URD can be characterized by discharge from the nose and eyes, coughing, and loud sounds while breathing. Signs of pneumonia include fever, depression, lack of appetite, increased breathing rate, coughing, and death.

BEST MANAGEMENT PRACTICES

- *Improve ventilation*: Cows out on pasture are at a lower risk for respiratory disease than cows in a poorly ventilated barn. Fans improve airflow inside barns and you can retrofit many older barns with tunnel ventilation. In tunnel ventilation systems, fresh air is drawn through openings in one endwall by exhaust fans mounted in the opposite endwall.
- *Vaccination*: If respiratory disease is a problem on the farm, or if new cattle are introduced to the farm, consider vaccinating both the home herd and new cattle. Vaccinations should include: IBR, PI3, BVD (Types I and II), BRSV, *Pasteurella*, and *Haemophilus somnus*. Early in a herd outbreak, vaccination or boostering with an intranasal vaccine may be helpful.
- *Feed quality*: Dust from poor-quality hay can increase the risk of respiratory disease in a number of ways. Dust (possibly including mold spores) can act as a physical irritant on the respiratory system and may trigger allergic reactions and noninfectious pneumonia (interstitial pneumonia).

TREATMENT PRACTICES

- *Passive antibodies*: Although not effective for viral pneumo-

nia, these are effective for *Pasteurella* pneumonia.
- *Vitamin B and C injections*: Both of these vitamins are good antioxidants. You can administer both under the skin or in the vein.
- *Anti-inflammatories (aspirin or flunixin)*: Used to reduce fever and prevent damage to lungs.
- *Antibiotics*: In cases where the calf does not respond to the above treatments, you **must** give antibiotics to prevent suffering. You must **permanently remove** animals treated with antibiotics from organic production.

Alternative therapies for adult respiratory disease
These may not be scientifically evaluated or appropriate for all farms. Make sure you consult a reference recommended in the Resources list at the end of this chapter for specific instructions.
- herbal antibiotic tinctures
- garlic
- homeopathy determined by cow's presentation and symptoms
- essential oils (eucalyptus)

BASIC CATTLE IMMUNOLOGY

A good organic system and a holistic health system will naturally enhance the cow's immune system. A fundamental understanding of bovine immunity will help you troubleshoot problems when disease arises and will help you make educated management decisions to enhance immunity on your farm.

COMPONENTS OF THE IMMUNE SYSTEM THAT OFFER A LAYER OF IMMUNITY

- *Skin and mucous membranes*: The strength and health of the epithelium (skin of the body) is the first line of defense against disease by providing a physical barrier. Mucous membranes (nasal passages, mouth, vulva, etc.) contain skin-specific antibodies (immunoglobulin A) in secretions on their surface that are first alert immunity for diseases entering via that route.
- *White blood cells*: These cells are produced in the bone marrow, spleen, liver, and lymph nodes and are constantly circulating in the bloodstream, patrolling for disease agents. When they encounter disease, they engulf the agent and send out warning chemicals (to recruit other white blood cells from other parts of the body.
- *Lymphocytes*: Lymphocytes are specialized white blood cells that produce antibodies to disease agents.
- *Passive immunity*: Passive immunity is achieved when antibodies are passed from dam to calf in the colostrum. Active immunity comes through antibodies created in response to a vaccination or an illness; it is generally much stronger than passive.

STRATEGIC VACCINATION PROGRAMS

The NOP allows vaccination. It is strongly recommended for open herds (e.g., when animals are bought in or sold, animals are shown in fairs, or other animals are occasionally boarded on the farm). It is always cheaper and healthier for the animal and farm budget if the farm can prevent diseases.

HOW VACCINATIONS WORK

Vaccinations stimulate the cow's immune system against specific diseases that may be present or later introduced into the herd. You should discuss vaccination with your herd veterinarian. There is no one-size-fits-all vaccine program appropriate for every dairy. Your farm's disease history, purchase of new animals, management, housing, breeding, cost versus benefits, and efficacy are all important considerations when evaluating whether a vaccination will benefit your farm.

No vaccination program can protect all the animals. The goal is to build a level of herd immunity so disease cannot spread to susceptible animals. If a disease is introduced into an unvaccinated herd, 100% of the animals are susceptible. Disease may spread from the initially sick animal throughout the herd. Some may get sick and die; others will survive the infection and develop an immunity that may last weeks or a lifetime. When a disease infects vaccinated herds, the herd already has a level of immunity that slows or even halts its spread. It may therefore never reach the susceptible animals. Figure 7.3 illustrates an unvaccinated herd. All cows in this herd are susceptible. If a disease is brought into the herd, it can spread rapidly from cow to cow since so many animals are at risk. On the other hand, Figure 7.4 illustrates a vaccinated herd where 17 of the 19 animals in the herd are immune, leaving only two susceptible animals. If a disease enters *this* herd, it will likely encounter an immune animal and be stopped in its tracks before reaching a susceptible one.

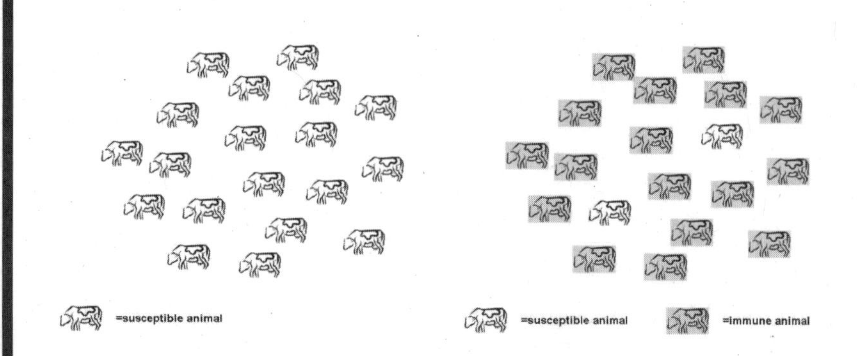

FIGURE 7.3
Unvaccinated herd

FIGURE 7.4
Vaccinated herd

TYPES OF VACCINES

There are four kinds of vaccines, and each has different benefits and drawbacks. These are described in Table 7.19.

ADMINISTERING VACCINES

Stress, other diseases, parasites, and poor nutrition can affect the body's ability to produce a good immune response to a vaccine and make it more susceptible to vaccine reactions. Failure to give booster vaccinations at the recommended intervals leaves animals unprotected. Poor handling of the vaccine, such as exposing it to temperature extremes, may reduce the quality and fail to stimulate a response.

Vaccinating young calves is tricky. Antibodies absorbed from colostrum often interfere with the vaccines during the first few months of life. Also, newborn calves' immune systems are usually not mature enough to respond to vaccination.

Some people are concerned with the risks of vaccination. Some vaccines, particularly killed ones, are more likely to stimulate an allergic reaction. If a cow receives too many vaccines at one time, her immune system may be overwhelmed responding to all the vaccines and may not respond completely to one or more of them,

TABLE 7.19 VACCINE TYPES, NEEDED ACTION, BENEFITS, AND CAUTIONS

TYPE OF VACCINE	ACTION	BENEFITS	CAUTIONS
Modified Live (MLV)	Multiplies in animal and stimulates white blood cells to produce antibodies	Create a very strong immune response and usually just one vaccination is necessary	Often not given to pregnant animals for fear of causing abortion or birth defects (IBR, BVD)
Killed	Inactivated virus or bacterium, so it does not reproduce in body	Safe for pregnant animals	Immunity is not as strong or long-lasting; requires frequent boosters; delayed development of immunity
Autogenous	Bacterial or viral vaccine produced from organism or tissue of sick animal	Used for farm-specific diseases where commercial vaccines are not available or effective	May contain bacterial toxins, so may create allergic reaction
Nosodes	Homeopathic remedy produced from diseased tissue or culture	Used as preventive to stimulate natural immunity	Not effective in outbreaks: not legally recognized

leaving her susceptible. You should also take care using MLV vaccines. Since they are modified live viruses, animals may shed the virus for a brief time after vaccination. This may cause complications in immuno-suppressed or young animals if they are in contact with the MLV-vaccinated animals. Table 7.20 outlines common vaccines administered on organic dairy farms.

BIOSECURITY

A biosecurity program is a set of management practices used to minimize the introduction to disease, spread of disease within the herd, and transport of disease off the farm. Using biosecure best management practices reduces clinical disease, improves production, and increases profitability. As always, it is easier to prevent disease than treat it.

TABLE 7.20 DISEASES TO BE CONSIDERED IN DESIGNING A STRATEGIC VACCINATION PROGRAM

DISEASE	CONDITIONS	VACCINATION SCHEDULE
Infectious bovine rhinotracheitis (IBR)	Respiratory disease, abortion	4–6 months of age; prebreeding and yearly booster
Parainfluenza-3 (PI3)	Respiratory disease, abortion	
Bovine viral diarrhea (BVD)	Diarrhea, abortion, fever, respiratory disease	
Bovine respiratory syncytial virus (BRSV)	Respiratory disease	
Bovine respiratory complex (*Pasteurella, Mannheimia, Haemophilus*)	Respiratory disease, fever	
Leptospirosis	Fever, abortion, blood in urine	
Brucellosis	Abortion, reproductive diseases, repeat breeding	Calfhood vaccination at 4–8 months (once)
Clostridium	Sudden death	Start at 2–6 months and booster in 2 weeks, yearly booster
Scours (E.coli, Corona and Rotaviruses)	Calf diarrhea	Give at 6–7 months of pregnancy
Pinkeye	Watery eyes, clouding, swelling	Two doses at least 30 days prior to pinkeye season
Coliform mastitis	Acute mastitis, fever, depression, watery milk, death	Three doses: at dry off, two weeks later, and after freshening for herds with coliform mastitis problems

BIOSECURE BEST MANAGEMENT PRACTICES TO REDUCE OFF-FARM RISKS

Off-farm risks are diseases or problems that may occur from introducing animals to the farm, moving animals, or transferring biological hazards on or in objects (truck tires, boots, and water sources). The questions below address the chances that a new disease will be introduced on the farm.

When purchasing animals, you buy more than just the cow. Healthy animals may be carriers of contagious mastitis, Johne's disease, BVD, and other risks. Keeping a closed herd (where no new animals are introduced) is ideal. However, farms interested in expanding their herd may feel the need to bring in outside animals. There are steps you can take to minimize risks. Think about the following.

- What is the origin of cattle you are buying? Buying from a private, closed herd with rigorous testing is preferable to buying from an auction or cattle dealer where animals from many different farms are mixed.
- Ask for the current herd health records. Has the farm screened for contagious disease within the past year?
- Can you quarantine animals when they arrive on your farm for 3 to 4 weeks so that testing, vaccination, or treatment may be accomplished?
- Are the cattle already on the farm vaccinated? (See the section, Strategic Vaccination Programs on page 226).

There are other off-farm risks that can introduce diseases to the farm. For the following situations in open herds, taking precautionary biosecure measures lessens the risk of disease.

- *Boarding and exhibiting*: Boarding animals off the farm (e.g., sending heifers to a custom raiser) or exhibiting animals in fairs and shows: protect by vaccination and quarantine for 3–4 weeks when they return home.
- *Wildlife*: Noncattle wildlife (particularly birds and rodents) can introduce *Salmonella*, *Cryptosporidium*, *Leptospirosis*, rabies, and other diseases. Biodiversity is important for organic farms, but it is important to create control programs for pests and to keep feed in areas or containers where rodents and birds have minimal access.
- *Manure from noncertifed herds*: Use of manure from noncertified herds is allowed in organic production, but it carries some risks. Before spreading manure from the neighboring dairy on your fields, ask about the Johne's status of the herd. High-temperature composting of off-farm manure will reduce the level of infectious organisms (*E. coli*, *Listeria*, *Salmonella,* and Johne's bacteria) in the manure.
- *Vehicles and people*: Restrict access of vehicles that travel farm to farm (e.g., feed trucks) to only necessary parts of the farm and keep them out of contact with animal areas. Ask everyone to sanitize their boots on entry to your farm and wear clean coveralls. It would be good to have a boot wash station with disinfectant readily available. Request that your

hoof trimmer disinfect the table and tools before they come onto your farm.

BEST MANAGEMENT PRACTICES TO REDUCE WITHIN-FARM RISKS

Disease can also spread among susceptible groups on the farm. The most common scenario is milking the cows and then traveling down to the calf barn to feed the calves, wearing the same coverall and boots. Ideally, cattle should be worked in order of increasing age, with adjustments made for sick animals. Calves are the most immunologically immature animals on the farm, so make contact with this group before any others on the farm. Sick calves should be separated from healthy calves and fed and treated after the healthy group. Maternity and sick pens should be separate and sick cows should be the last group worked on the farm. Washing boots, hands, and changing coveralls between groups creates optimal biosecurity.

BEST MANAGEMENT PRACTICES FOR SPECIFIC DISEASES

Johne's disease

Johne's disease, caused by the bacterium *Mycobacterium paratuberculosis* subspecies *avium*, is a chronic wasting disease of cattle and other ruminants, characterized by weight loss and diarrhea. Infected cattle shed the bacteria in manure and may classify as light, moderate-high, or super shedders depending on the number of bacteria per gram of manure. It may take only a thimbleful of contaminated manure from a heavy shedder to infect a calf.

Young calves up to a year old are most susceptible to oral infection with Johne's bacteria. Symptoms, however, rarely appear until years after the initial infection occurs. Johne's disease can also be spread in the uterus and in milk by heavily infected mothers, although this route of transmission is less common.

Best management practices are designed to prevent susceptible calves from coming in contact with adult manure. Most states have a Johne's Disease Control Program that assists farmers in designing plans to identify Johne's disease on the farm and to work to reduce it over time. Your state veterinarian can provide additional information.

The basic steps of all Johne's disease control programs

1. Establish a Johne's testing and removal program for your farm.
    - This plan varies farm to farm and should be developed with your veterinarian and the state Johne's disease coordinator.
    - Small herds should fecal-culture all cattle older than two years of age; larger herds may test a subset of cattle that have been statistically selected.
    - Environmental screening (culture of pooled feces from concentrated cattle areas) can be performed to assess if

Johne's is present on a farm.
- Cull heavily infected animals.

2. Prevent fecal to oral spread of Johne's bacteria.
   - Use separate tools to scrape manure. Do not use manure contaminated tools or equipment to handle feed.
   - Prevent manure contamination of the drinking water.
   - Prevent runoff from adult areas to calf areas.
   - Dedicate pastures for youngstock use only. Avoid leader-follower grazing schemes.
   - Do not walk in the feed.
   - Feed and care for calves before working with adult cattle.

3. Manage calves to break the cycle of transmission.
   - Know the Johne's status of the dam before calving.
   - Calve cows in separate pens or expansive pasture. Pens should be dedicated to calvings and cleaned between uses.
   - In herds with Johne's disease, calves should be removed from birth mother as soon as possible.
   - Feed calves colostrum and milk from test-negative cows only.

You can access additional information on Johne's disease at: www.johnes.org and www.nyschap.vet.cornell.edu.

Bovine viral diarrhea (BVD)
Bovine viral diarrhea is a problem in dairy and beef cattle and other ruminants. As with Johne's disease, infected cattle may not show any outward symptoms. Although BVD was initially associated with diarrhea, we now know that the virus can cause a range of symptoms. BVD affects the cow's immune system, making it more likely to contract other respiratory and intestinal diseases. Other symptoms include ulcers in the mouth and nose. If calves are infected later in gestation they may be born with birth defects.

Cattle may have acute infections or be persistently infected. Acute animals may have a fever, depression, diarrhea, or respiratory disease. BVD may cause abortions during the first trimester. Acutely infected cattle shed the virus for about two weeks and either recover completely or die. Persistently infected (PI) animals contract the disease *in utero*. PI animals are infected for life, constantly shedding the virus. The offspring of PI cows are also persistently infected. These infected animals may not exhibit any symptoms at all or they may be sickly and die in the first year of life. Bulls can also be infected and shed BVD in semen.

Best management practices for BVD
- Screen the herd for BVD in bulk milk. Your veterinarian can submit bulk milk samples to your state diagnostic laboratory for screening. Although this is an accurate test, it does not check the dry cows and calves.
- Test and remove persistently infected animals. Nasal swabs,

blood (serum) samples, and ear notches can identify infected animals.
- Vaccinate to prevent additional spread.
- Screen bulls for BVD before they come to the farm or be sure your semen company uses BVD-free semen.

Bovine leukemia virus (BLV)
Bovine leukemia is another contagious viral disease of cattle also known as bovine leukosis or lymphosarcoma. BLV is spread among cows by blood contact. Multiple-use needles and rectal sleeves, dehorning equipment, and biting flies can all spread this disease. Dams can also spread it to calves in colostrum and milk. Many cattle in the U.S. are infected with BLV, but only a small percentage (less than 5%) develop cancer of the lymph nodes (lymphosarcoma). Finding lymphosarcoma will result in the carcass being condemned at slaughter.

Best management practices for BLV
- Identify and remove positive cattle by blood testing.
- Prevent the further spread in the herd.
  - Use single-use needles.
  - Change rectal sleeves between palpations.
  - Sanitize dehorning instruments between animals, or use butane or electric dehorners.
  - Control flies.

CONCLUSION
Animals managed in organic systems have the opportunity to enjoy tremendous health and longevity, but herd health on organic farms requires a proactive, holistic approach, and not merely substituting alternative medicine for conventional synthetic treatments. As the farmer, you need to assess the animal's condition on a daily basis so that problems are detected early. Once an abnormality is detected, you must act and treat not only the problem, but also reflect on what could be wrong in the organic management plan that is the root of the problem.

Your herd health toolbox (see Appendix P) will contain both alternative and allowed synthetic treatments. Review the contents with your veterinarian and develop standard operating procedures for common herd health problems. Work with established organic dairy producers to educate yourself on effective alternative therapies. Remember, however, that all products may not work equally well in all situations and you may need to tailor these therapies for your farm. Finally, when alternative treatments are not effective or the severity of a disease warrants it, prohibited conventional medications **must** be used to relieve animal suffering and that animal must be removed from organic production forever.

"Treat the cow kindly, boys; remember she's a lady—and a mother."—Theophilus Hacker, 1938

RESOURCES

Animal Welfare Resources

Fraser, A. F. 1997. *Farm Animal Behaviour and Welfare*. CABI Publishing: New York, NY.

**Animal Welfare Information Systems:
http://www.nal.usda.gov/awic/pubs/dairy/dairy.htm**

**New York State Cattle Health Assurance Program Animal Welfare Module:
http://www.nyschap.vet.cornell.edu/module/welfare/welfare.asp**

**Outline of Animal Welfare Critical Control Points for Dairies:
http://www.grandin.com/cow.welfare.ccp.html**

Alternative and Complementary Treatment Resources

de Bairacli Levy, Juliette. 1991. *The Complete Herbal Handbook for Farm and Stable*. Faber and Faber: London.

Dettloff, Paul. 2004. *Alternative Treatments for Ruminant Animals*. Acres U.S.A.: Austin, TX.

Karreman, Hubert. 2006. *Treating Dairy Cows Naturally: Thoughts and Strategies*. Acres U.S.A.: Austin, TX.

Sheaffer, C. Edgar. 2003. *Homeopathy for the Herd: A Farmers Guide to Low-Cost, Non-Toxic Veterinary Care for Cattle*. Acres U.S.A.: Austin, TX.

Verkade, Tineke. 2001. *Homoeopathic Handbook for Dairy Farming*. HFS: New Zealand.

Wynn, Susan and Barbara Fougere. 2007. *Veterinary Herbal Medicine*. MOSBY-Elvesier: St. Louis, MO.

NOTES

1 Pasteurized Milk Ordinance of 2003: http://www.cfsan.fda.gov/~ear/pmo03toc.html.

2 There are many versions of the Homeopathic Materia Medica, but one helpful resource is: George Macleod. 2004. A Veterinary Materia Medica and Clinical Repertory: With Materia Medica of the Nosodes. Random House: UK.

Rick, Kathie, Bob, and Miriam Arnold.

BOB, RICK, AND KATHIE ARNOLD, TWIN OAKS DAIRY, LLC

LINDA TIKOFSKY AND BETHANY WALLIS

Twin Oaks Dairy, LLC is a picturesque hillside dairy of 130 Holstein, Scandinavian Red, and Brown Swiss cows munching pasture along river-bottom lands in central New York. The farm partnership began in 1980 among Bob, Rick, and Kathie Arnold. Parts of the farm, however, have been in the family since the 1930s. Originally a conventional dairy, it became management intensive to increase herd average using a variety of approaches that were currently in vogue. In the early 1990s, the Arnolds confined high and fresh cows to increase production. However, feed costs became extremely high and to remain profitable, they made the choice to move to management intensive rotational grazing in 1993. The best land on the farm was gradually added to their pasture system (prior to that only the hillsides were permanent pasture). They farm more than 700 total acres: 50 acres of corn, 30–50 acres of small grains, soybeans, about 200 acres in pasture, and the remainder in hay.

Kathie had long had an interest in sustainable and organic farming methods, but their transition was delayed until a milk market came to their region. Although Rick was reluctant at first, when he saw the price differential in a newspaper ad he realized there was an economic incentive to transition. By 1997 there was a stable market for organic milk, so they started the process of certifying land and in January 1998 began transitioning the herd. At that point it was only a three-month herd transition, so they began shipping organic milk in May 1998. They would never go back to conventional methods. Organic farming has become a way of life for them.

In the early 1990s when their cows were confined, health problems abounded. One of the biggest bonuses when they switched to intensive grazing was the tremendous increase in herd health. Displaced abomasums decreased with grazing, and without corn silage in the ration, they no longer had problems with ketosis. Cows also now have fewer dystocias, which Kathie attributes to grazing and outdoor access. Their cows are strong and vigorous; they are athletes.

Udder health and milk quality are important to the Arnolds. They prefer to see bacteria counts under 5,000, and they would love to be under 150,000 cells/ml for bulk milk Somatic Cell Count (SCC) consistently—that is when they achieve their highest premiums—but realistically, they are satisfied with under 200,000 cells/ml.

In the 1980s, they received some milk quality awards while going through caseloads of intramammary antibiotics. Mastitis settled down such that by later in the 1980s, they stopped using antibiotics cold turkey for udder treatments after reading about other farms doing the same. After several years of lax mastitis management, elevated SCC raised its head as a problem and selective dry and lactational antibiotic treatments were used again. At the time of transition, they struggled to keep bulk milk SCC under 400,000 cells/ml. This was a concern since organic processors are reluctant to accept bulk milk with cell counts of 400,000 or above. Cornell University's Quality Milk Production Services (QMPS) cultured all the lactating cows and found that one-third of the animals were infected with contagious mastitis due to *Staphylococcus aureus*. The Arnolds identified all *S. aureus* infected cows with leg bands, moved them to one end of the barn so that they were milked last, and gradually culled them from the herd. They have not eradicated *S. aureus* from the herd, as new cases continue to pop up, but it is a matter of constant management.

KATHIE'S ADVICE FOR TRANSITIONING FARMERS WHO WANT TO MAINTAIN OR IMPROVE MILK QUALITY

1. It is important to know the status of your herd in terms of mastitis pathogens. Culture your herd on a regular basis, or at least culture your bulk tank to determine if you have *S. aureus* or *S. agalactiae*.
2. Maintain your milking equipment. Twin Oaks Dairy maintains its system twice a year and changes inflations monthly. QMPS personnel also evaluate the system independently twice a year.
3. Follow good sanitary and consistent milking procedures.
4. Enroll your herd in DHIA's monthly SCC testing and use the results. Most of the mastitis in our herd is subclinical; the highest SCC cows usually do not have abnormal milk or udders.
5. Be ever vigilant. Experience has shown us that getting complacent will come back to bite us.

Early identification of new contagious mastitis problems is important, and Kathie aims to be vigilant. If a cow has a high SCC on Dairy Herd Improvement Association (DHIA) testing, Kathie performs a California Mastitis Test on the cow to identify the problem quarter. She takes a sample for culture and may treat her, depending on the cell count. They have also purchased a Delaval DCC somatic cell counter for testing fresh and treated cows. They test milk and withhold any milk that has an SCC of 1 million or greater. For the treatment of mastitis, they give an injection of a colostrum-whey product accompanied by an herbal intramammary tube, a homeopathic topic cream or pellets, and/or a topical essential oil mix.

The farm's milking procedures consist of predipping with a 1% iodine teat dip that stays on for 30 seconds before being wiped off with individual paper towels after prestripping. After the unit comes off, teats are post-dipped with 1% iodine in a separate bottle. Milkers wear gloves. Because of problems with Herpes mammilitis in winter, they use a chlorhexidine-based postdip during colder months.

As when they were conventional, mastitis remains the Arnold's greatest challenge. Under organic management, they must be much more vigilant and management intensive. However, the payoff is great. They are able to have a healthy herd without the use of antibiotics and other synthetic materials and a large acreage without pesticides and herbicides.

TOM AND SALLY BROWN, SHIPHRAH FARM

LINDA TIKOFSKY AND BETHANY WALLIS

Located in central New York, Shiphrah Farm is nearly a century old. Tom and Sally Brown have been farming there for 30 years and made the transition to organic management in 2000. Their decision to transition was economically driven: they needed to find a more cost-effective way for the dairy to remain in business. When the Browns transitioned in 2000, only a 90-day transition was required. They had already been grazing rotationally for two years and most of their land was certifiable right away. They only needed to transition the animals. The transition helped improve their overall farm management and created a paradigm shift for their personal lifestyle as well.

Tom and Sally Brown.

Their main management foci are to build the health of the land and improve herd health. They currently milk 100 animals and farm 325 acres, growing hay—mostly grasses (orchard grass/white clover and reed canary/red and white clover and timothy). They have roughly 43 acres of pasture for the milking herd. The Browns do not plant row crops; they concentrate on managing the soil to provide high nutrient values through perennial grass crops. They also routinely test their soil for pH and trace elements. Sally explains, "We constantly go back to the soil. We take soil tests to monitor how healthy the soil is and work on building the soil."

Johne's disease has been their biggest herd health challenge. They first realized this disease was a problem several years ago with a clinical animal. When they first tested for Johne's, 62% of the herd was positive on fecal samples that summer. They had recently purchased test-negative cattle from another farmer, and those were positive as well. One of the cows tested was a "super shedder" who was passing millions of infective bacteria in her manure. Unfortunately, this super shedder was a cow they were using for calf milk at the time. Most of their youngstock, the most susceptible population on the farm, had been exposed to this deadly disease. As a result, 80% of their two-year old heifers never made it to freshening. They died of clinical Johne's disease and wasting.

Under organic management, calves **must** receive whole milk, so Sally thinks it is essential that organic farmers know the Johne's status of their herd. In normal farm practices, you feed milk from high SCC cows or ones with other problems to the calves, saving the better-quality milk for the bulk tank.

When they go to organic meetings, it is difficult for them to admit

Shiphrah Farm's calves in the greenhouse.

that they have a herd with Johne's, but since it nearly put them out of business they feel obligated to educate their fellow farmers. They had the challenge of "buying the farm" over again. It cost them over $600,000 when they just factor the youngstock that were lost.

They joined the NYS Cattle Health Assurance Program (NYSCHAP) so that they could test for Johne's more economically. Controlling exposure of young animals to manure was key. They watched everything they did on the farm. They created protocols such as no walking in manure and then in the feed alley and reduced tractor traffic in and around the barns. They now have individual calving pens that are cleaned after every use, and they know the Johne's status of every cow before she calves so she can be properly segregated. Johne's cows may calve anywhere but in the designated calving pens. If a calf is born in manure, this is noted in their herd records and that animal is closely observed as she grows. Heifers from known Johne's-positive dams have the letter 'J' added to their ID so they know at a glance the heifer's history. Tracking animals is essential to their battle with Johne's.

Foot traffic on the farm is limited, and individuals who have been to multiple farms are not allowed in the calf greenhouse. They do not allow the cattle trucker in the greenhouse or the barns. Salesmen and truckers are the least concerned with biosecurity in the Browns' opinion. Veterinarians and AI technicians wash their boots; truckers and salesmen do not, and the Browns therefore view them as big risks. When Tom comes from the freestall to the feed room, he stops in the milkhouse first to wash his boots so that contamination of feed is prevented.

Unfortunately the Johne's group is next to their breeding-age heifers and so there are concerns about manure splatter, but that is the best the Browns can do. Calves that come from the greenhouse are housed near the holding area, so the Browns have installed large sheets of plywood on that side of the holding area to eliminate splatter into the calf pen. Being obsessive about the details has brought the Browns from a prevalence of 65% Johne's to 5–7% Johne's. Johne's-positive cows are culled as soon as they become moderate or heavy shedders.

Selling all the heifer calves from Johne's cows is not economically feasible for the Browns. Strict recordkeeping allows them to review their progress, and they have not necessarily seen a correlation between a Johne's dam and the Johne's status of the daughter. "When managing Johne's you have to watch the details on a day-to-day basis," explains Sally.

The organic milk price is what saved their farm when the Johne's problem arose. If they were shipping conventionally, given that milk price, they would not be in business today. They also think that the emphasis on organic management aided in their recovery from the crisis. Bringing the soils and crops into balance with the emphasis on trace minerals had an impact on cow health and immune systems. In the Browns' opinion, even though the cows may have some minimal exposure to the Johne's bacteria, their high level of health allows them to fight that exposure and not become infected.

Sally testifies that organic management has absolutely changed and enhanced their overall farm management. She states, "Actually, going organic did a paradigm shift in our entire life—not just on the farm. When we saw how well it worked for the animals, we figured that it would work just as well for us." One of their missions, as organic farmers, is to pass on what they have learned and help others.

VAUGHN AND SUSAN SHERMAN, JERRY-DELL FARM

LINDA TIKOFSKY AND BETHANY WALLIS

Vaughn and Susan Sherman, with the help of their sons and nephews, farm Jerry-Dell Farm—about 1,300 acres of central New York farmland. Vaughn's parents, Jerry and Ardella, started the farm in 1946 with 100 acres and 20 cows. Vaughn and Susan purchased the farm in 1976 and added cows to a herd size of 150 cows. They modernized the facilities with the addition of a new barn and several Harvestore silos. Grazing however, was still an integral part of cow management.

Susan and Vaughn Sherman.

Disaster struck in 1985 with a barn fire that destroyed the old facilities. Following university recommendations at the time, the Shermans rebuilt and expanded the herd to 350 cows. They gradually reduced grazing until the herd was 100% confinement. It was a vicious cycle at that point, pushing the cows for production so that they could pay the bills. As Susan relates: "By 1997, the cows were sick, they were dying, and everybody was stressed out." Realizing that some changes were necessary, they met with a consultant whose sole recommendation was: sell. Liquidation was not an option, so they decided to return to what they knew—grazing.

Working with Graze-NY, they formulated a grazing plan and put cows out on pasture in the fall of 1997. The cows got healthier every day. A few years later, Vaughn and Susan were approached by an organic milk processor and began to investigate organic production through NOFA-NY. Since they had never applied many synthetic fertilizers or amendments on their soils, most of their land was immediately certifiable. At that time, only a three-month transition was required for animals, so they were ready to ship organic milk by January 2001. In their opinion, this is the best move they ever made for themselves and for the cows. "Farming is a lot more fun now," says Susan. "Our kids and nephews have come back to the farm."

Their total farm acreage is about 1,300 acres, and 340 animals graze on approximately 250 acres. They would like to develop more pasture in the future to meet the grazing needs during the dry years. "If you have good pasture you can really reduce the grain you are feeding. During lush pasture season, we get down to 10–15 pounds of grain. In the wintertime it's 15–20 pounds." They rotationally graze their cows and give them new pasture every 12 hours. There are three grazing groups: the high-production group, medium-production group, and low-production group. They move water wagons with the cows so that fresh water is always accessible. Their pasture season runs from May 1 to November 1. During pasture season, the cows are in the barn for about 4 hours during milking and feeding TMR.

During the winter months, the Shermans feed a completely TMR diet, but in summer months, they reduce it to 25% TMR. The TMR components are minerals, salt, haylage, and grain mix (high-moisture corn, some soy, and small grains when they are available). They grow most of their own small grains (triticale and spelt) on 173 acres and grow about 200 acres of corn. With each succeeding year, they purchase less grain. They also grow nearly all the hay they use. The Shermans use several reference rations depending on pasture conditions and grain availability. Managing nutrition on an organic farm means planning ahead. When they were conventional and needed grain, they picked up the telephone and it was delivered that afternoon. Now they call and need organic grain, it may take a week or two. "When we were feeding cows conventionally it was pretty uniform. Every day was the same, especially when they were total confinement; now it's always changing."

HERD HEALTH

At the close of their conventional years, the herd was under total confinement with highly intensive feeding, milking three times a day, and using rBST. Their herd average at that time was 100 lb./cow/day, but that production came at a high cost: many sick cows, laminitis, DAs, and a cull rate that pushed 50%. "We weren't too impressed managing that high-intensive style of farming," says Vaughn. Calf mortality was high, roughly around 30%. The biggest herd health problem during those final years was laminitis although, in reality, there were many others. Bulk milk somatic cell counts (SCC) reached 400,000 cells/ml and the Shermans stopped drinking their own milk because of the bad flavor.

Herd health under organic management is 90% better in Vaughn's opinion. Now the cull rate is 30%—cows they *want* to cull, rather than previously when they culled because of illness. The cows that leave now are healthy cows that leave for dairy purposes rather than beef. When they want to build cow numbers, they reduce their cull rate to 20%. Currently, perhaps there is one calf death a month, which translates to a mortality rate that is less than 5%. Laminitis problems are gone, but they still battle heel warts, which were a problem under conventional management as well. Reproductive problems (retained placenta, metritis, etc.) have also decreased. They manage the occasional metritis with organic uterine boluses and garlic/aloe infusions. Changing to an organic nutrition program higher in forage has drastically reduced the number of DAs. In the past, they may have had up to five a day; nowadays they treat five a year and they expect to achieve the goal of zero in the near future.

Calf management has changed as well. Calves are on whole milk to about 3 months of age. The Shermans introduce grain and TMR after that. They vaccinate calves twice (two weeks apart) with a 9-way vaccine and for pinkeye. Pinkeye is rare now; previously they

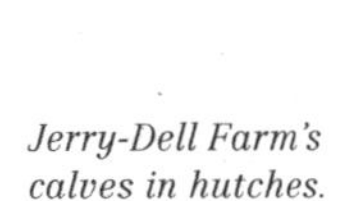

Jerry-Dell Farm's calves in hutches.

had it almost year-round. They breed the heifers at between 14 and 16 months of age so they will freshen around two years of age. Coccidiosis is a concern in older calves. They keep the environment extremely clean and monitor the calves through fecal samples. If they notice a problem, they add ground black walnut hulls to the feed.

Their milk quality is quite good and they enroll the herd in monthly DHIA testing. Average bulk milk SCC is between 150,000 and 200,000 cells/ml, typically closer to 150,000. Their goal for SCC is 100,000 cells/ml. The biggest milk-quality concern at this point is keeping a handle on *Staph aureus* contagious mastitis. Quality Milk Production Services cultures cows quarterly to identify new cases; the Shermans milk these last. In between herd cultures, they submit individual samples from fresh cows and cows with clinical mastitis. They strip out clinical mastitis cases and treat them with a whey product under the skin and an organically approved udder liniment topically on the quarter. Hot, hard quarters are treated homeopathically and the cows receive supportive treatment (fluids, dextrose, and vitamin C). When they were conventional, they would vaccinate for coliform mastitis, but they have discontinued this practice as coliform mastitis cases are now rare. Initially after transition, clinical mastitis increased but now few cases exist. They abruptly dry off cows by limiting their feed for 24 hours in an attempt to drastically decrease production. The average dry period is 45 days.

Johne's disease is a problem in their herd. They submit cows' fecal samples for Johne's culture two months before dry-off so that results are back before the cows freshen again. They immediately cull moderate and heavy shedders and are currently at less than 1% in their herd with the goal of total eradication. Low shedders are identified and retested.

Their herd veterinarian is generally supportive of their organic management, although he is not familiar with alternative or complementary treatments. To learn about alternative treatments for

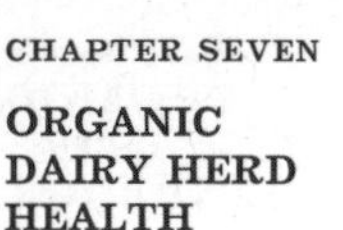

dairy cattle, they rely on books and attend local meetings on the topics. Their medicine cabinet contains homeopathic remedies, botanical tinctures and herbs, whey, and allowed synthetics (electrolytes, aspirin, etc.). Using the alternatives requires more vigilance to detect problems sooner, and cows are often treated longer than they would be with conventional medications.

When choosing products to use on their cattle, they like to see proof that it is effective and that it is an accepted product in the organic world. They confer with other farmers and ask about their experiences. "We have learned that if someone says it's an organic product, it isn't necessarily so. You must look into it and check with your certifier," remarks Susan.

Their greatest herd health challenge is respiratory disease in calves, and the alternative treatments do not seem to work very well in these cases. Calves with respiratory disease are given garlic; occasionally they need an antibiotic and must be removed from the herd. Respiratory disease is by far the most common reason cows or calves are given an antibiotic treatment. It is always a challenge to decide when to draw the line with alternative treatments and treat them with antibiotics. There is a conventional farm that will buy treated animals from them. For the most part, there are many fewer sick animals than there were before, so overall the Shermans think organic management has made the farm more animal and environmentally friendly.

Even though Vaughn says herd health is his biggest challenge, he thinks his biggest success under organic management is that the animals are healthy. "Managing for health is a daily thing. It's all animal health, animal comfort, animal well-being," says Vaughn.

Life is good and a recent mission statement they developed for their farm sums up their priorities: "Our goal is to have a successful farm with healthy animals and to work efficiently and cost effectively."

CHAPTER EIGHT

ORGANIC RECORDKEEPING

BETHANY WALLIS

Recordkeeping is a required part of organic certification and helps verify the integrity of the organic label. According to the National Organic Program (NOP) rule §205.103, recordkeeping by certified operations must follow these protocols.

> (a) A certified operation must maintain records concerning the production, harvesting, and handling of agricultural products that are or that are intended to be sold, labeled, or represented as "100 percent organic," "organic," or "made with organic (specified ingredients or food group(s))."
> (b) Such records must:
> (1) Be adapted to the particular business that the certified operation is conducting;
> (2) Fully disclose all activities and transactions of the certified operation in sufficient detail as to be readily understood and audited;
> (3) Be maintained for not less than 5 years beyond their creation; and
> (4) Be sufficient to demonstrate compliance with the Act and the regulations in this part.
> (c) The certified operation must make such records available for inspection and copying during normal business hours by authorized representatives of the Secretary, the applicable State program's governing State official, and the certifying agent.[1]

Well, there it is plain and simple: Recordkeeping defined by the government. But what does it really mean? What are you required to document? How should your records be maintained? Who has the right to view them? This chapter will attempt to simplify recordkeeping by outlining the requirements, showing different recordkeeping methods, and providing real-farm record management and compliance examples.

KEY ORGANIC TRANSITION POINTS

- Choose a recordkeeping system that you and your farm staff can maintain easily.
- All records must be accessible and easily audited by your inspector.
- Document that everything brought onto the farm operation is certified organic or approved for use on your organic operation.
- Always check with your certifier before using a new product.
- Maintain records of all products that leave the farm.

- The integrity of organic milk is dependent on all of the producers, certifiers, and the federal government. To market your products as certified organic, your operation must meet *all* of the requirements. The levels of inspection and review serve to check and maintain the integrity of the organic community.

KEEPING ORGANIC RECORDS

Documenting farm records for organic certification provides proof that an organic operation is following its organic system plan. Organic dairy farmer and Northeast Organic Dairy Producers' Association (NODPA) board member Kathie Arnold explains that, "The basis for the whole certification process is detailed recordkeeping on just about everything that comes onto the farm, leaves the farm, or is done on the farm."[2]

Although organic management may require you to maintain significantly more records than you once did, most farmers find that these additional records are an asset to their overall farm management. Detailed records provide farmers with a method for tracking patterns in soil, crop, and herd health, allowing them to more readily identify problem areas. This is essential to organic management because it is much easier to correct the problem at its source rather than treat the symptoms repeatedly.

Organic dairy farmer Les Miller recommends making your records compatible with your certifier's records. He suggests, "If you can use their forms, do so. Most certifier forms are pretty thorough; you don't need to reinvent the wheel."

EVALUATE YOUR CURRENT RECORDKEEPING SYSTEM

Evaluate your current recordkeeping system for any existing faults and determine how well it meets the farm's organic recordkeeping needs.

- *Is the system user-friendly?* If the current recordkeeping system is difficult, consider adopting a new system that is more user-friendly. All employees required to record certification information should have a complete understanding of what information must be recorded, why it is needed, and how to maintain the records properly.
- *Are past records available for the farm operation?* Organic certification requires that farms maintain their records for at least five years.
- *Is the system easy to access for inspections and audits?* The farm recordkeeping system should make it easy to present both current and past years' files and information to inspectors.
- *Does the system include all certification record requirements?* In order to comply with the NOP, all certified farms must maintain the required records for organic management. If the current farm records do not include all the organic required records, the current system must be updated.
- *Does the system include verifiable data sources for the audit trail?* The recordkeeping system must keep labels, receipts, bills of sale, and all other records and copies needed to verify all farm practices outlined in the organic system plan.

Les Miller has developed a highly organized and simple system for maintaining his organic certification documentation. He has clearly labeled an area of shelving to store all of the farm's certification information. His certification file folders are color-coded in relation to the color of the form provided by his certifier. His system is easy for everyone working on the farm to learn and use. Color-coding makes it easy to find records during inspections or if there is a certifier question. He keeps all product labels in file boxes and organizes all receipts in file folders by month.

CHOOSING A RECORDKEEPING SYSTEM

Farm managers have the flexibility to design their own recordkeeping systems. The NOP requires many detailed records but does not regulate the form in which they are kept. However, some certifiers do prefer farms to use their recordkeeping forms. When developing a recordkeeping system, keep in mind the individuals who will be maintaining the records and what type of system will work best for them.

COMPUTERIZED RECORDKEEPING

When developing a computerized recordkeeping system for your organic dairy, first check if your certifier has electronic application forms. Some certifiers prefer that their certified farms submit their information on these forms, so incorporating them into a farm recordkeeping system from the start will save some time! There are many electronic recordkeeping formats available in the marketplace, or you can create your own using a spreadsheet program.

Any computerized system must include a backup system on disks, external hard drives, and/or paper to ensure a secure audit trail. A hard copy of all records must also be available for the inspector to review during the farm inspection. Computer system failures or errors can lose valuable documentation and cause noncompliance problems. Certification documentation can be difficult or impossible to reproduce. A computer crash is not an acceptable excuse for not being able to produce the required records.

PAPER FILE RECORDKEEPING

There are many types of paper recordkeeping systems. Often producers simply use the certification application forms provided by their certifier. Check with your certifier for a list of the specific records required. All documentation must be clear, accessible, and well organized for inspection purposes. Some farmers create their own system including field and barn notebooks, calendars, lists, file boxes, binders, folders, or other methods to keep the information organized. Keep all paper records in a safe, dry area—weather damage is not an acceptable excuse for not being able to produce required records.

RECORDS SUPPORT THE ORGANIC SYSTEM PLAN

As you learned in the first chapter of this handbook, the organic system plan is a detailed description of the farm resources, soil management, crop production, livestock management, and all purchased inputs and sales of products from the farm system. For most certifiers, the organic certification application completes the organic system plan requirement. The following section outlines the records needed to support the information included in the organic system plan.

FARM MAPS

The farm map provides a picture of the farm for the inspector, certifiers, and the certification review committee. When developing a farm map, keep the certifier's requirements in mind. Most agencies suggest that you use hand-drawn maps, tracings from Agricultural Stabilization and Conservation Service (ASCS) maps, or outlines of aerial photographs that show the farmland and buildings. Copies of aerial photographs and ASCS maps are available at your local Soil and Water Conservation District office. Figure 8.1 shows two examples of hand-drawn farm maps.

Farm maps must clearly identify and label the following.

- Individual fields with either a number or letter that corresponds to the number or letter used on the application forms.
- The use of adjoining lands, whether owned by the certified farm or another individual (e.g., conventional farmland, woods, swamp, organic farmland, neglected, or residential).
- All buffer zones between organic and nonorganic land (hedgerow, 50-foot buffer, etc.).
- The farmstead, including location of barns, paddock areas, manure storage areas, water supplies (wells, ponds, and streams), and feed storage areas (grain bins, bunks, and silos).

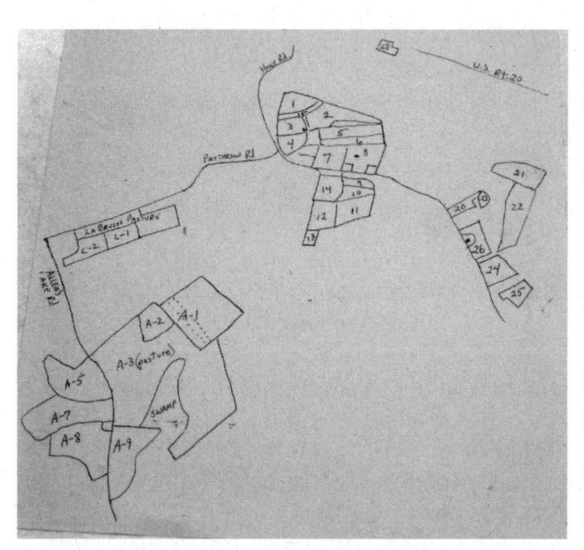

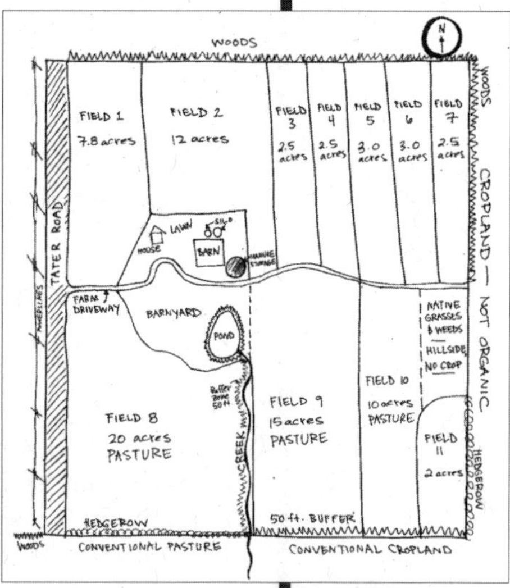

FIGURE 8.1
Farm map examples for organic records

FIELD HISTORY RECORDS

All fields that a farmer intends to certify must be included in the application. The NOP permits farmers to certify rented land, providing that they have daily management control of the fields. These fields require signed documentation from the previous owners or tenants that outline its past uses.

The NOP requires farmers to create a complete soil management plan that documents how they intend to maintain and improve soil health. The field history records prove that the farm is completing the soil management plan and outline crop rotations, applied amendments, green manures, cover crops, and erosion control methods. These records must include dates, seeds or amendments used, and quantities applied, when appropriate. Soil tests can also be used and recorded in the field history records to demonstrate soil improvements.

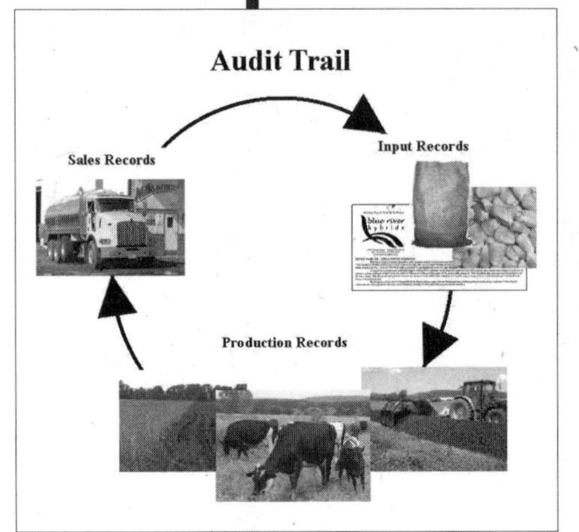

FIGURE 8.2 *Illustration of a complete audit trail*

THE AUDIT TRAIL

An audit trail is a chronological sequence of records that tracks all products used on the farm from their origins to their points of sale. It is the evidence that backs up the organic integrity of the information outlined in the organic system plan. The NOP requires specific records for the audit trail, outlined in Table 8.1. Some certifiers may require additional audit trail records.

Input records

Input records document that everything brought onto the farm for use in the organic operation is certified organic or approved for use. They also document that everything used on the farm was included in the organic system plan.

Off-farm inputs

Purchasing off-farm inputs requires many records to verify their organic status. Always check with your certifier before using a new product. There are different record requirements for single- and multiple-ingredient products.

- *Single-ingredient products*: Single-ingredient products, such as feed or seed from an off-farm source, require records to prove that it is organic. For these products, file a current copy of the source's organic certification certificate.
- *Multiple-ingredient products*: Multi-ingredient products (i.e., mixed feed, healthcare remedies, cleaning products, etc.) require that all ingredients are approved by the NOP. Many certifiers also require that farms keep a product label on file for all inputs. Before using a product on the farm, the certifier may have to review the ingredients to verify that they meet the NOP standards. Some certifiers publish their own lists of approved products or follow OMRI's review. Always check with the certifier before using a new product on the farm. These product labels are your record of approved inputs.
- *Transported and/or custom-harvested products*: When a farm purchases a transported and/or custom-harvested product, it must document that the organic product was not contaminated by commingling with a conventional product. To verify its clean transport, many certifiers require a "clean truck affidavit" from the driver and/or a "clean-out log" for the vehicle.

Harvest records
Organic farmers maintain harvest records for all certified and non-certified crops. All of these records must identify field numbers, letters, or names that match the farm maps in the organic system plan. The harvest records include the following documentation.

- field identification
- planting dates
- production records (harvest date and yield)
- post-harvest handling records
- storage records (see Figure 8.3)
- buffer zones (If crops are harvested from buffer zones they must be treated as a conventional crop.)

Split operations and custom-harvest: Split operations are farms that have both organic and nonorganic products for sale. The producer is responsible for submitting management plans that prevent

BOX 8.1 FROM THE INSPECTOR

Organic inspectors are important to the organic certification process and help verify that certified organic operations uphold the organic integrity of the NOP standards. Good complete on-farm records are crucial to a successful inspection. Below are two examples of how good and poor recordkeeping practices affected a farm's certification.

Farmer's Audit Trail Verifies Farm Input
Due to complaints from neighbors that a conventional feed truck was making regular stops at an organic farm, an inspector was sent out to investigate the allegations. When questioned, the farmer explained that they had been purchasing lime and shavings from the company. The farmer was able to provide purchase receipts to verify his statement. When the inspector followed up with the feed company, the mill provided sales receipts from the previous year for all of the farmer's purchases—which only included lime and shavings. This story is just one example of how a well-maintained audit trail proves the authenticity of an organic product.

Due to Falsified Records, Farm Loses Certification Status
After an inspection, the certification agency's organic review committee reviewed the inspection report. While calculating the herd's feed requirements in relation to milk production and herd size, a reviewer noticed a discrepancy. The amount of purchased and harvested feeds on the farm could not sustain the volume of milk the operation produced. The reviewer advised the certification agency to investigate further. An inspector made an unannounced spot inspection and found that the farm had been purchasing conventional commodities and had recorded them as approved minerals. This became evident by the unreasonably high quantity of minerals purchased in a short time frame. The inspector took milk samples from the entire herd. He also took soil and plant tissue samples. Each of these samples produced traces of prohibited materials. The certifier gave the producer a written notice of noncompliance that included a suspension and mandatory steps to rectify the noncompliance. The farmer appealed the certifier's decision, but ultimately lost the organic certification.

product commingling. In additional to the harvest records outlined for organic crops, split operations or those that use custom-harvest must also include equipment clean-out logs, clean truck affidavits, custom-work receipts, and sales records of certified and noncertified crops. Some certifiers require an equipment list.

Organic Farm Documentation Series — Provided courtesy of NCAT's ATTRA Project, 1-800-346-9140

Organic Feed Storage Record

Use this form to document your harvest of farm-grown feed crops, storage of purchased feeds, and proper handling of organic feeds, feed supplements, and feed additives in split storage to prevent commingling or contamination. [§205.237(a)]

Farm Name or Unit: — Production Year:

Storage Unit I.D.: — Organic only?

Date crop/feed stored	Type of crop/feed	Field # if farm-grown	Lot # if purchased	Quantity In	Quantity Out	Date of cleanout / by whom*

FIGURE 8.3
Example of feed storage record from ATTRA

Source: George Kuepper, Lisa Cone, and Joyce Ford. 2003. "Organic Livestock Documentation Forms," ATTRA, National Center for Appropriate Technology (NCAT), http://attra.ncat.org/attra-pub/PDF/livestockforms.pdf.

Livestock records

Basic livestock records track each individual animal's birth, health, production, sale, or death. Certification inspectors use the animal records to verify herd size, feed needs, and production levels. The following records are required components of the livestock records.

Animal list: Basic livestock records begin with an animal list. All certified organic animals must have a form of identification (i.e., nametags, ear tags, identification numbers, etc.). Many certifiers require animal lists and provide guidelines for these records. Figure 8.4 illustrates how an organic farmer records his individual animal list.

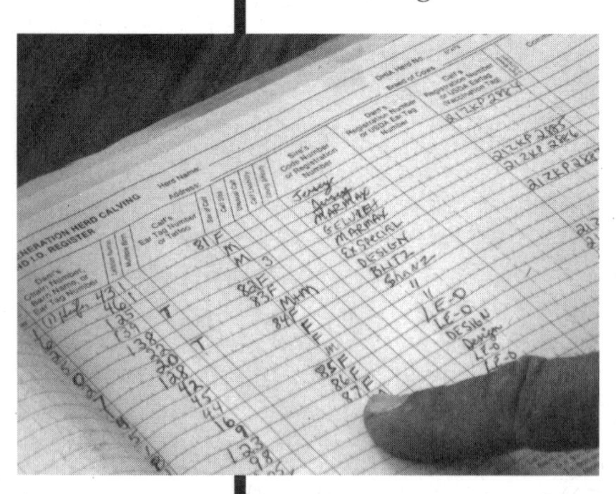

FIGURE 8.4
Animal identification records

Origin of livestock: These records must track all animals from the time they arrive on the farm (birth or purchase) until the time they leave the farm (death or sale from the herd), and log the purchase of animals, breeding, freshening, and sale information (see Figure 8.5). All purchased certified organic animals must include written proof of their organic certification. Required records include a purchase receipt or other transaction document that shows both the seller's and buyer's name and contact information, the date of the transaction, the animal identification number, breed and birth date (if available) of each animal, as well as a valid proof of certification from the seller. The records should also identify whether the organic dairy animal qualifies for organic slaughter stock.

Feed records: All certified operations must document that they grew or purchased enough feed to provide their organic livestock with 100% organic feed. Certifiers use a combination of records to determine if the feed grown and purchased is sufficient to meet feed requirements for all stages and age groups of livestock.

Required feed records include:

- feed ration records for each type of animal during each stage of growth and production (see Figure 8.6),
- pasture intake records: check with your certifier and milk handler for specific pasture requirements,
- purchased feeds (including forages and grains),
- purchased feed supplements (vitamins and minerals),
- feed additives (preservatives, coloring agents, and appetite stimulants), and
- feed-storage records detailing the exact storage areas with correlation to feed type, quantity, sources, and dates of purchase.

FIGURE 8.5
Breeding wheel used for reproduction records

To back up this documentation, files must include copies of receipts and labels from all purchased feeds, supplements, and additives.

Heath and medication records: The NOP requires a detailed medication inventory (including everything in the medicine cabinet or refrigerator) with product names, ingredients, manufacturer names, and regulatory status. Health management records include administered medications and vaccinations, the reason for use, date of use, and animal identification (see Figure 8.7). As for all farm inputs, the product labels and purchase receipts for all medications used must be included in the recordkeeping system.

Pest management for animals: The management practices (i.e., ventilation, etc.) and materials used (i.e., biological controls, essential oils, etc.) for managing pests on your animals must be recorded with the appropriate receipts and product labels.

Milk quality: These records track milk quality including somatic cell counts (SCC) and standard plate counts (SPC). Milk handlers use these milk quality counts to monitor milking and sanitation practices and to identify potential dairy health and management problems.

Sales records

Certified organic operations must maintain records of all products sold off the farm, including but not limited to milk, animals, crops, compost, seeds, and all other organic and nonorganic products. If the farm is a split operation, the certifier may track nonorganic product sales to make sure that none are sold as organic. Record the product, date of sale, field harvested in, animal ID, and purchaser of product. If the farm sells milk wholesale to a processor, keep milk slips to verify milk sales. On-farm processing operations must provide milk production records and production records for all organic products produced from that milk. Recording product sales completes the audit trail required for maintaining organic certification. » *continued on page 254*

Organic Farm Documentation Series — Provided courtesy of NCAT's ATTRA Project, 1-800-346-9140

Organic Feed Ration Record

All feed must be organic or on the National List (see §205.603(c) of the National Organic Regulations) and meet the nutritional requirements of the animal, including vitamins, minerals, protein and/or amino acids, fatty acids, energy sources, and fiber. Feed requirements change as animals age, and the nutritional needs of different types of stock (e.g. breeders, slaughter animals) vary. Seasonal changes in ration formulation may also occur. [§205.237, §205.238(a)(2), §205.603]

Farm Name or Unit:						Production Year:	
Ration #: **Purpose*:**		**Ration #:** **Purpose*:**		**Ration #:** **Purpose*:**		**Ration #:** **Purpose*:**	
Feed Ingredient:	% Total Ration:	Feed Ingredient:	% Total Ration:	Feed Ingredient:	% Total Ration:	Feed Ingredient:	% Total Ration:

FIGURE 8.6
Example of feed ration record form from ATTRA

Source: George Kuepper, Lisa Cone, and Joyce Ford. 2003. "Organic Livestock Documentation Forms," ATTRA, National Center for Appropriate Technology (NCAT), http://attra.ncat.org/attra-pub/PDF/livestockforms.pdf.

Individual Organic Animal Health Record: Large Livestock

Preventive health care practices must be used, including appropriate species selection, complete feed ration, good housing, outdoor access, exercise, needed physical alterations, and vaccines. Synthetic medications, including parasiticides, must be on the National List. You must not withhold medical treatment to preserve organic status; all livestock treated with prohibited substances must be clearly identified. [§205.103, §205.238, §205.603]

Farm Name or Unit: ______________ Production Year: ______

Tag#, ID#, or Name of Animal ______ *and/or* Description of Animal ______

Birth Date ______ Dam I.D. ______ Sire I.D. ______ Sale Date ______ Buyer Name ______

If the animal died, what was the cause of death? ______ Date of Death ______

Vaccinations / Date Administered	Vaccinations / Date Administered

Physical Alteration / Date	Physical Alteration / Date

Health care problem	Practice(s) used to treat health care problem	Product(s) used to treat health care problem	Date used / Effectiveness

FIGURE 8.7
Example of individual animal health record form from ATTRA

Source: George Kuepper, Lisa Cone, and Joyce Ford. 2003.
"Organic Livestock Documentation Forms," ATTRA, National Center for Appropriate Technology (NCAT), http://attra.ncat.org/attra-pub/PDF/ livestockforms.pdf.

TABLE 8.1 REQUIRED ORGANIC DAIRY CERTIFICATION RECORDS

RECORD	REQUIRED	REQUIRED IF . . .
INPUT RECORDS		
Expense receipts	yes	
Expense ledger, checkbooks, checks	yes	
Feed and grain receipts	yes, if . . .	Purchasing feed/forage or other approved inputs
Certified organic certificate for purchased inputs	yes, if . . .	Purchasing organic seed, grains, or forages
Transaction receipt or certificate	yes	
Weigh slips	yes, if . . .	Purchasing bulk inputs
Clean truck affidavits	yes, if . . .	Purchasing feed/forage or other approved inputs in bulk
PRODUCTION RECORDS		
Field maps	yes	
Field identification for each given crop	yes	
Field history forms	yes	
Purchase receipts for seeds	yes	
Documentation of organic seed search (check with certifier for specific records)	yes, if . . .	Purchasing nonorganic seed that is untreated and GMO-free. (Must document contacting at least 3 viable sources of organic seeds for one specific variety.)
Custom-work records and receipts	yes, if . . .	Hiring custom field work done on your farm or performing custom field work for another farm
Machinery clean-out logs	yes, if . . .	Hiring custom field work done/ maintaining a split operation
Applied amendments/spray records	yes	
Field activity log	yes	
Harvest records	yes	
Records of post-harvest handling or processing	yes	
Compost production records	yes, if . . .	Producing certified organic compost
Storage records	yes	
Monitoring records (soil, water, and crop tests)	yes, if . . .	Requested by certifier
Animal identification records	yes	
Breeding and birthing records	yes	
Health and medication records	yes	
Loss/cull records	yes	
Milk quality records	yes, if . . .	Requested by certifier
Pasture records (see new pasture rules)	yes	
SALES RECORDS		
Records of what is sold as organic and as nonorganic	yes, if . . .	The farm has a split operation
Sales invoices	yes	
Income ledger	yes	
Milk sales records	yes	
Transaction receipt or certificate	yes	
Lot numbers	yes	
Weigh slips	yes, if . . .	Selling bulk commodities
Clean truck affidavit	yes, if . . .	Selling bulk commodities

continued from page 252 »

Processing and facility records

Processing, as defined by the NOP includes cooking, baking, curing, heating, drying, mixing, grinding, churning, separating, extracting, slaughtering, cutting, fermenting, distilling, eviscerating, preserving, dehydrating, freezing, and chilling. It also includes packaging, canning, jarring, or otherwise enclosing food in a container. If an operation conducts any of the preceding processes, it will be required to certify as a processing facility. Processing facilities must keep a complete audit trail of any product sold or labeled as organic. The NOP rule §205.103 outlines the required processing facility records.

EVALUATE YOUR ORGANIC RECORDKEEPING SYSTEM

As this chapter demonstrates, organic recordkeeping is extensive, but it is also one of the most important aspects of organic management. Accurate records support and uphold organic integrity—the basis of the organic market and future of organic farming. Table 8.1 reviews all the required records detailed in this chapter. Use the table to determine if your farm maintains all the required records. Some certifiers require additional records, so keep the line of communication open to make sure everyone is on the same page about the recordkeeping expectations.

RESOURCES

The NOP and National Organic standards are available online: (http://www.ams.usda.gov/nop/) and through the USDA NOP office at: 202–720–3252.

These websites also include helpful recordkeeping information.

ATTRA Recordkeeping Resources

- ***Forms, documents, and sample letters for organic producers*: (http://www.attra.ncat.org/attra-pub/PDF/producerforms.pdf)**
- ***Organic field crops documentation forms*: (http://www.attra.ncat.org/attra-pub/PDF/cropforms.pdf)**
- ***Organic livestock documentation forms*: (http://www.attra.ncat.org/attra-pub/PDF/livestockforms.pdf).**

Rodale Institute. *The New Farm, Organic System Plan Tool.* (http://www.tritrainingcenter.org/code/osp_index.php).

NOTES

1 The National Organic Program, Rule §205.103 (http://www.ams.usda.gov/nop/).

2 Kathie Arnold. 2007. "Making the Leap to Organic Dairy Production." NODPA website (http://www.nodpa.com/MakingTheLeapKathieArnold.html). Site accessed 2/28/2008.

LES MILLER, MILLER BROTHERS' FARM

BETHANY WALLIS

Just around the bend from scenic Young's Lake is the Miller Brothers' Farm. Les and Jim Miller have been farming since 1972 with the help and guidance of their father, Warner. The Millers started backing away from chemicals in the late 1980s. They became interested in organic production when bST was introduced, a practice they staunchly opposed. They certified organic in 1997.

Today, with the addition of Jim's son Ken, the dairy grows crops on about 250 acres of land with supplementary acres in pasture. More than 100 acres of grain, along with a variety of forages and pasture, keep 50 cows milking year-round on a pipeline system. The management focus is to concentrate on soil improvement and growing all inputs. The Millers feel strongly about the organic industry putting its best foot forward, securing the market with product quality and integrity.

Warner, Les, Ken and Jim Miller.

Since the Millers used almost no chemicals for years, they were close to organic management when the standards were first developed. They were able to grow along with the program, allowing for an easier herd health transition than some farmers now experience. Overall somatic cell counts (SCC) have decreased, in part by an improved milking system and corrected husbandry practices. Current practices include dry-washing, postdipping, improved laneways to pasture, and clean, ample bedding. If needed, the entire herd is sampled for SCC, allowing for individual culture samples, identifying and culling chronic cases, or drying off problem quarters. A nutritionist balances their total mix ration (TMR) for 60 pounds of production. The TMR includes haylage, silage, sorghum silage, whole grains (15–17 lb. per cow), and minerals, with free-choice dry hay. They supplement pastures with a little dry hay, corn silage, and some grain (11–14 lb. per cow). Kelp is fed year-round. The Millers have added land and storage over the years, as another farm goal is grain independence.

The milking herd enjoys an intensive pasture management system, with about ¾ acre per head. The milking herd has a new paddock daily on a 30-day rotation. Temporary posts support a single-strand aluminum electric fence wire. The water supply is

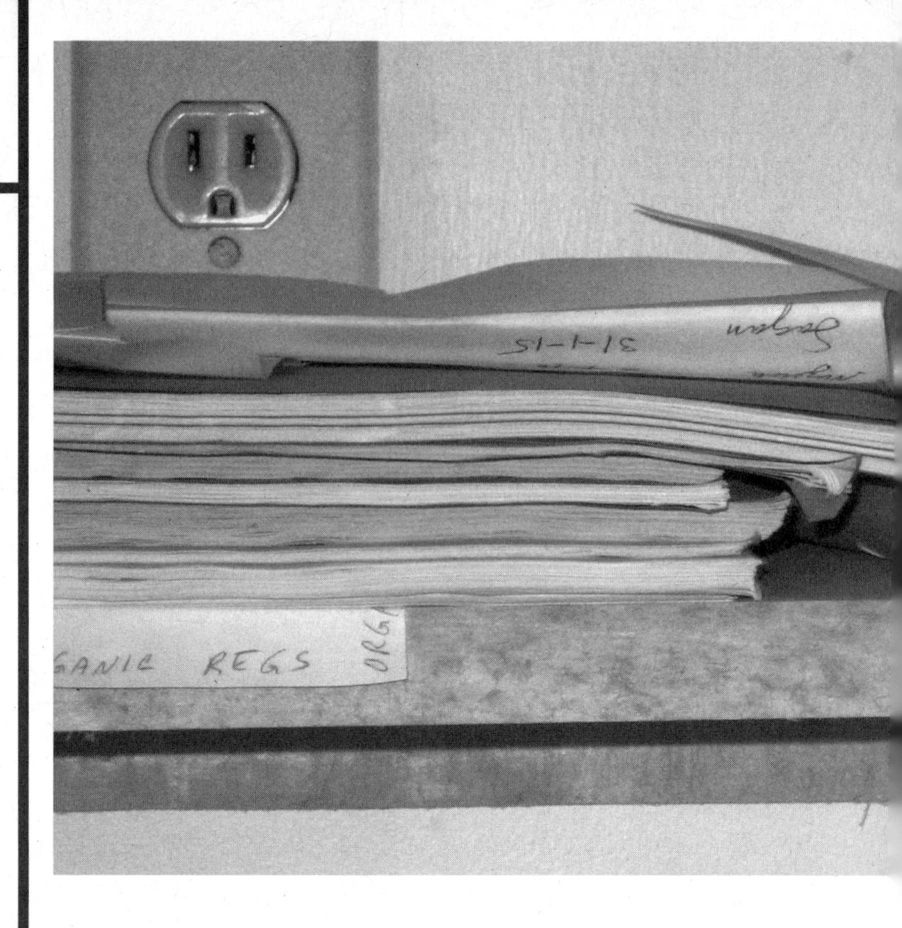

Labeled filing shelves.

moveable stock tanks supplied by a water wagon. Heifers have a very large pasture, clipped in sections, and dry cows have 1.5 to 2 acres of rotated pasture.

The results of continually working to improve soil health with crop rotation and manure management are evaluated by observing soil tilth, earthworms, weeds, and soil samples. They compost the manure on the farm, contributing to soil biological activity and soil health.

Recordkeeping is simple, consisting of files of required documents for organic certification. The Millers use labeled shelves, color-coded files, and file boxes. They organize the records to match their certifier's required records, having separate folders for each record category. For example, their certifier provides blue forms for herd health records; the Millers file them in a blue folder. Everything else, including receipts, is kept with the financial and tax records. The recordkeeping system is compatible with the requirements of the certifier, so they have not reinvented the wheel. Les likes paper records because copies are easy to make and share with the people he is working with. He records information right onto the certification forms, rather than keeping extensive notes that he would later need to transcribe onto the forms. A dry-erase board, herd record book, and breeding wheel help track breeding, freshening, and health problems.

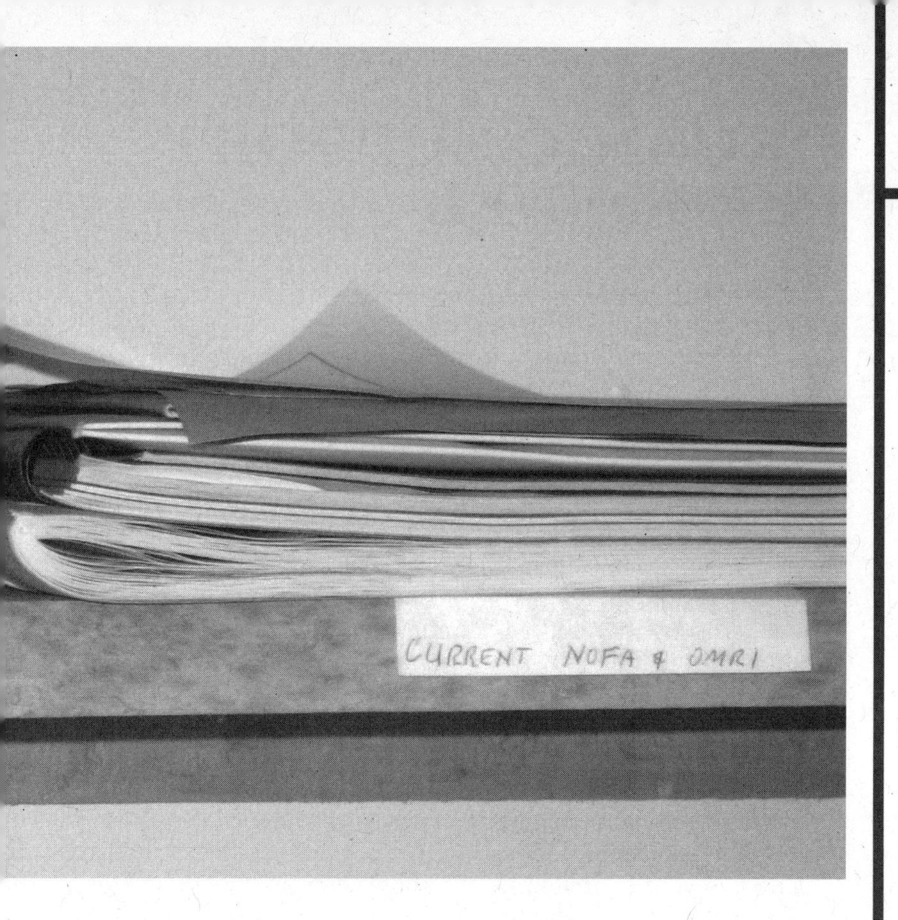

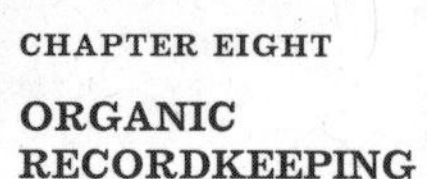

The Millers hand draw farm maps that indicate their field identification numbers. They keep copies of the maps in files such as harvest records for easy reference when they are completing certification forms that require them to indicate activity in a specific field. Farm maps are copied onto green paper, so they are easy to spot.

Although the many records required in organic certification can at times seem overwhelming, Les believes that once you get your system down they are actually quite beneficial. "After you've been in it for a couple of years, you get a pattern down. You know what is necessary, what's required, and you try to keep those records where you can get at them. You take your labels and keep them in a box so you can find them quickly, or your seed information in a folder for seed purchases of organic and nontreated seeds, now that there are more options." Records help you see more clearly what works and what does not. Les reflects that farm management used to be a "toolbox full of chemicals" to get you out of trouble. Organic management forces you to solve the problem and not just treat it.

CHAPTER NINE

ORGANIC SUPPORT PERSONNEL

STEVE RICHARDS

Learning something new is not always easy—whether you are transitioning to organic dairy farming or an experienced farmer, there are times that you will need to look for some additional help. One thing that is always advisable is to form a support network. Just like a family, a business support network can be there for you when times are good or times are tough. This chapter analyzes how organic dairy farmers collect knowledge from their support personnel, which personnel they find most helpful, and where other sources of additional information reside. In addition, it will speak to the topic of mentorship, because most farmers tend to learn their practices from other farmers.

KEY ORGANIC TRANSITION POINTS

- Find the right sources of information for you and your farm.
- Assemble a team of competent advisors knowledgeable in organic principles.
- Educate your farm employees about their new responsibilities under the organic standards.
- Locate an organic farmer mentor.
- Constantly evaluate the services and information you receive.

WORK AS A TEAM WITH YOUR FAMILY AND FARM STAFF

The first part of organizing your support personnel is to make sure that all the people involved with your farm are on the same page. Family members and business partners should be your closest advisors. Judging from interviews with farmers who have navigated the transition successfully, there are two key ingredients for success: attitude and communication. A positive outlook on the challenges ahead helps one visualize success. And, just as it was important to talk with your family and staff about your decision to transition to organic dairy, keeping them informed during the process will be just as important.

SEEK HELP AND INFORMATION FROM MANY SOURCES

Organic dairy farming has many facets, so your pool of support personnel needs to be deep. Some of the sources of information that you will use as an organic dairy farmer may differ from those you used when you were operating conventionally. Remember that you are trying something that is a different way of farming, and some advisors may be locked into a more conventional mindset. Regardless, off-farm advisors are still critical to the success of your transition. You should consider the service providers you currently work with and judge their willingness to work with your new production methods. Some regions have good service providers and sources of organic dairy information and others do not. Ask some pointed questions of your advisors before you assume that they will or will not be helpful. In addition, always poll a number of different resources for the same question in order to see how others approach a common challenge.

SURVEY RESULTS: HOW FARMERS RANK SOURCES OF INFORMATION

If you are just beginning your transition research, it is helpful to know where organic dairy farmers obtain their information and how useful this information might be. NOFA-NY conducted a survey that asked these very questions of both experienced dairy farmers and farmers still in their transition. You will see from the survey results that there are a number of different sources of information that organic dairy farmers have used for their transition research. The survey asked the same question in two different manners to enrich the results. First, what source of information had the farmer tried? Second, what was the value of that information? The top four in each category was as follows.

Transition information used by experienced organic dairy farmers

1. Other farmers and farm organizations
2. Certification agency
3. Conferences
4. Magazines, books, and websites

Transition information used by transitioning organic dairy farmers

1. Magazines, books, and websites
2. Milk buyers
3. Certification agency
4. Other farmers

Ranked value of information by experienced organic dairy farmers

1. Private consultants (nutritionist, financial advisor, crop advisor)
2. Certification agency
3. Other farmers and farm organizations
4. Cooperative Extension

Ranked value of information by transitioning organic dairy farmers

1. Milk buyer
2. Magazines, books, and websites
3. Certification agency
4. Conferences

There were some significant differences in the responses between the experienced group and the transitioning group. Some may have to do with the timing of the transition: today there are more websites, books, and magazine articles about organic dairy farming than there were just a few years ago. In addition, when many of the experienced organic farmers transitioned, there were not many organic milk buyers—especially milk buyers offering technical assistance. It also makes sense that the results show the experienced group trying many more sources of information than the transitioning group. Transitioning farmers may still be building their support network and may not have found the consultants, service providers, conferences, or farmer mentors that are particularly valuable to them. Regional differences in service providers might have something to do with this, as well as explaining the perceived value of Cooperative Extension information between the two groups.

Farmers and farmer organizations were consistently used and ranked highly among both groups. Networking with other farmers and participating in peer groups is the most common source of information for organic dairy farmers. Conferences, field days, seminars, and farm tours are routinely organized by organic educational organizations like NOFA, producer groups, certifiers, milk marketing organizations, product and service vendors, and public agencies. Attendance at formal educational functions provides a perfect opportunity to engage in informal networking and to meet potential mentors.

Certification agencies are another top source of information. Most certification agencies work closely with an educational organization that provides programs and personnel to help you through the transition process. For example, certifier Vermont Organic Farmers (VOF) is associated with NOFA Vermont which has a Dairy and Livestock Technical Assistance Program that assists Vermont dairy farmers to:

- determine if certified organic management is suited for their farm resources and goals,
- design economically viable and ecologically sound whole farm plans to transition to and sustain certified organic management,
- learn skills, tools, and management strategies that will sustain and improve certified organic management goals, and
- provide management information, training, and consulting to certified organic farmers.[1]

Pennsylvania Certified Organic (PCO) coordinates a mentoring program of producer volunteers who have agreed to answer transitioning farmers' questions about organic production. While your certifier cannot give you direct advice on how you manage your farm, he or she can answer concrete questions about the certification process and tell you whether certain products or processes are allowed under organic management. Ask whether your certifier works with an educational organization that offers more direct technical assistance.

Do not underestimate the value of service providers, especially the ones who have a financial stake in assuring that your farm succeeds. Milk buyers want to help make sure that your milk makes it onto their truck and meets their quality standards. Private consultants are paid for their advice, so they strive to provide advice that is timely and useful for their clients. Books, websites, magazines, and farmer organizations also want to provide quality information. Each chapter in this handbook includes a list of information resources that explore the various aspects of organic dairy farming in more depth. This is a great start to begin creating your organic dairy library.

TIPS ON FORMING A GOOD MENTOR-MENTEE RELATIONSHIP

It is a good idea for transitioning farmers to establish an informal mentorship relationship with at least one seasoned organic farmer. Formal mentorship opportunities exist in some regions; seriously consider participating if one is available to you. Mentoring and advisory relationships are about communication and the exchange of useful information. Organic agriculture features a strong culture of mentorship. As a rule, organic farmers genuinely enjoy discussing their vocation and are eager to educate and coach one another in their methods. This presents a unique opportunity for guidance and ease in the transition to certified organic production.

Consistent with results from the producer survey described earlier in the chapter, the ideal mentor farmer is one who has become very knowledgeable through study and experience and who can derive satisfaction from helping you. Milk buyers, organic educational organizations, or the local Cooperative Extension may help you find other experienced organic farmers who are willing to serve as mentors to transitioning farmers.

A good mentor will possess many of the following characteristics.

- understands that what has worked on his/her farm may not be directly applicable to what will work on your farm
- enthusiastic, learned, skilled, and experienced
- realistic and willing to alert you to risks and pitfalls
- understands and explains "why" as well as "how"
- genuinely interested in helping you succeed
- communicates well and is a great listener
- located nearby, close enough to make regular farm visits
- organized, takes mentoring seriously, and incorporates a good sense of humor

Having identified a willing mentor farmer, it is up to you, the transitioning farmer, to assume responsibility for your own mentoring experience. A few points of mentoring etiquette are worthy of consideration.

Organic dairy farmer Sally Brown says, "As an organic farmer it is really important to find a mentor. It's critical. One of the things I appreciate about the organic community is that we all help each other. If someone finds something that works well, we let everyone else know. There are enough challenges in farming as a whole, and then you are coming into organics, which is whole new mindset. You need to attach yourself to someone who knows what is going on."

Mentoring etiquette

- Agree on frequency of contact and how much time they can give.
- Schedule appointments and keep them without fail.
- Prepare ahead of your visits and conversations to make the most effective use of the mentor's time.
- Maintain realistic expectations. The mentor will not have all of the answers.
- Be open-minded. Accept guidance and occasional criticism cheerfully.
- The mentor should not simply perform tasks for you, but should challenge you to perform them yourself, no matter how inconvenient.
- Follow through on what you agree to do – give the mentor's advice a fair try by implementing it precisely as described or demonstrated.
- Understand that the mentor's advice may not work out for you quite as it seems to work for him or her.
- Be diligent and accurate in evaluating the results of changes you make on the advice or your mentor.
- Include the mentor in your successes as well as your setbacks.
- Resist becoming too dependent upon the mentor. Mentors often report that much of the satisfaction and reward for them is in watching the trainee build self-confidence.
- Recognize when the mentoring relationship arrives at its natural conclusion.
- Thank your mentor for his or her time.

EVALUATING THE COSTS AND BENEFITS OF PAID CONSULTANTS

Selecting the right advisors in the first place will save you much time and energy.

For many farmers, having a competent team of farm advisors who help them keep the farm business running smoothly is an important part of their farm business. However, there are some questions farmers must ask themselves to evaluate the effectiveness of paid advisors.

Tips for selecting the right advisors

- Always investigate beyond sales brochures and testimonials.
- Ask for references from other organic dairy farmers or your farming mentor.
- Make sure your advisor has a grasp of whole-farm concepts and how his or her role contributes to the success of your farm operation.
- Make sure that your advisor can work in tandem with your other advisors.
- Make sure that your advisor is qualified to help you. This may be difficult in organic dairy. In addition to enthusiasm and familiarity with organic production, in every case the advisor should be professionally trained within the field of expertise.

- A veterinarian, homeopathic veterinarian, or alternative medicine veterinarian should be licensed to practice in your state and should have strong skills in herd health management, especially preventative practices.
- A nutritionist should have knowledge of working with intensive grazing and pasture-based rations, organic feedstuffs, and have extensive training in ruminant nutrition.
- A seed salesperson or agronomist needs to know what agronomic production practices will work best on your farm and have knowledge of soil health and organic cropping strategies such as cover cropping and crop rotation. They should also be knowledgeable in the NOP approved and restricted substances.
- A financial advisor should specialize in whole-farm business analysis and planning.
- All advisors should be familiar with the National Organic Program, the National Organic Standards, and the fundamental principles of organic agriculture.

THE BUCK STOPS WITH YOU

Evaluation and progress monitoring are critical for all aspects of organic dairy farming, and organic support personnel are no different. Make sure you periodically evaluate your farm employees and paid advisors' performances. Your farm staff will appreciate hearing positive feedback on their contribution to your farm. Offering constructive criticism will give them concrete ideas on how to improve. The organic transition may require your farm staff to take on more responsibilities and learn new methods of farming. Taking the time to perform personnel evaluations will let them know you acknowledge that you are all learning together and appreciate their effort in making the transition a success.

Evaluating paid farm advisors is also an important management task to make sure that you are getting your money's worth. Your expectations of competent advisors should be that they know their industry and their products, know your business' needs, and know when to apply their products or knowledge to your benefit. The current status of your farm operation reflects, to a degree, how you have managed your advisors' efforts. You are the manager—you make the decisions and are accountable for the results. Clear communication among you, your farm employees, and paid advisors is key to managing a successful transition and profitable organic farm.

To recap what this chapter has discussed, please evaluate your progress in developing the following support systems for your organic dairy transition and/or success as a new organic dairy farm.

- I have explored many different sources of information regarding organic dairy farming.
- I have interviewed my current farm advisors as to their inter-

est and expertise in helping me with my transition to organic dairy farming and continuing build a successful organic dairy.
- To the extent that my current advisors are insufficient, I have picked new advisors that are more in line with organic dairy production methods.
- I have created networks with other organic farmers and expect that at least one will agree to mentor me through my transition and in the early years of organic certification.
- I plan to evaluate the information that I receive (and the advisors who give the information) periodically to ensure that the right decisions are implemented on my farm.

CONCLUSION

Be sure to do your research, pick the right advisors, and critically evaluate their advice. In addition, you will find yourself rebuilding your team every now and then, as team members will change due to retirements, career changes, and differences in opinion. Make it a priority to spend an adequate amount of time whenever you choose an advisor, as this will make a big difference in your transition to organic dairy farming. Remember, although your success ultimately rests upon the decisions that you make, the surveys of your colleagues show that your advisors will have a large influence on your decisions.

NOTES

1 Northeast Organic Farming Association of Vermont. (http://www.nofavt.org). Site accessed 2/18/2009.

Sarah, Jamie, and James Huftalen.

JAMIE AND SARAH HUFTALEN

STEVE RICHARDS AND BETHANY WALLIS

Jamie and Sarah both attended SUNY Morrisville where Jamie studied dairy science and Sarah studied horticulture. After college, Jamie worked on a conventional dairy, but did not like the farm's policy to push cows to higher and higher milk production, resulting in many cows having to leave the farm for health reasons. After taking a few years off from farming, Sarah and Jamie decided to pursue dairy farming again, but this time with organic management and marketing.

Jamie began his organic transition research by speaking with neighboring farmers about their experiences. A local organic dairy farmer advised the Huftalens to begin their transition with heifers, rather than purchasing organic cows. On this advice, Sarah and Jamie certified their family farm first, and then purchased heifers that would freshen into a milking herd when the transition was complete. This allowed them to transition without the burden of the high organic feed costs required for maintaining a milking herd. Jamie recommends transitioning the way they did if possible. He believes their transition was easier than they had initially anticipated and the advice from the neighboring farmer saved them thousands of dollars.

Huftalens' dairy herd grazing.

The Huftalens certified their farm on June 1, 2006, and started shipping organic milk from their 17 cows 9 days later. Jamie heeded additional advice from other farmers who recommended that he wait to set up his permanent pasture infrastructure until he figured out the best fit for his grazing system. This advice was beneficial, as infrastructure is expensive to implement and even more expensive to change. His current plans are to provide two acres of pasture per cow. He intends to install a permanent water system when he and Sarah figure out their permanent pasture design.

Jamie and Sarah's biggest concern with transitioning was feeling comfortable with alternative health treatments. They had difficulty finding information and instructions on how to implement alternative healthcare for their dairy herd. Currently, they use homeopathics and seek advice from other organic dairy farmers. Fortunately, every problem has been an opportunity to learn. They have already gained a lot of herd health knowledge by choosing not

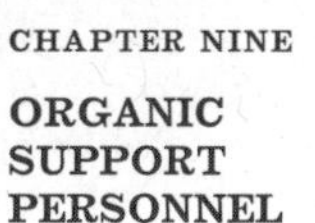

to send sick animals off the farm, but rather to ask another farmer for advice on treatments and learn from treating the animal. Jamie has also implemented a herd health monitoring system of his own. Their milking system includes four floor pails. Jamie has found that, "the weight of each individual cow's pail is a good measure of cow health. Since you are actually carrying the pail it's hard to miss a change in milk production."

Jamie and Sarah realize that their transition could not have been so successful without help from others. They credit their families' support for their success—Jamie's brother even helped them with their initial business plan. Finally, the organic recordkeeping requirements have forced them to be better farm managers because they now have a detailed record trail to track their progress over the years to come.

CHAPTER TEN

FULL FARM CASE STUDY: SIOBHAN GRIFFIN, RAINDANCE FARM

BETHANY WALLIS

In the northern foothills of the Catskill Mountains, dairy cows graze over 200 acres of pasture on hillsides at Raindance Farm. Siobhan Griffin began her first farm in Vermont in 1991 with 20 cows, 10 owned and 10 lent. She recalls, "The farmers created a support group. It was just a neighbor who lent us the other 10 cows and we'd help him with hay as a trade off. There was community push to get us started." At the time, she was not managing her farm organically. She was, however "a fanatical grazer."

Siobhan Griffin.

In 1994, members of the UVM grazing program (Bill Murphy, Sarah Flack, and Lisa McCrory) visited her farm and provided guidance on everything from paddock rotations to barn ventilation. "Vermont made it a friendly place to start farming."

Siobhan contemplated transitioning to organic in Vermont. She believed organics would be a good fit for her farming style and focus on grazing. She did not grow corn and was not interested in using chemicals. The only thing that she would have had to change was discontinuing the use of conventional mastitis treatments on the cows, which she felt was no longer effective anyway. While researching organic management, she visited Butterworks Farm to see how Jack and Annie Lazor managed their herd; Siobhan still considers them mentors today.

Because she rented her Vermont farm, Siobhan was unable to transition the milking herd to organic. When her herd size outgrew the dairy facility there, she began looking for a new farm to rent or purchase. An ad in *Country Folks* led her to her current farm in Westville, New York, and she moved in May 1997.

Relocating to New York created the opportunity to purchase the farm and have control of how the land was managed. At the time, Elmhurst Creamery was developing an organic market. The creamery lent her a portion of the funds to transition to organic. This helped cover the cost of feeding 100% organic feed for the three-month transition. Siobhan recalls that the transition period was very difficult financially, and even though she had received a loan, she still had to sell her pick-up truck to meet expenses. The farm was certified in late fall 1997.

Raindance Farm was "organic by neglect." Although she had moved in May, there was no grass for pasture and the old hay fields were mostly thatch from not having been mowed. Siobhan spent the first season running ahead of the cows putting up low-tensile wire and taking down barbed wire. Her laneways were a mess and it was a challenge to keep the cows out of the mud. Siobhan says, "It is critical to have good laneways for a grazing program." At that time, the NRCS EQIP program provided a cost-share program for pasture improvement. Siobhan participated in the program to put in lanes and create a way for the cows to get out of the barn and over three waterways.

Now Raindance Farm milks 87 cows, employs three people full-time, and harvests its own forages. Siobhan explains that her herd size has increased over the years due to improved herd health. Under organic management her herd average is 12,000 pounds. Siobhan says, "The cows are not pushed. I have very few feet problems and the herd just multiplies. I think it's the biggest advantage of a grazing farm, because you will have excess animals to sell. In fact, that is probably where the profit center is." Siobhan feels that as much as she likes organic management, it is the grazing that really makes her herd size increase.

PASTURE MANAGEMENT

The grazing system at Raindance Farm includes more than 200 acres, and they add more every year. The grazing season begins with annual rye pasture the last week in April, then perennial pastures beginning in May. Siobhan likes the annual rye grass because "I can pasture it early and it will come back really well, and I get a lot of green manure to plow down out of it." Beginning in October, she supplements the cows with round bales on pasture.

The grazing acreage is always increasing. Siobhan says, "In drought years we are forced to increase the pasture acreage, which is a good thing because it is making the farm more sustainable." Last year she added 44 pasture acres, and this coming year she will create an additional 24 acres with another rented farm. Siobhan's goal is to have everything grazeable. "It's just a matter of figuring out how to do it," she says.

The grazing system is set up as a management intensive system designed to be flexible. Siobhan uses high tensile wire for permanent boundaries and polywire for temporary divisions. The water system provides water to every paddock and consists of piped well water from the barn and the house. As the pasture system grows, the water system expands with it. She also has a homemade portable solar-powered water pump for watering cows in pastures by the river.

Milking cows receive a new piece of pasture every 12 hours or more, depending on the amount of grass and clover available in the pasture, and the heifers follow them the next day. The paddocks are roughly 2 acres in size and support 80 to 90 cows. Less desirable pasture, such as wet fields or wooded areas with poorer grasses, are designated heifer pastures when they are adequately dried up. Siobhan reseeds 20 acres every year. Her rotation includes plowing up perennial sod pasture after the first grazing and planting sorghum, grazing the sorghum twice, and then spinning on annual rye for the winter. She plows the rye down before seeding perennial rye grass (BG34) and clover (Alice White) in the spring.

SOIL MANAGEMENT

To foster good soil health, Siobhan spreads composted manure in the fall and spreads lime according to soil test results. The farm uses limestone dust for bedding, which helps maintain a soil pH of around 6.5. She maintains records for fertility on all of the farm fields and uses them to monitor soil improvements. On fields that have a low pH and do not receive manure, she applies 2 tons of lime to the acre. Siobhan also monitors the fields' organic matter ranges as a benchmark for soil health. She sees a marked improvement in the soil organic matter under her current management practices and attributes these improvements to grazing and manure applications. She plans to begin monitoring trace minerals because her fields have great natural fertility. "The farm soils are gravel and silty loam on top of gravel, so without rain things won't grow, which is why improving organic matter is so important."

To manage weed pressure, Siobhan rotates the fields harvested for stored feed and sometimes mows them after grazing. She has found that giving the pasture a clean slate for growing cuts back the weed pressure. She has not experienced a problem with refusal of forages from pastures that have been grazed prior to chopping or haying and feels this is because she waits for the manure to be washed in by rain. She notes that grazing cows have very liquid manure to begin with.

In the winter, her pasture season does not truly end; it just changes. She rotates pastures closest to the barn, which helps spread the manure. She feeds the cows round bales in pasture to create an outside bedded pack. Siobhan notes that the round bales keep the cows out of the mud and are a great carbon source for composting. Their winter ration includes grain at a rate of 4 pounds per cow. Siobhan purchases whole grains, shell corn, and whole barley; she grinds her own grain and adds minerals to the feed. The grain ratio is 2 parts corn to 1 part barley, ground fresh daily. She prefers to

Raindance Farm dairy herd grazing sorghum.

feed minerals with her grain so she knows that all the cows are getting some, as opposed to a free-choice feeding system.

DAIRY NUTRITION

The summer ration consists of up to 6 pounds of grain per cow. If the season is dry, she supplements the pasture by increasing grain according to how much stored feeds they receive. However, if they receive sorghum pasture, she cuts the grain back by 2 pounds. Siobhan states, "The pasture comes first. I won't put the cows out to pasture until it's ready. It's not a complicated formula. The grass has to be one hand high. I will feed them stored feed if I have to and make them wait until the pastures are tall enough to go in."

One of the things Siobhan liked the most about going organic was that she knew where her grains were coming from and that they were grown in New York state. Although the organic market is bigger now, she still tries to get local grains when available. She has a long-standing relationship with her feed company, Lake View Organic Grain, and trusts them to source local grains. Siobhan has never had a problem sourcing grain, but she predicts it will become a challenge with the ethanol industry's pressure for conventional corn acres and the organic chicken market pulling grains to the South. With increasing grain prices, she says, "I don't think you can have a successful organic dairy operation without grazing being the center of your nutrition program."

On her best perennial rye grass and white clover pastures, she gets almost 7 tons of dry matter per acre from 9–10 grazings. "Grazing doesn't reduce the amount of feed you pull off a piece of land. If I were a confinement system harvesting all of my forages and bringing them to the cow, the only way that I would do better with tonnage is with corn silage." Siobhan uses her experience and training from grazing in Vermont to determine the dry matter.

HERD HEALTH

Her herd's health has improved through organic management. Prior to transitioning, she lost cows at the tail end of winter (March/April) due to swollen hocks and feet problems. Since transitioning, foot health has improved and she does not lose those animals. One of the best investments she made was a hoof rack. She states, "The root of all evil in herd health is the foot problems. If you don't get the feet under control, the next thing you know you have a retained placenta or they have trouble calving or mastitis. If she is not getting the right nutrition and her immune system is compromised, it becomes the domino effect. If her feet aren't perfect, it snowballs." She also thinks there is a direct correlation between the amount of grain cows eat and hoof problems. Her personal experience has shown that with less grain, her cows have fewer hairy hoof warts.

When Siobhan initially transitioned to organic management, she had a problem with high somatic cell counts. Now Raindance Farm participates in the Cornell Quality Milk Program, testing the entire herd twice a year. She believes that the key to controlling mastitis is in prevention, and that using a monitoring program is a good tool for farmers. Siobhan also emphasizes the importance of clean cows. The limestone dust she uses for bedding is ideal for keeping her cows clean, and the herd's somatic cell count is usually under 100,000. "If you don't have clean cows, it doesn't matter what you do for mastitis, because you're going to have it." She pre- and postdips with iodine, but stresses that clean stalls and clean pasture are most important.

Garlic tincture.

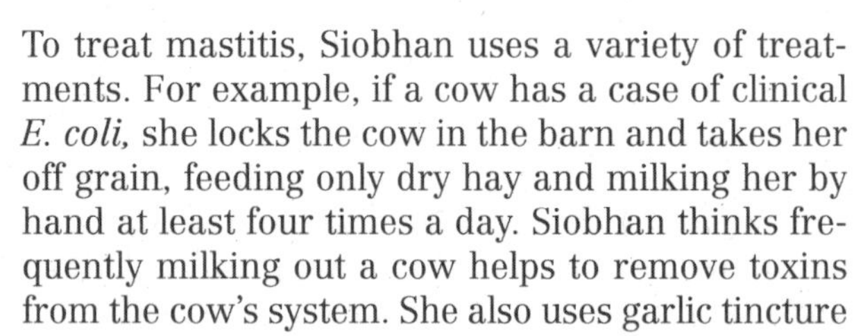

To treat mastitis, Siobhan uses a variety of treatments. For example, if a cow has a case of clinical *E. coli,* she locks the cow in the barn and takes her off grain, feeding only dry hay and milking her by hand at least four times a day. Siobhan thinks frequently milking out a cow helps to remove toxins from the cow's system. She also uses garlic tincture and aspirin orally and vitamin C intramuscularly. For white flakey mastitis, she treats with garlic tincture and aspirin. Siobhan never infuses anything into the udder.

Siobhan vaccinates the herd with Triangle 9 every fall. She also vaccinates for rabies because it is prevalent in her area, and she vaccinates with Scour Guard and Endovac during the dry period. Siobhan has never had a clinical case of Johne's and does not currently test for it. She maintains a closed herd and has only purchased animals once in the last ten years. To clean fresh cows more quickly, she allows newborn calves to suckle on the dams and does not take the calf away until the cow has cleaned. Also, fresh cows receive all the warm water they can drink after calving.

Raindance Farm runs a seasonal herd that begins calving in February and ends in May. Each pasture season, the calves receive a new clean environment to prevent coccidiosis and worms. The calves are housed in group pens until they are weaned, between 2 and 6 months. Siobhan uses mob feeders for calf feeding in groups, which she builds from 15-gallon drums and McCarvil nipples and straws. She regularly replaces the hoses to reduce bacteria. Calves are put into pasture with double wires. The older yearlings being bred follow the milking cows in the pasture rotation.

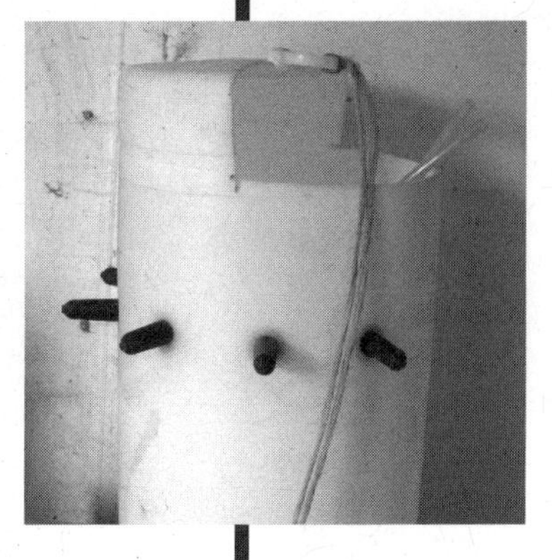

Mob calf feeder.

Siobhan uses both artificial insemination and a bull for breeding. Generally, she AIs the cows for 6 weeks and puts the bull in at 6 to 8 weeks for cleanup. She runs the bull with the heifers prior to the cows and cleans up any unbred heifers with AI.

When she has herd health questions, Siobhan relies on her mentor for homeopathic advise and books by Drs. Sheaffer, Dettloff, and Karreman. She shares successful treatments with other farmers. Siobhan has found, "You have to change your mind's way of thinking. You can't just plug in organic treatments to conventional practices." Siobhan still uses a vet for chronic cases and if a cow gets pneumonia. She is fortunate because her vet works with her and uses organic practices if he can. Her biggest health challenge under organic management is pneumonia, especially in calves. Siobhan no longer tries to treat full-blown-with-fever pneumonia organically. She treats it conventionally and sells the animal to a conventional dairy as soon as it is well. This helps her to select for a genetically stronger herd.

RECORD KEEPING

Siobhan's advice for recordkeeping is to keep it simple. She keeps all of her information on the computer, using spreadsheets she develops and electronic forms from her certifying agency. She maintains her financial records in Quicken and uses Farm Service Agency forms. The financial trends on her farm help her determine if her current management practices are working. The required records for organic certification are handy tools for monitoring her farm. Siobhan has taken a holistic farm-planning course and has participated in a financial planning workshop to improve her farm management. Planning is important for her organic farm management.

Siobhan's advice for transitioning farmers is to, "have a strong grazing program in place first before you even think about transitioning. . . . Farmers have to have a good grazing system in place to make money at organics because if you don't have a good pasture system you won't see the benefits of reduced grain feeding per pound of milk."

APPENDIX A

ORGANIC CERTIFICATION AGENCIES

Baystate Organic Certifiers
683 River Street
Winchendon, MA 01475
Contact: Don Franczyk
PHONE: (978) 297-4171
EMAIL: baystateorganic@earth-link.net
www.baystateorganic.org

Demeter USA/Stellar Certification Services, Inc.
Biodynamic and Organic Certification
PO Box 1390
Philomath, OR 97370
Contact: Jim Fullmer
PHONE: (541) 929-7148
EMAIL: demeter@peak.org
www.demeter-usa.org

Global Organic Alliance, Inc.
PO Box 530
Bellefontaine, OH 43311
Contact: Betty J. Kananen
PHONE: (937) 593-1232
EMAIL: kananen@logan.net
www.goa-online.org

Maryland Department of Agriculture
Food Quality Assurance Division
Organic Certification Program
50 Harry S Truman Pkwy
Annapolis, MD 21401
Contact: Deanna Baldwin
PHONE: (410) 841-5769
EMAIL:
baldwidl@mda.state.md.us
www.mda.state.md.us

MOFGA Certification Services, LLC
294 Crosby Brook Road
PO Box 170
Unity, ME 04988
Contact: Mary Yurlina
PHONE: (207) 568-4142
EMAIL: certification@mofga.org
www.mofga.org

New Hampshire Department of Agriculture, Division of Regulatory Services
25 Capitol Street
PO Box 2042
Concord, NH 03302-2042
Contact: Victoria M. Smith
PHONE: (603) 271-3685
EMAIL: vsmith@agr.state.nh.us
www.nh.gov/agric/divisions/markets

New Jersey Department of Agriculture
369 S. Warren Street
PO Box 330
Trenton, NJ 08625-0330
Contact: Erich Bremer
PHONE: (609) 292-5536
EMAIL:
erich.bremer@ag.state.nj.us
www.state.nj.us/agriculture/divisions/md

NOFA-NY Certified Organic, LLC
840 Upper Front Street
Binghamton, NY 13905
Contact: Marilyn Murray
PHONE: (607) 724-9851
EMAIL: certifiedorganic@no-fany.org
www.nofany.org

Ohio Ecological Food and Farm Administration
41 Croswell Road
Columbus, OH 43214
Contact: Janie Marr Werum
PHONE: (614) 262-2022
EMAIL: Janiemarr@oeffa.org
www.oeffa.org

Pennsylvania Certified Organic
106 School Street, Suite 201
Spring Mills, PA 16875
Contact: Leslie Zuck
PHONE: (814) 422-0251
EMAIL: leslie@paorganic.org
www.paorganic.org

Rhode Island Department of Environmental Management
Division of Agricultural and Resource Marketing
235 Promenade Street
Providence, RI 02908
Contact: Matt Green
PHONE: (401) 222-2781
EMAIL: matt.green@dem.ri.gov
www.dem.ri.gov/programs/bnatres/ agricult/orgcert.htm

Vermont Organic Farmers, LLC
PO Box 697
Richmond, VT 05477
Contact: Nicole Dehne
PHONE: (802) 434-4122
EMAIL: nicdehne@hotmail.com
www.nofavt.org

APPENDIX B
RESOURCE ORGANIZATIONS

ORGANIC EDUCATIONAL ORGANIZATIONS

Maine Organic Farmers & Gardeners Association (MOFGA)
294 Crosby Road
PO Box 170
Unity, ME 04988
PHONE: (207) 568-4142
EMAIL: mofga@mofga.org
www.mofga.org

The New England Small Farm Institute
275 Jackson Street
Belchertown, MA 01007
PHONE: (413) 323-4531
EMAIL: info@smallfarm.org
www.smallfarm.org

New York Certified Organic (NYCO)
1443 Ridge Road
Penn Yan, NY 14527
PHONE: (315) 536-9879
EMAIL: kandmfarm@sprint-mail.com

Northeast Organic Dairy Producers Alliance (NODPA)
O-Dairy Listserve
30 Keets Road
Deerfield, MA 01342
EMAIL: info@organicmilk.org
www.nodpa.com

NOFA Connecticut
PO Box 164
Stevenson, CT 06491
PHONE: (203) 888-5146
EMAIL: ctnofa@ctnofa.org
www.ctnofa.org

NOFA Massachusetts
411 Sheldon Road
Barre, MA 01005
PHONE: (978) 355-0853
EMAIL: info@nofamass.org
www.nofamass.org

NOFA New Hampshire
4 Park Street, Suite 208
Concord, NH 03301
PHONE: (603) 224-5022
EMAIL: info@nofanh.org
www.nofanh.org

NOFA New Jersey
60 S. Main Street
PO Box 886
Pennington, NJ 08534
PHONE: (609) 737-6848
EMAIL: nofainfo@nofanj.org
www.nofanj.org

NOFA New York
PO Box 880
Cobleskill, NY 12043
PHONE: (607) 652-6632
EMAIL: office@nofany.org
www.nofany.org

NOFA Rhode Island
51 Edwards Lane
Charlestown, RI 02813
PHONE: (401) 464-7557
EMAIL: nofari@nofari.org
www.nofari.org

NOFA Vermont
PO Box 697
Richmond, VT 05477
PHONE: (802) 434-4122
EMAIL: info@nofavt.org
www.nofavt.org

Pennsylvania Association for Sustainable Agriculture (PASA)
114 West Main Street
PO Box 419
Millheim, PA 16854
PHONE: (814) 349-9856
EMAIL: info@pasafarming.org
www.pasafarming.org

Rodale Institute
611 Siegfriedale Road
Kutztown, PA 19530-9320
PHONE: (610) 683-1400
www.rodaleinstitute.org
www.newfarm.org

ADDITIONAL ORGANIZATIONS

Acres USA
PO Box 91299
Austin, TX 78709
PHONE: (800) 355-5313
EMAIL: info@acresusa.com
www.acresusa.com

Alternative Farming Systems Information Center (AFSIC)
National Agricultural Library
10301 Baltimore Ave. Rm 132
Beltsville, MD 20705
PHONE: (301) 504-6559
www.nal.usda.gov/afsic

ATTRA - National Sustainable Agriculture Information Service
PO Box 3657
Fayetteville, AR 72702
PHONE: (800) 346-9140
www.attra.ncat.org

Dairy One
730 Warren Road
Ithaca, NY 14850
PHONE: (800) 496-3344
www.dairyone.com

Farmers Legal Action Group, Inc.
360 Robert Street N, Suite 500
Saint Paul, MN 55101-1589
PHONE: (651) 223-5400
www.flaginc.org

National Mastitis Control
421 S. Nine Mound Road
Verona, WI 53593
PHONE: (608) 848-4615
www.nmconline.org

National Organic Program (NOP)
1400 Independence Ave. SW, Rm 4008
Washington, DC 20250-0020
PHONE: (202) 720-3252
EMAIL: nopaqss@usda.gov
www.ams.usda.gov/nop

Natural Resource Conservation Service (NRCS)
PO Box 2890
Washington, DC 20013
PHONE: (202) 720-3921
www.nrcs.usda.gov

Organic Materials Review Institute (OMRI)
PO Box 11558
Eugene, OR 97440
PHONE: (541) 343-7600
www.omri.org

Organic Trade Association (OTA)
PO Box 547
Greenfield, MA 01302-0547
PHONE: (413) 774-7511 Ext. 19
EMAIL: llutz@ota.com
www.howtogoorganic.com

Quality Milk Production Services (QMPS)
Central Laboratory/Park View Tech Center
22 Thornwood Drive
Ithaca, NY14850
PHONE: (607) 255-8202
www.qmps.vet.cornell.edu

Vermont Dairy Herd Improvement Association
226 Holiday Drive, Suite 3
White River Junction, VT 05001
PHONE: (800) 639-8068
EMAIL: mnault@vtdhia.org
www.vtdhia.org

APPENDIX C

ORGANIC SEED AND SOIL AMENDMENT SOURCES

MAINE

Fedco Seeds
PO Box 520
Waterville, ME 04903
PHONE: (207) 873-7333
www.fedcoseeds.com

Johnny's Selected Seeds
310 Foss Hill Road
PO Box 2580
Albion, ME 04910
PHONE: (800) 437-4290
www.johnnyseeds.com

MASSACHUSETTS

Great Cape
2624 Main Street
PO Box 1206
Brewster, MA 02631
PHONE: (800) 427-7144
EMAIL: ginkgo@greatcape.com
www.greatcape.com

Neptune's Harvest Organic Fertilizers
88 Commercial Street
Gloucester, MA 01930
PHONE: (800) 259-4769
EMAIL:
ann@neptunesharvest.com
www.neptunesharvest.com

NEW YORK

Agriculver Seeds
2059 Route 96
Trumansburg, NY 14886
PHONE: (800) 836-3701
(607) 387-5788

Cold Spring Farm
379 Slate Hill Road
Sharon Spring, NY 13459
PHONE: 518-234-8320
EMAIL: csfarm@wildblue.net
www.coldspringsfarm.net

Green Haven Open-Pollinated Seed
8225 Wessels Road
Avoca, NY 14809
PHONE: (800) 582-0952
www.openpollinated.com

Lakeview Organic Grain
119 Hamilton Place
Penn Tan, NY 14527
PHONE: (315) 531-1038
EMAIL: kandmhfarm@sprint-mail.com

Lightning Tree Farm
132 Andrew Haight Road
Millbrook, NY 12545
PHONE: (914) 677-9507

Mara Seeds
4451 Route 221
Marathon, NY 13803
PHONE: (607) 849-7871

Rocky Top Acres
1659 Quaker Hill Road
Hubbardsville, NY 13355
PHONE: (315) 899-8907

Schuster Seeds
1833 Route 89
Seneca Falls, NY 13148
PHONE: (315) 568-9337

Turtle Tree Seed
Camphill Village
Copake, NY 12516
PHONE: (888) 516-7797
www.turtletreeseed.org

West Wind Farm
196 Hoyer Road
Cherry Valley, NY 13320
PHONE: (607) 264-3635
EMAIL: bill@westwindfarm.com
www.westwindfarm.com

OHIO

Merit Seeds
PO Box 205
Berlin, OH 44610
PHONE: (800) 553-4713 /
(330) 893-2338

PENNSYLVANIA

Doeblers Seed
202 Tiadaghton Ave.
Jersey Shore, PA 17740
PHONE: (800) 853-2676
EMAIL: jmorse@doeblers.com
www.doeblers.com

The Fertrell Company
PO Box 265
Bainbridge, PA 17502
PHONE: (717) 367-1566
www.fertrell.com

Homestead Nutrition, Inc.
245 White Oak Road
New Holland, PA 17557
PHONE: (888) 336-7878 /
(717)354-4398
www.homesteadnutrition.com

King's Agriseeds
96 Paradise Lane
Ronks, PA 17572
PHONE: (866) 687-6224 /
(717)687-6224

Lancaster Ag Products
60 North Ronks Road
Ronks PA 17572
PHONE: (717) 687-9222
www.lancasterag.com

Seedway
1225 Zeager Rd
Elizabethtown, PA 17022
PHONE: (800) 952-7333
EMAIL: info@seedway.com
www.seedway.com

VERMONT

G. Boucher Fertilizer
2343 Gore Rd
Highgate Center, VT 05459
PHONE: (802) 868-3939

Bourdeau Brothers
25 Severance Rd
Sheldon, VT 05483
PHONE: (800) 499-2128 /
(806) 933-4581
www.bourdeaubros.com

Brookfield Ag Service
366 West St
Brookfield, VT 05036
PHONE: (888) 293-1200

Butterworks Farm
421 Trumpass Rd
Westfield, VT 05874
PHONE: (802) 774-6855
www.butterworksfarm.com

Green Mountain Feeds
65 Maine Street
Bethel, VT 05032
PHONE: (877) 234-6278 / (802) 234-6278
www.greenmountainfeeds.com

High Mowing Seed Co.
76 Quarry Road
Wolcott, VT 05680
PHONE: (802) 472-6174
www.highmowingseeds.com

J. R. Farm Service
189 Summer Street
Morrisville, VT 05661
PHONE: (802) 888-3522

LD Oliver Seed Co., Inc
29 Sunset Drive
PO Box 156
Milton, VT 05468
PHONE: (800) 624-2952 / (802) 777-7177
EMAIL: oliverseeds@aol.com

Morrison's Custom Feeds, Inc.
1140 Old Country Road
Barnet, VT 05821
PHONE: (802) 633-4387
www.morrisonsfeeds.com

North Country Organics
Depot St / PO Box 372
Bradford, VT 05033
PHONE: (802) 222-4277
www.norganics.com

Seedway
3442 Route 22A
Shoreham, VT 05770
PHONE: (888) 863-9099 / (802) 897-2881
EMAIL: info@seedway.com
www.seedway.com

CANADA

Homestead Organics
1 Union Street
Berwick, Ontario KOC 1GO
PHONE: (877) 984-0480
www.homesteadorganics.ca

Wehrmann Grain and Seed Ltd.
RR#1 460 Sideroad 20
Ripley, Ontario NOG 2RO
PHONE: (519) 395-3126
EMAIL: ingasven@hurontel.on.ca

OTHER

ILLINOIS

American Organics
PO Box 385
Warren, IL 61087
PHONE: (815) 266-4010
www.american-organics.com

INDIANA

Great Harvest Organics
6803 E 276 Street
Atlanta, IN 46031
PHONE: (317) 984-6685
www.greatharvestorganics.com

Prairie Hybrid Seeds
27445 Hurd Road
Deer Grove, IL 61243
PHONE: (800) 368-0124

IOWA

Welter Seeds & Honey
17724 Hwy 136
Onslow, IA 52321-7549
PHONE: (800) 728-8450 / (563) 485-2762
www.welterseed.com

MINNESOTA

Albert Lea Seed House
1414 W. Main Street
Albert Lea, MN 56007
PHONE: (800) 352-5247 / (507) 373-3161
www.alseed.com

Gold Country Seeds
16506 Hwy 15N
Hutchinson, MN 55350
PHONE: (800) 795-8544
EMAIL: gcs@goldcountryseed.com
www.goldcountryseed.com

NEBRASKA

Blue River Hybrids
PO Box 327
Seward, NE 68434
PHONE: (800) 370-7979
EMAIL: info@blueriverorgseed.com
www.blueriverorgseed.com

NORTH DAKOTA

Agassiz Seed & Supply
445 7th Street NW
West Fargo, ND 58078
PHONE: (701) 282-8118
EMAIL: tisha@agassizseed.com
www.agassizseed.com

Blaine's Best Seeds
6020 22nd Ave
Rugby, ND 58368
PHONE: (701) 776-6023
EMAIL: bbest-seeds@stellarnet.com
www.localfoods.umn.edu/blainesbest

WISCONSIN

Superior Organic Grains, Ltd.
N 7076 Hwy C
Seymour, WI 54165
PHONE: (920) 833-6953

APPENDIX D
ORGANIC GRAIN AND MINERAL SOURCES

CONNECTICUT

Highland Thistle Farm
289 North Society Road
Canterbury, CT 06331
PHONE: (860) 546-3960
EMAIL: htf@charterinternet.com

MAINE

Blue Seal Feeds, Inc.
10 Dalton Road
Augusta, ME 04330
PHONE: (800) 734-5778
EMAIL: services@blueseal.com
www.blueseal.com

The Farmer's Shed
243 Knight's Pond Rd
South Berwick, ME 03908
PHONE: (207) 384-5090
(207) 384-2154
EMAIL: thefarmer'sshed@meedamfarm.com
www.thefarmersshed.com

Northern-Most Feeds, LLC
155 Gagnon Road
Madawaska, ME 04756
PHONE: (207) 728-3150
EMAIL:
northernmostfeeds@yahoo.com

MASSACHUSETTS

Greenfield Farmers Coop. Exchange
269 High Street
Greenfield, MA 01301
PHONE: (413) 773-9639
www.greenfieldfarmerscoop.com

Horse and Buggy Feeds
Route 12 North
Winchendon, MA 01475
PHONE: (978) 297-2518
www.horseandbuggyfeeds.com

Moon in the Pond Farm
816 Barnum Street
Sheffield, MA 01257
PHONE: (413) 229-309
EMAIL:
dom@mooninthepond.com
www.mooninthepond.com

United Cooperative Farmers, Inc.
22 Kimball Place
Fitchburg, MA 01420
PHONE: (978) 345-4103 Ext. 219
www.ucf-inc.com

Waquoit Feed and Grain
411 Waquoit Road
Falmouth, MA 02536
PHONE: (508) 457-9400

NEW HAMPSHIRE

Blue Seal Feed
PO Box 8000
Londonderry, NH 03053
PHONE: (800) 367-2730
EMAIL: services@blueseal.com
www.blueseal.com

Horse and Buggy Feeds
Dunbar Street
Keene, NH 03431
PHONE: (603) 352-0328
www.horseandbuggyfeeds.com

NEW YORK

Blue Seal Feeds
PO Box 127
Sangerfield, NY 13455
PHONE: (800) 228-2709
www.richerfeeds.com

Buffalo Molasses, LLC
5133 Robinson Road
Fillmore, NY 14735
PHONE: (585) 567-2106
www.buffalomolasses.com

Cold Spring Farm
379 Slate Hill Road
Sharon Springs, NY 13459
PHONE: (518) 234-8320
EMAIL: coldspringsfarm@verizon.net
www.coldspringsfarm.net

Cornerstone Farm Ventures
242 Dan Main Road
Norwich, NY 13815
PHONE: (315) 334-2833
EMAIL: info@cornerstone-farm.com
www.cornerstone-farm.com

Lakeview Organic Grain
119 Hamilton Place
Penn Yan, NY, 14527
PHONE: (315) 536-9879
EMAIL: kandmhfarm@sprint-mail.com

Lightning Tree Farm
132 Andrew Haight Road
Millbrook, NY 12545
PHONE: (845) 677-9507
Richer Feeds /

Walkers Feed Mill/MGK Enterprises
5565 Route 4
Fort Ann, NY 12827
PHONE: (800) 480-5223

Wight and Patterson, Inc.
8 Miller Street
Canton, NY 13617
PHONE: (315) 386-2751

OHIO

Dale Filbrun Farms Inc.
3993 State Route 503 S.
West Alexandria, OH 45381-9355
PHONE: (937) 787-4885

Ft. Recovery Equity, Inc.
2351 Wabash Road
Ft. Recovery, OH 45846
PHONE: (419) 375-4119
EMAIL: ftequity@bright.net

Highland Naturals
3878 County Road 135
Millersburg, OH 44654
PHONE: (330) 893-2016
www.highlandnaturalproducts.co.uk

Holmes Ag Service
301 South Market St./PO Box 67
Holmesville, OH 44633
PHONE: (330) 279-2501

Maysville Elevator
10583 Harrison Road
Apple Creek, OH 44606
PHONE: (330) 695-4413

Mt Eaton Elevator
15911 East Berry Road
PO Box 195
Mt Eaton, OH 44659-0195
PHONE: (330) 359-5028

Mt Hope Agri Serv Center
SR 241
Mt Hope, OH 44660
PHONE: (330) 674-0416

Ohio Earth Food
5488 Swamp Street NE
Hartville, OH 44632
PHONE: (330) 877-9356
www.ohioearthfood.com

PENNSYLVANIA

Agri-Dynamics
PO Box 267
Martins Creek, PA 18063
PHONE: (610) 250-9280
(877) 393-4484
www.agri-dynamics.com

The Fertrell Company
PO Box 265
Bainbridge, PA 17502
PHONE: (717) 367-1566
www.fertrell.com

Homestead Nutrition Inc.
245 White Oak Road
New Holland, PA 17557
PHONE: (717) 354-4398 / (888) 336-7878
www.homesteadnutrition.com

Kreamer Feed/Nature's Best
215 Kreamer Ave.
PO Box 38
Kreamer, PA 17833
PHONE: (800) 767-4537
EMAIL: info@kreamerfeed.com
www.kreamerfeed.com

Lancaster Ag Products
60 North Ronks Road
Ronks, PA 17572
PHONE: (717) 687-9222
www.lancasterag.com

McGeary Organics
PO Box 299
Lancaster, PA 17608-0299
PHONE: (800) 624-3279
EMAIL: sales@mcgearygrain.com
www.mcgearyorganics.com

MS Bio-Ag
5802 Torrance Drive
Export, PA 15632
PHONE: (724) 733-2594
EMAIL: msinan@surfnet.net

Organic Unlimited/Vintage Feed
PO Box 238
Atglen, PA 19310
PHONE: (610) 593-2995
EMAIL: organicunlimited@one-mane.com
www.organicunlimited.com

Pennfield Corporation/Horn Co. Organic Feeds
711 Rohrerstown Road
PO Box 4366
Lancaster PA, 17604
PHONE: (800) 732-0467
www.pennfield.com

Zook Molasses Company
4960 Horseshoe Pike
Honey Brook, PA 19344
PHONE: (800) 327-4406

VERMONT

Blue Seal Feeds, Inc.
38 Union Street
Brandon, VT 05733
PHONE: (800) 766-6391
www.blueseal.com

Green Mountain Feeds, Inc.
65 Maine Street
PO Box 505
Bethel, VT 05032
PHONE: (802) 234-6278
EMAIL: gmferic@verizon.net
www.greenmountainfeeds.com

Morrison's Custom Feeds, Inc.
1140 Old Country Road
Barnet, VT 05821
PHONE: (802) 633-4387
www.morrisonsfeeds.com

North Country Organics
Depot Street
PO Box 372
Bradford, VT 05033
PHONE: (802) 222-4277
www.norganics.com

North Star Farm Supply
PO Box 299
S. Royalton, VT 05068
PHONE: (802) 763-8870

Poulin Grain
24 Depot Street
Swanton, VT 05488
PHONE: (802) 868-3323
www.poulingrain.com

Renaissance Nutrition
70 Grand Ave.
Swanton, VT 05488
PHONE: (802) 868-2675
EMAIL: formoremilk@verizon.net

CANADA

Bio-AG Consultants & Distributors, Inc.
PO Box 189
Wellesley, Ontario NOB 2TO
PHONE: (800) 363-5278
www.bio-ag.com

Great Lakes Organic
RR2 Parkhill
Petrolia, Ontario NON 2K0,
PHONE: (519) 232-9458
www.greatlakesorganic.com

Homestead Organics
1 Union Street
Berwick, Ontario K0C 1G0
PHONE: (613) 984-0480
(877) 984-0480
www.homesteadorganics.ca

Jirah Farms and Mills
2780 North River Road
Ormstown, Quebec J0S 1K0
PHONE: (450) 829-1025

P. A. Lessard
745 93E Rue St. Georges de Beauce
Quebec G5Y 3K4
PHONE: (418) 228-8918
EMAIL: pa@palessard.qc.ca
www.palessard.qc.ca

Robert Mosher Transport, Inc.
1736 Rte 225
Noyan, Quebec J0J 1B0
PHONE: (450) 294-2757

Robitaille Feeds
190 Rue Comeau Nord
Farnham, Quebec J2N 2N4
PHONE: (800) 363-6323

OTHER

ILLINOIS

Helfter Feeds, Inc
PO Box 266
Osco, IL 61274
PHONE: (866) 435-3837
www.helfterfeeds.com

UTAH

Redmond Minerals, Inc
6005 North 100 West
PO Box 219
Redmond, UT 84652
PHONE: (866) 735-7258
www.redmondnatural.com

VIRGINIA

Thorvin Kelp, USA
PO Box 458
New Castle, VA 24127
PHONE: (800)464-0417
www.thorvin.com

WISCONSIN

Crystal Creek Products
1600 Round House Road
Spooner, WI 54801
PHONE: (888) 376-6777
www.crystalcreeknatural.com

Midwestern Bio-Ag
PO Box 160
Blue Mounds, WI 53517
PHONE: (800) 327-6012
www.midwesternbioag.com

APPENDIX E
SOIL, FORAGE, AND MANURE TESTING SERVICES

SOIL TESTING

AGPDTS/SoilTest.htm
University of Vermont
Soil Testing Lab
219 Hills Bldg, UVM
Burlington, VT 05405
PHONE: (802) 656-3030
www.uvm.edu/pss/ag_testing/

Analytical Laboratory and Maine Soil Testing Service
5722 Deering Hall
Orono, ME 04469-5722
PHONE: (207) 581-3591
www.anlab.umesci.maine.edu

A&L Analytical Laboratories
411 North Third Street
Memphis, TN 38105-2723
PHONE: (800) 264-4522
www.al-labs.com

Cornell Soil Health Assessment
1003 Bradfield Hall
Ithaca, NY 14853
PHONE: (607) 255-1706
www.soilhealth.cals.cornell.edu

Crop Services International Inc.
1718 Madison SE
Grand Rapids, MI 49507-2518
PHONE: (616) 246-7933
www.cropservicesintl.com

Kinsey's Agricultural Services
297 County Hwy 357
Charleston, MO 63834
PHONE: (573) 683-3880
www.kinseyag.com

Midwestern Bio-Ag
10955 Blackhawk Drive
PO Box 160
Blue Mounds, WI 53517
PHONE: (608) 437-4994
www.midwesternbioag.com

North Country Organics
Depot Street
PO Box 372
Bradford, VT 05033
PHONE: (802) 222-4277
www.norganics.com

Rutgers Soil Testing Laboratory
PO Box 902
Milltown, NJ 08850
PHONE: (732) 932-9295
www.njaes.rutgers.edu/soil-testinglab/

Soil Foodweb, Inc.
980 NW Circle Blvd
Corvallis, OR 97330
PHONE: (541) 752-5066
www.soilfoodweb.com

Soil Nutrient Analysis Laboratory
6 Sherman Place, U-102
University of Connecticut
Storrs, CT 06269-5102
PHONE: (860) 486-4274
www.soiltest.uconn.edu/#testing

University of Massachusetts Soil Testing Lab
682 N. Pleasant St
Amherst, MA 01003-8021
PHONE: (413) 545-2311
www.umass.edu/plsoils/soil-test

University of New Hampshire Coop. Extension
38 Academic Way, Rm G28A
Durham, NH 03824
PHONE: (603) 862-3200
www.extension.unh.edu/Agric/

Woods End Laboratories, Inc.
290 Belgrade Road
PO Box 297
Mt Vernon, ME 04352
PHONE: (207) 293-2457
www.woodsend.org

FORAGE TESTING

Cumberland Valley Analytical Services
Batavia Forage Lab
5049 Clinton Street Road
Batavia, NY 14020-3357
PHONE: (800)282-7522
EMAIL: mail@foragelab.com
www.foragelab.com

Dairy One Forage Lab
730 Warren Road
Ithaca, NY 14850
PHONE: (800) 496-3344
www.dairyone.com/Forage/services/ Forage/forage.htm

Green Mountain Feed Testing Laboratory
24 Railroad Square
Newport, VT 05855
PHONE: (802) 334-6731

National Forage Testing Association
PO Box 451115
Omaha, NE 68145-6115
PHONE: (402) 333-7485
www.foragetesting.org
Provides a complete national list of forage laboratories

William H. Miner Institute
586 Ridge Road
Chazy, NY 12921
PHONE: (518) 846-7121 x123
EMAIL: cotanch@whminer.com
www.whminer.com/facilities_lab.html

MANURE TESTING

Agricultural Analytical Services Laboratory
Penn State University
Tower Road
University Park, PA 16802
PHONE: (814) 863-0841
www.aasl.psu.edu/ manureprgnew.html

Analytical Laboratory University of Maine
5722 Deering Hall
Orono, ME 04469-5722
PHONE: (207) 581-3591
www.anlab.umesci.maine.edu

Dairy One
730 Warren Road
Ithaca, NY 14850
PHONE: (800) 496-3344
www.dairyone.com/Forage/services/ Manure/manure.htm

APPENDIX F

NORTHEAST ORGANIC MILK MARKETS

AgriMark Inc.
PO Box 5800
Lawrence, MA 01842
www.agrimark.net
CONTACT: Bob Stoddard
PHONE: (978) 689-4442
EMAIL: info@agrimark.net

CROPP Cooperative / Organic Valley Family of Farms
One Organic Way
LaFarge, WI 54639
PHONE: (888) 809-9297
www.organicvalley.coop
NORTHEAST (NY, PA, AND MD)
CONTACT: Peter Miller
PHONE: (888) 444-6955
Ext: 3407
EMAIL:
peter.miller@organicvalley.coop

NEW ENGLAND CONTACT:
John Cleary
PHONE: (888) 444-6544 ext. 3330
EMAIL:
john.cleary@organicvalley.coop

Dairy Marketing Services (DMS)
5001 Brittonfield Pkwy
PO Box 4844
Syracuse, NY 13221
PHONE: (888) 589-6455
www.dairymarketingservices.com
CONTACT: Chris Cardner
PHONE: (800) 582-2552 Ext: 5403

Horizon Organic Dairy/ The Organic Cow
PO Box 17577
Boulder, CO 80308
PHONE: (800) 769-9693
www.horizonorganic.com

NEW ENGLAND CONTACT:
Cindy Masterman
PHONE: (888) 648-8377
EMAIL:
CindyMaster-man@whitewave.com

NORTHEAST CONTACT (EASTERN NY AND PA):
Peter Slaunwhite
PHONE: (800) 381-0980
Cell: 315 420-3293
EMAIL: Peter.Slaunwhite@whitewave.com
MIDEAST CONTACT (WNY, PA, OH, IN, MI):
Steve Rinehart
PHONE: (866) 268-4665
EMAIL: Steve.Rinehart@white-wave.com

MID-ATLANTIC CONTACT (SOUTHERN PA, MD, VA, NJ):
Michelle Sandy
PHONE: (866) 268-4665
EMAIL:
Michelle.Sandy@whitewave.com

HP Hood, LLC
Six Kimball Lane
Lynnfield, MA 01940
PHONE: (800) 383-6592
www.hphood.com
Contact: Karen Cole
PHONE: (866) 383-1026
EMAIL: karen.cole@hphood.com

Lancaster Organic Farmer Cooperative (LOFCO)
101 S. Lime Street, Suite A
Quarryville, PA 17566
PHONE: (800) 996-0383
www.lancasterfarmfresh.com
CONTACT: Levi Miller
PHONE: (717) 661-8682
Contact: Jerry McCleary
PHONE: (717) 577-8809

United Ag Services/ Organic Dairy Farmers Cooperative, Inc.
12 North Park Street
Seneca Falls, NY 13148
CONTACT: Mima Kaiser
PHONE: (800) 326-4251
EMAIL: unitedag@flare.net

Upstate Niagara Cooperative, Inc.
7115 West Main Road
Leroy, NY 14482
PHONE: (800) 724-6455
www.upstatefarmscoop.com
CONTACT: Bill Young
PHONE: (585) 768-2247 Ext. 6225
EMAIL:
byoung@upstatefarms.com

APPENDIX G

DAILY MILK PRODUCTION PER COW

MILK PRODUCTION (LB.)									
DATE:									
1									
2									
3									
4									
5									
6									
7									
8									
9									
10									
11									
12									
13									
14									
15									
16									
17									
18									
19									

 Source: Graze NY Program, Cornell Cooperative Extension.

APPENDIX H
BODY CONDITION SCORING

First view the pelvic area from the side. Check line from hooks, to the thurl, to the pins.

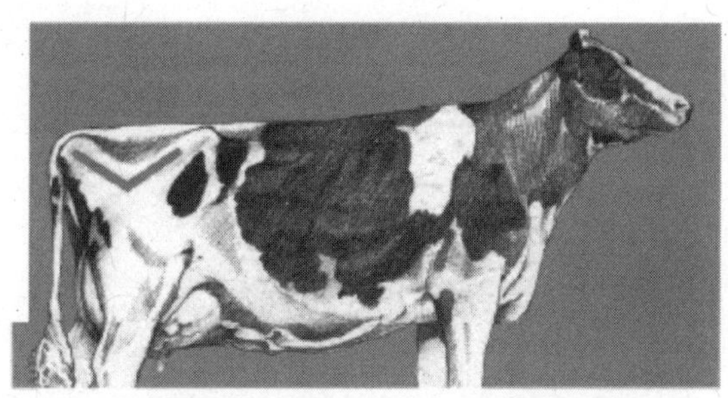

If the line forms a flattened V then **BCS ≤ 3.0.**

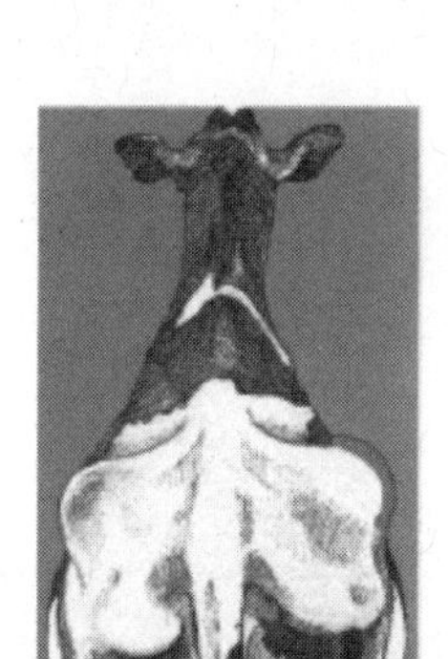

1 If hooks rounded BCS = 3.0.

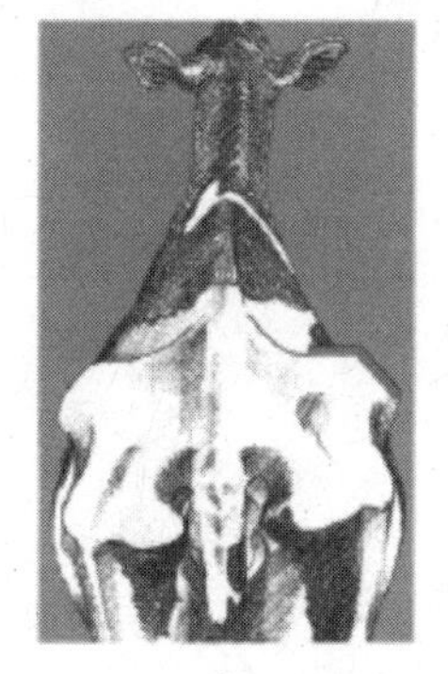

2 If hooks angular **BCS ≤ 2.75.** Check pins. If pins padded **BCS = 2.75.**

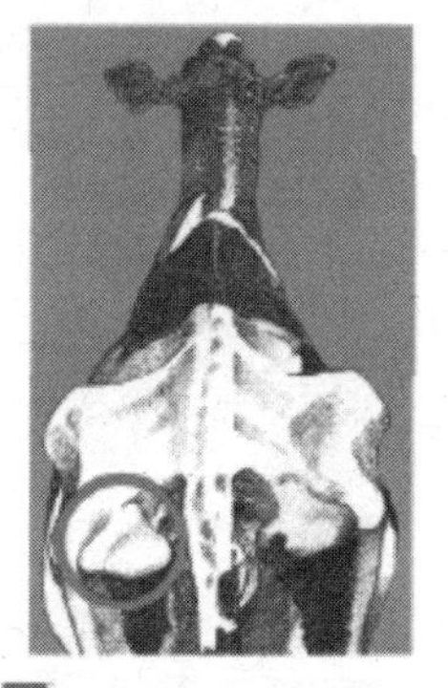

3 If pins angular **BCS < 2.75.** If palpable fat pad on point of pins **BCS = 2.50.**

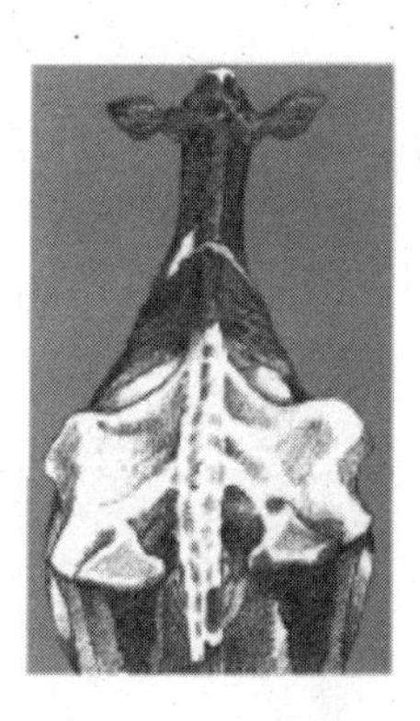

—

4 If no fat pad on pins BCS < 2.50. View the short ribs. Look for corrugations along the top of short ribs as fat covering disappears. If corrugations visible 1/2 way between tip and spine of short ribs, BCS = 2.25. If corrugations visible 3/4 way from tip to spine BCS = 2.0. If thurl prominent and saw-toothed spine BCS < 2.0.

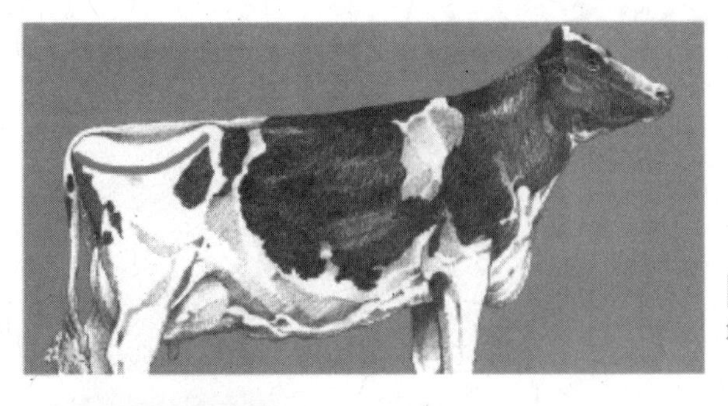

U If the line forms a cresent or flattened U consider BCS $\geq$ 3.25.

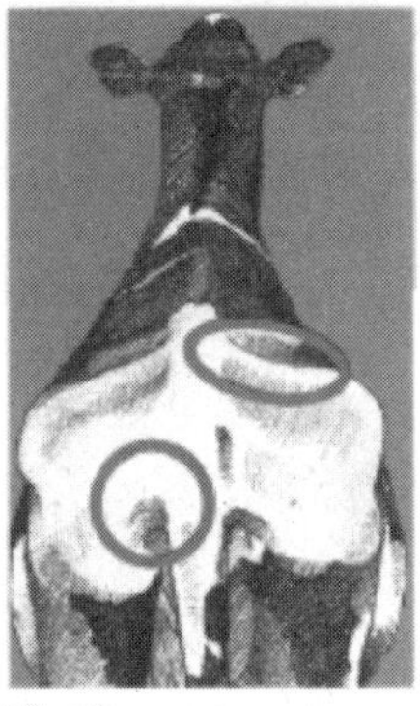

1 If sacral and tailhead ligaments visible BCS = 3.25.

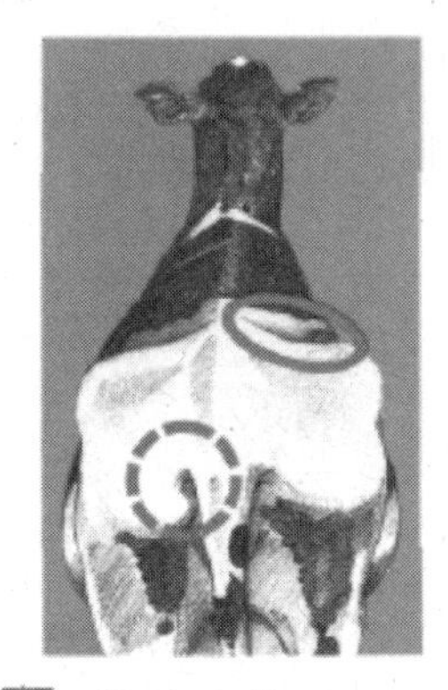

2 If sacral ligament visible and tailhead ligament barely visible **BCS = 3.50.**

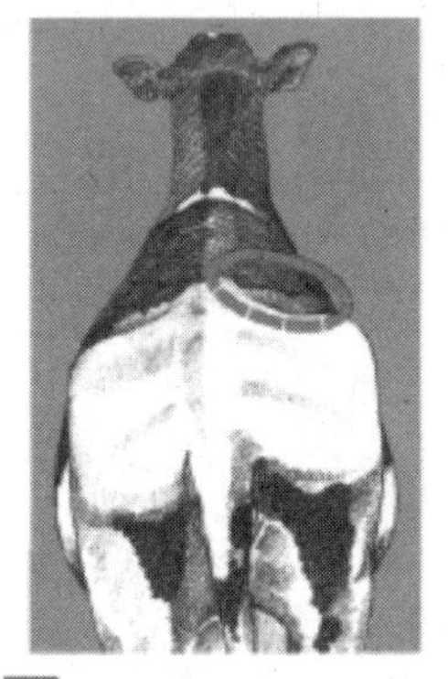

3 If sacral ligament barely visible and tailhead ligament not visible **BCS** = 3.75. If sacral and tailhead ligament not visible **BCS** $\geq$ **4.0.**

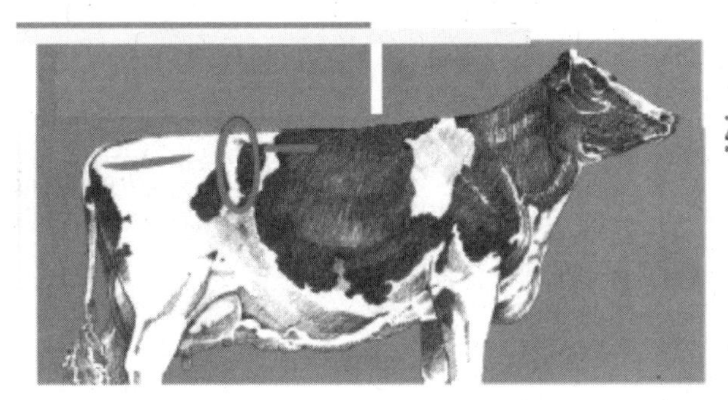

4 If thurl flat BCS > 4.0. If tip of short bs barely visible BCS = 4.25. If thurl at and pins buried BCS = 4.5. If hooks arely visible BCS = 4.75. If all boney rominences well rounded **BCS = 5.0.**

 Source: Elanco Animal Health. 2007. Greenfield, IN.

APPENDIX I
LOCOMOTION SCORING GUIDE

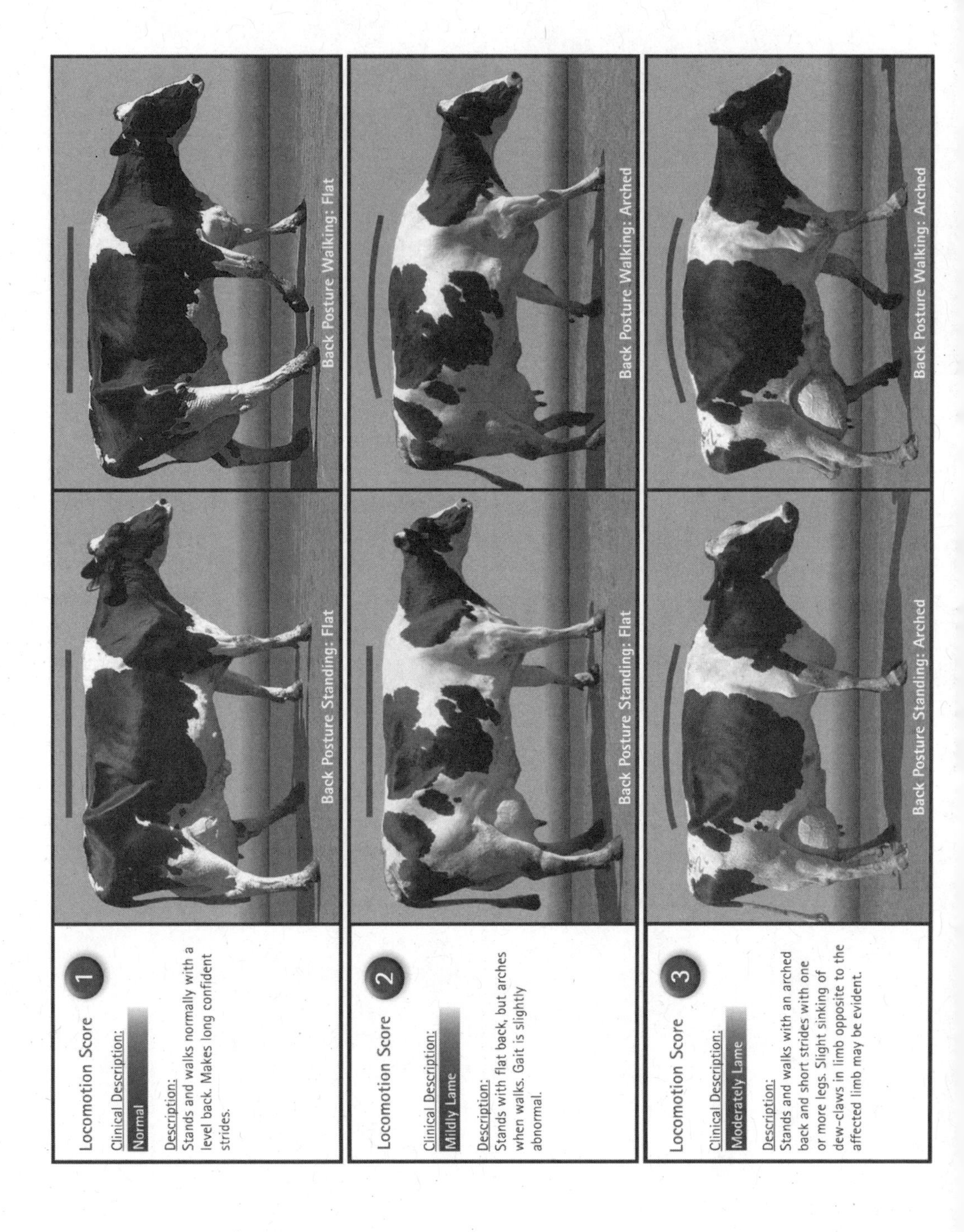

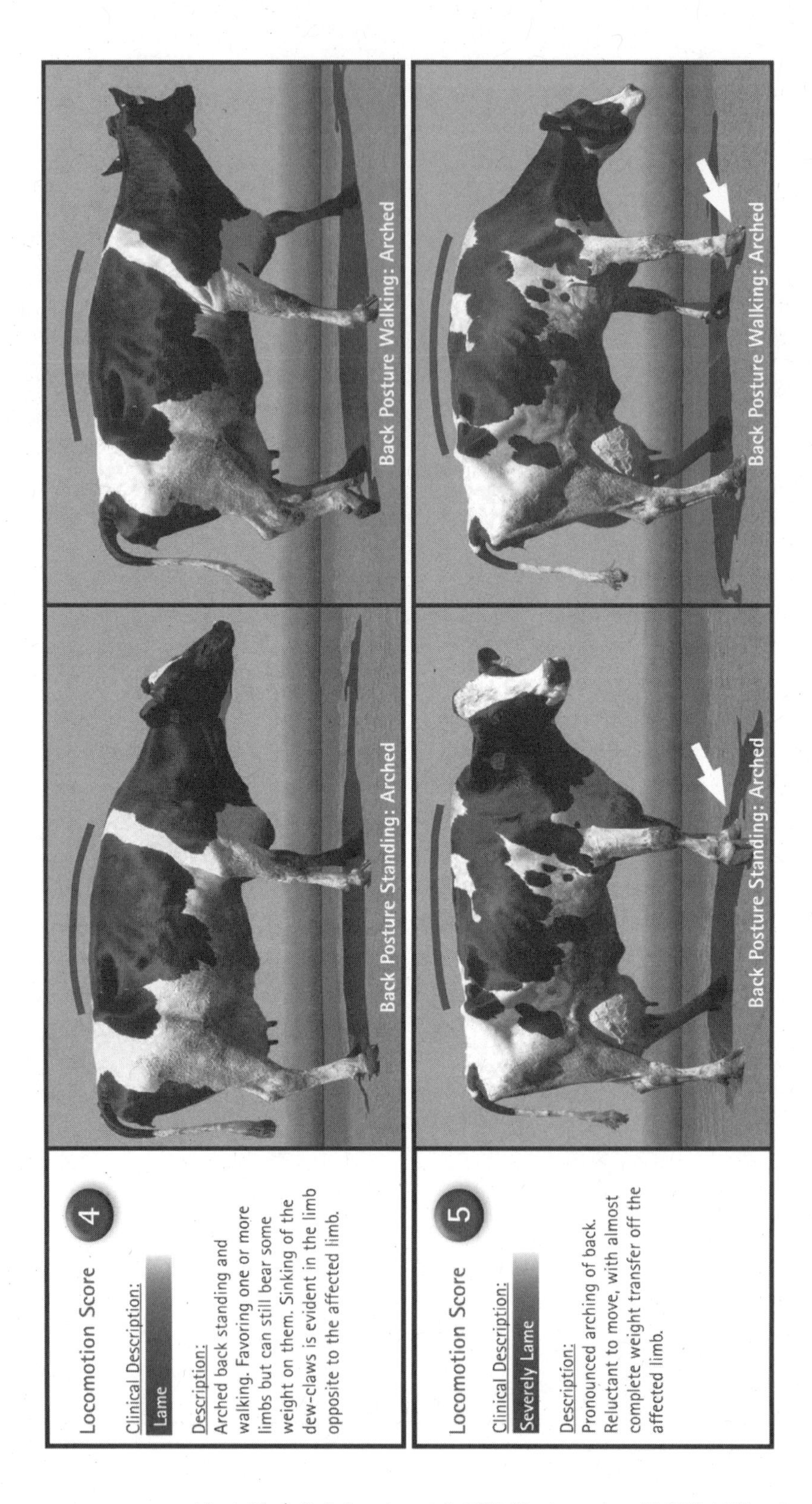

Adapted from D. J. Sprecher et al. 1997. Theriogenology 47:1178-1197 and contributions from N. B. Cook, University of Wisconsin. Reprinted courtesy of Zinpro Corporation, http://www.zinpro.com.

APPENDIX J

COW ASSESSMENT

INDIVIDUAL ANIMAL ASSESSMENT AND OBSERVATION

Attitude, alertness, interactions
Apart from group?
Eating? Standing? Lying down?
Carriage of head, ears, tail
Response to stimuli
Expression—eye brightness
Unwilling to move or rise?

Abnormalities of body, gait, or conformation
General body outline, swellings
Body condition relative to stage of lactation
Abnormal discharges (eyes, nose, mouth)
Vaginal discharge (presence and character)
Skin and hair coat
Signs of estrus
Lameness: lesions, swelling, smell
Stance (tucked up, straining)

Feed intake & outputs
Is there feed and manure?
Is feed fresh? Smell?
Eating? Chewing cud?
Belly full or distended?
Udder full or slack? Leaking?
Manure color and character
Urine: frequency and character

Respiratory function
Consider air temperature and humidity
10–30 breaths/minute in normal animal
Breathing regular, deep, even
Nasal discharge and character
Labored breathing?
Open mouth breathing? Frothing?
Grunting, groaning?

POINTS OF PHYSICAL EXAM OF THE DAIRY COW

Temperature
Rectal 100.5–102.5 ^{0}F
Consider air temp and activity
Milk fever: decreased temp

Digestive system
Appetite
Mouth (caution: RABIES)
Sores on lips, pad, tongue
Abdomen: normal shape?
Rumination: listen on left side
Gut sounds: listen on right
"Ping" on left or right side

Udder and teats
Hot, cold, swelling, redness?
Injuries or warts?
Teat end lesions?
Strip test: appearance
CMT test
Milk culture

Hydration
Skin pinch test
Eye appearance & position in socket
Mucous membranes (moist or tacky)?

Heart and lungs
Heart rate at rest: 40–70 beats/min
Heart sounds: muffled, splashing, squeaking?
Pulse: strong and regular?
Mammary and jugular veins: distended?
Color of mucous membranes: blue, red, pale?
Lung sounds: fluid, crackle, wheeze, grunt?

Reproductive & urinary systems
Retained placenta: visible, smell, history?
Vulva: color, swelling, bruised, torn?
Vaginal discharge: amount and character?
Urine: pH, color, clarity, smell, amount
Ketosis test (urine or milk)

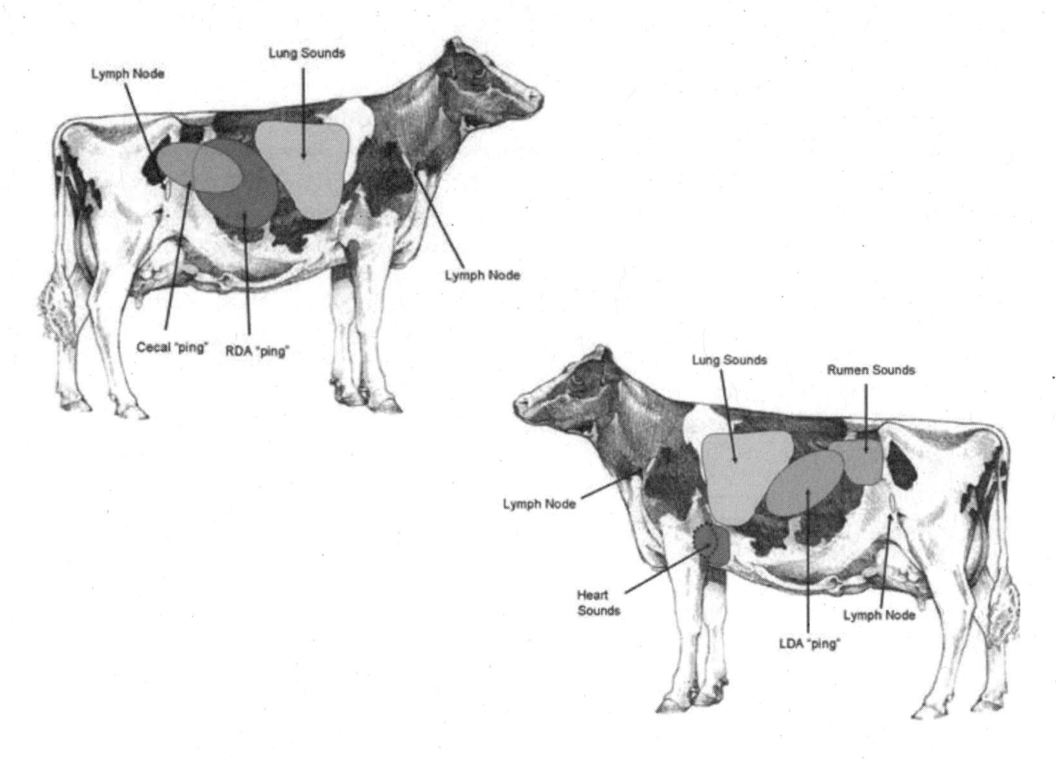

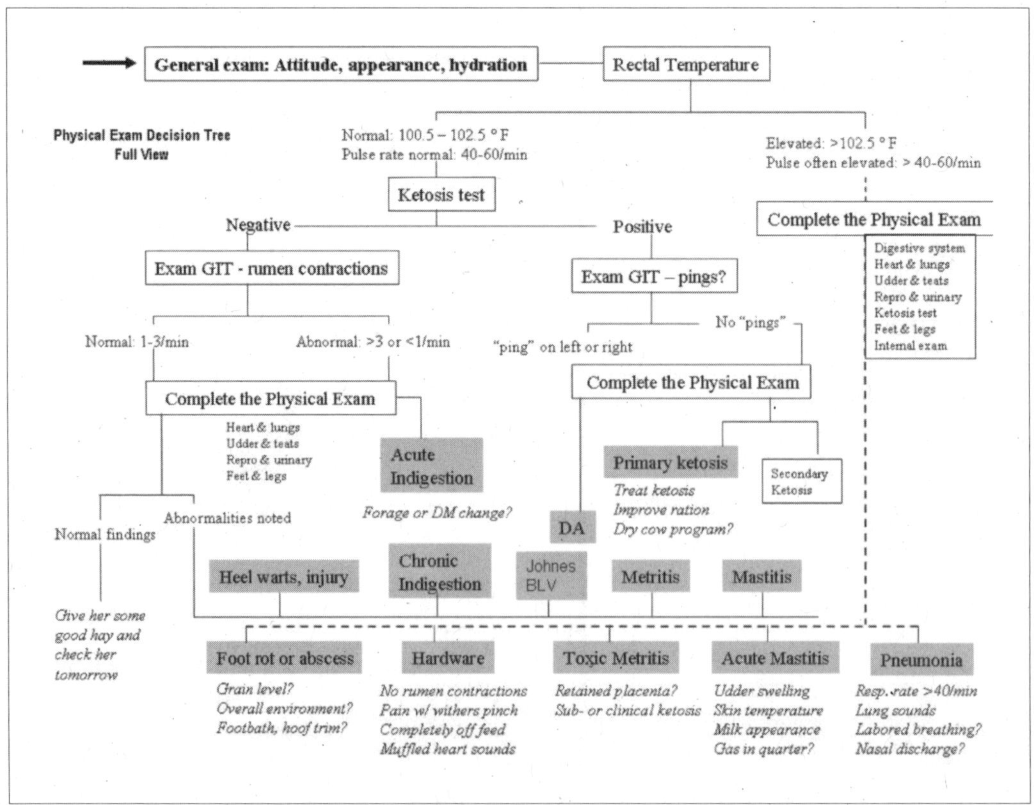

Source: Used with the permission of Pro-Dairy, http://www.ansci.cornell.edu/prodairy.

APPENDIX K

INDIVIDUAL ANIMAL HEALTH RECORD

ID/Name:	Registration No:	Birth date:
Sire:	Dam:	

Vaccinations	Date

Dehorning date:

Weaning date:

Johne's test status:

Comments:

Breeding history for	1st service		2nd service		3rd service		Date due to calving	Calving date	Calf sex
	Sire	Date	Sire	Date	Sire	Date			
1st calving									
2nd calving									
3rd calving									
4th calving									
5th calving									
6th calving									
7th calving									
8th calving									

HEALTH INCIDENT RECORD

DATE	SYMPTOM, INJURY, OR PROCEDURE	TREATMENT	OUTCOME (lived, died, culled)	COMMENTS

APPENDIX L

RECOMMENDED DIMENSIONS FOR TIESTALL AND FREESTALL FACILITIES

STALL TYPE	DIMENSION	RATIO AND REFERENCE BODY DIMENSION	AVERAGE HOLSTEIN
Tiestall Facilities	Bed length	1.2 × rump height	1.2 × 60 = 72 inches
	Tie rail height above cow's feet	0.8 × rump height	0.8 × 60 = 48 inches
	Stall width	2.0 hook bone width	2.0 × 26 = 52 inches
Freestall Facilities	Stall length from curb to solid front	2.0 × rump height	2.0 × 60 = 120 inches
	Stall length for open front (head to head)	1.8 × rump height	1.8 × 60 = 108 inches
	Bed length (imprint length)	1.2 × rump height	1.2 × 60 = 72 inches
	Neck rail height above cow's feet	0.83 × rump height	0.83 × 60 = 50 inches
	Neck rail forward location (bed length)	1.2 × rump height	1.2 × 60 = 72 inches
	Stall width (loops on center)	2.0 hook bone width	2.0 × 26 = 52 inches
	Space between brisket board and loop	Foot width	5 inches
	Brisket board location (from curb)	1.1 × rump height	1.1 × 60 = 66 inches

Source: Courtesy OMAFRA, 2007. (http://www.omafra.gov.on.ca/english/livestock/dairy/facts/info_tsdimen.htm).

APPENDIX M

HOCK SCORING GUIDE

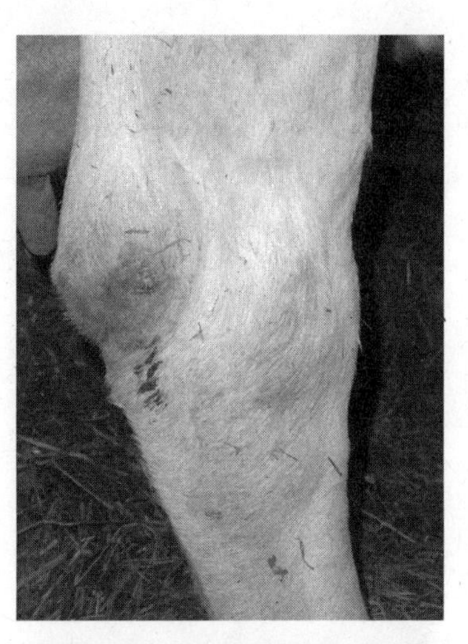

The normal healthy hock should have hair covering the surface and be free of scrapes, swellings, or sores. We can use hock health to assess cow comfort, if stall size is adequate, and to check if there is enough bedding.

WHAT PERCENTAGE OF YOUR COWS FALL INTO EACH OF THE THREE CATEGORIES?

__________% = SCORE 1

NO SWELLING. NO HAIR IS MISSING.

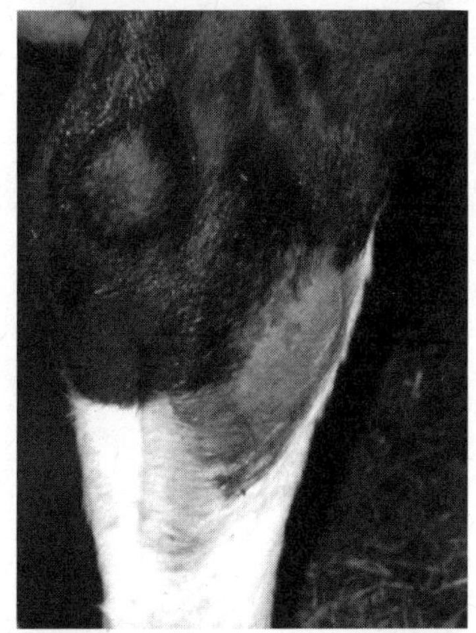

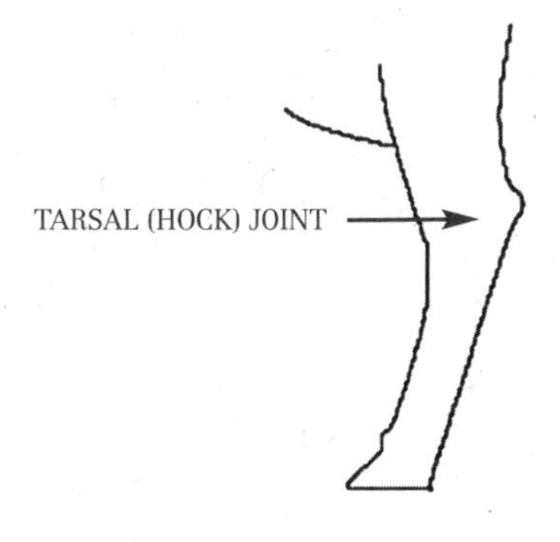

__________% = SCORE 2

NO SWELLING. BALD AREA ON THE HOCK.

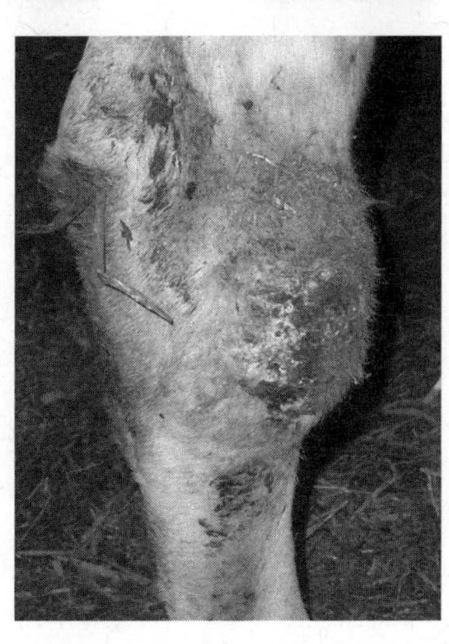

__________% = SCORE 3

SWELLING IS EVIDENT OR THERE IS A LESION THROUGH THE HIDE.

APPENDIX N
FEMALE COW ANATOMY AND ESTROUS CYCLE

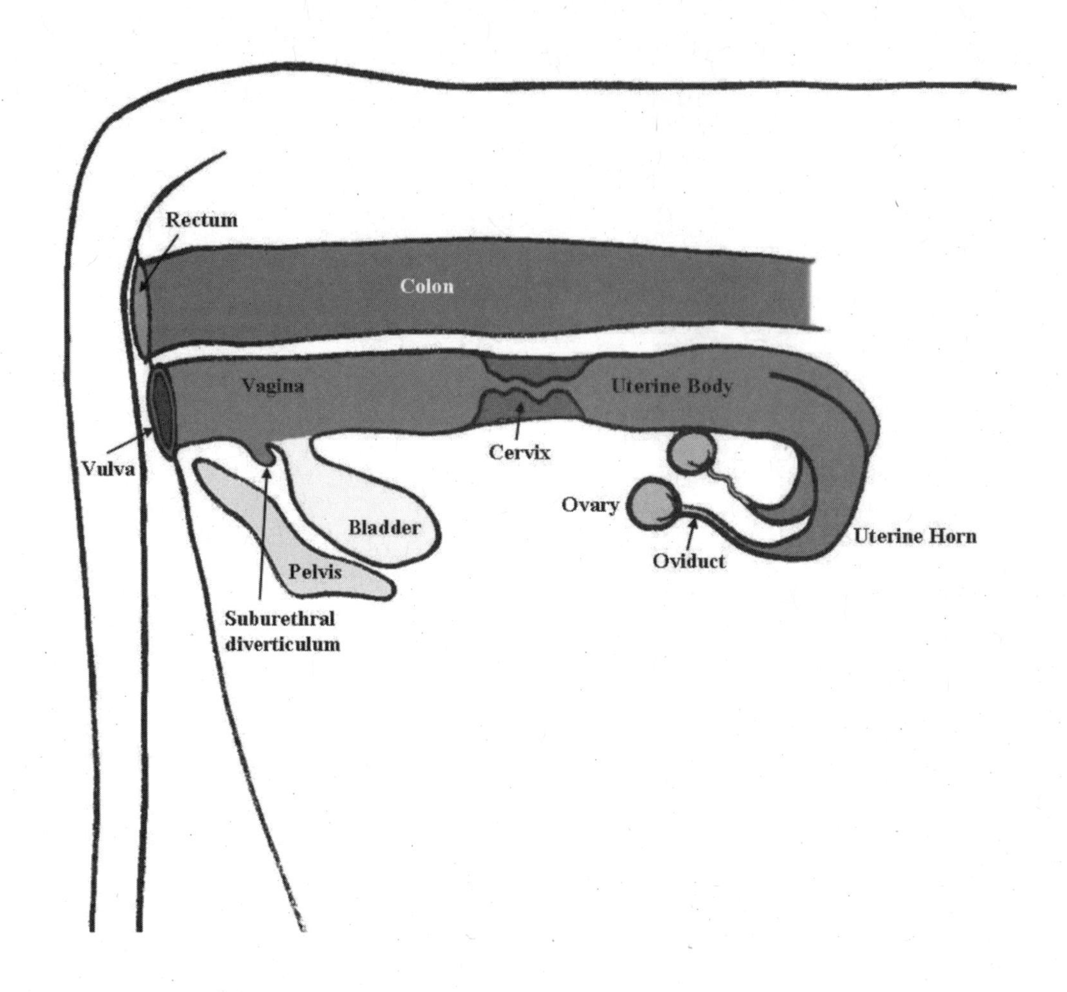

ANATOMY

The female reproductive anatomy is a fairly simple structure consisting of the following.

- *Ovaries (paired)*: secrete the natural reproductive hormones (estrogen and progesterone) and produce eggs, following the development of a follicle.
- *Oviducts*: two tubes that connect the ovaries to the uterus. Fertilization of the egg occurs here.
- *Uterus*: a muscular organ with two horns (left and right) that sustains the calf's needs during pregnancy. The muscular walls of the uterus contract during birth to deliver the calf.
- *Cervix*: muscular, ringed tube that connects the uterus to the vagina. It is sealed off with a mucus plug during pregnancy to keep the calf free of infection.
- *Vagina*: (birth canal) connects the cervix to the vulva.
- *Vulva*: exterior of the cow's reproductive tract.

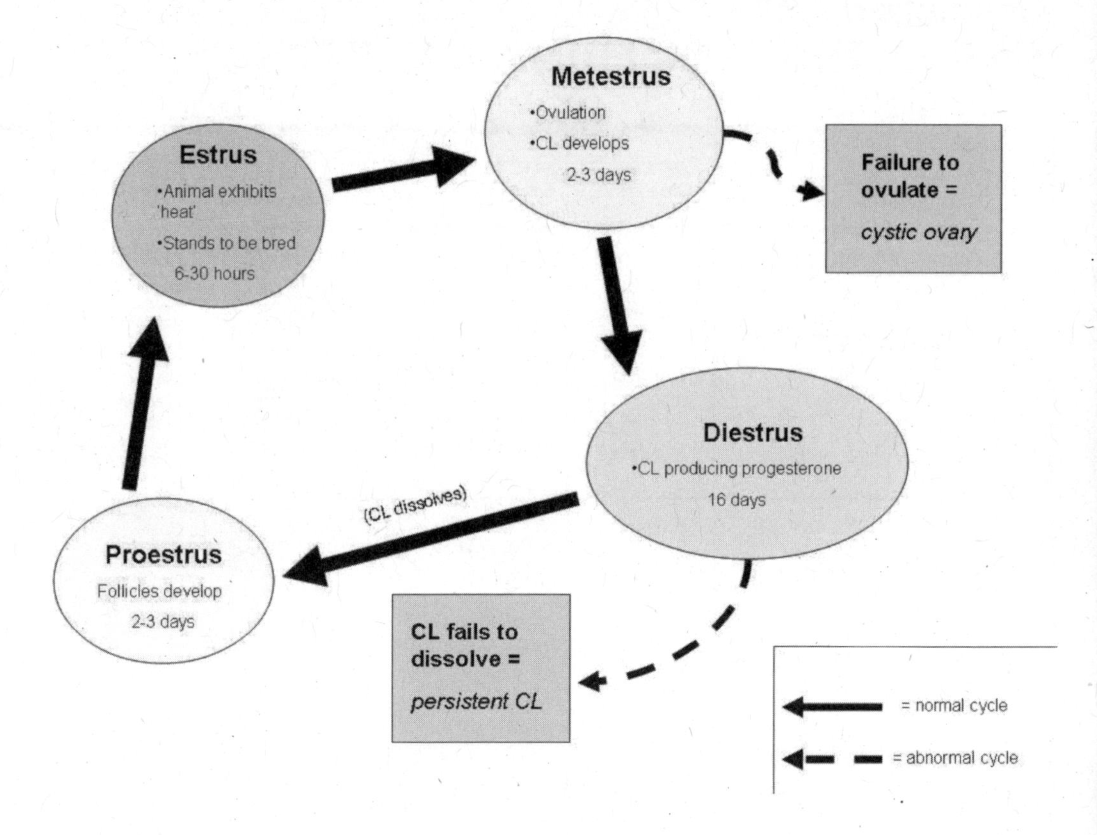

THE ESTROUS CYCLE

Cows typically have a 21-day estrous cycle (range: 17–24 days). Heifers may experience their first estrous cycle anytime between 5 and 12 months, depending on breed and level of nutrition.

The estrous cycle consists of four phases.

1. *Proestrous*: the 2–3 day phase occurring just before the cow shows signs of "heat." During this time the cow's estrogen levels are increasing and follicles are developing on the ovaries. A thin, clear mucus discharge may be noticed.
2. *Estrus*: the phase commonly referred to as "heat." The cow may bellow, be restless, and stand ("standing heat") to be mounted by a bull or other cows. This phase lasts about 18 hours.
3. *Metestrus*: the 2–3 day phase after "standing heat" and ovulation usually occurs in the first 12 hours of this phase. A corpus luteum (CL), which produces the pregnancy hormone *progesterone*, is formed on the ovary at the site of ovulation. Midway through metestrus a blood-tinged vaginal discharge may be seen.
4. *Diestrus*: the longest phase of the reproductive cycle, lasting about 16 days. The CL continues to secrete progesterone. If the cow is not pregnant, a CL will normally dissolve after 17 days and the estrous cycle will begin again. If fertilization results in pregnancy, the CL will remain through the length of gestation.

Anestrus may also occur. This is a period of reproductive failure when the cow's body does not cycle, no signs of heat are exhibited, and the cow cannot be bred.

APPENDIX O
RAW MILK BACTERIA COUNTS

BACTERIAL TEST	PROCEDURE	ACCEPTED LIMITS	SOURCES OF BACTERIA	TROUBLE-SHOOTING ON FARM	LIMITATIONS OF TEST
Standard Plate Count (SPC)	Milk is plated onto special agar and incubated for 48 hours @ 90°F	PMO: <100,000 legal limit Quality: <10,000	Mastitis Dirty equipment Dirty cows Poor cooling	Quantitative bulk tank culture Perform additional testing below	Only gives a raw number of bacteria Does not indicate what types of bacteria are causing the high counts
Lab Pasteurized Count (LPC)	Milk sample is heated to 145°F for 30 minutes to simulate pasteurization, then SPC is performed on the sample	<250 cfu	*Detects heat-stable bacteria* Mastitis bacteria should be killed by pasteurization High LPC come from dirty cows and dirty equipment	Review cow cleanliness Improve premilking teat preparation Check wash water temperatures Check thoroughness of cleaning cycle Sanitize before milking	
Preliminary Incubation Count (PIC)	Milk is held at 55°F for 18 hours, then SPC is performed on sample	Quality payments vary by processor; less than 50,000 cfu/ml or no more than 3–4X SPC	*Detects bacteria that grow at refrigeration temperatures* Poor cooling Poor cleaning Worn rubber parts Dirty cows	Check rubber parts Check cooling Evaluate wash system Sanitize before milking	Very sensitive test Can be affected by handling, storage conditions, and age of sample
Coliform Count (CC)	Milk is plated on special agar and incubated	<50 cfu/ml	*Detects coliform bacteria* Dirty cows Poor equipment cleaning Poor cooling Mastitis (rarely)	Review cow cleanliness Check system cooling Check thoroughness of wash cycle	

APPENDIX P

YOUR ANIMAL HEALTH TOOLBOX

Your organic system plan requires you to list any animal treatments that you may potentially use on your farm. It is also wise to assemble these products and basic equipment and store them properly so that they will be on hand if a health problem arises. Below is a list of recommended products. Please be sure to check with your certifier about specific brands and their allowed status *before* you use them.

ITEM	USE	ADDITIONAL INFORMATION
Aloe vera (juice or pellets)	Nonspecific immunostimulant	Colostral whey must come from organic cows
Aspirin	Anti-inflammatory	
Black walnut hulls	Internal parasites	
Calcium gluconate (23%) bottles	Treatment of milk fever	
California Mastitis Test	Subclinical mastitis detection	
Colostral whey	Nonspecific immunostimulant	
Dehorner	Dehorning calves	
Dextrose	Ketosis / Toxic mastitis	
Epinephrine	Severe allergic reactions	
Flunixin (Banamine®)	Anti-inflammatory / Fevers / Toxic infections	Administered by or on recommendation of a veterinarian
Garlic tincture	Antibiotic replacement	None
Homeopathic Remedy Kit	Disease treatment	None
Hypertonic saline	Dehydration / Toxic mastitis or other severe infections	Cow must drink after administration or be stomach-tubed with water
Immunoboost™	Nonspecific immunostimulant	None
Iodine	Disinfectant	
Lidocaine	Anesthetic: dehorning	
Magnesium oxide boluses	Indigestion	Check with your certifier about the specific brand
Magnets	Treatment / Prevention of hardware disease	None
Molasses	Ketosis / Appetite stimulant	
Oral electrolytes	Dehydration / Scours / Toxic mastitis	
Peppermint liniment	Clinical mastitis	Check with your certifiers about the specific brand
Poloxalene	Emergency treatment of bloat	None
Probiotic boluses	Scours / Indigestion / Off-feed	
Rope halter	Animal restraint	
Simplex (IV administration set) and needles	IV fluid administration	
Stomach tube	Relief of bloat / Administration of oral fluids	
Treating Dairy Cows Naturally (Hubert Karreman, VMD)	Reference book	
Vitamin B-12	Appetite stimulant	
Vitamin B-complex	Antioxidant / Toxic infections / Appetite stimulant	
Vitamin C	Antioxidant / Toxic infections	

APPENDIX Q

PASTURE MANAGEMENT TROUBLESHOOTING GUIDE

SYMPTOM	PROBLEM	SOLUTION	CHAPTER 4 "PASTURE MANAGEMENT" SECTIONS
COWS REJECTING PASTURE	Pasture is too tall/overmature.	Move cows to pasture with appropriate pregrazing height. If that is not available, strip graze the tall pasture.	What you need to know to set up a successful grazing system (69)
	Paddock is sized incorrectly.	Recalculate paddock size.	Calculating paddock sizes, DMI, and acreage needed (72)
	Supplemental feed in barn contains too much protein.	Reformulate ration to take out or reduce high protein forages such as haylage, and replace high protein grain with energy grain. Test MUN.	Ration balancing and pastures (88) Pasture pitfalls in poorly managed grazing systems (67) Foraging behavior and diet selection (86) See also Chapter 6, "Managing Dairy Nutrition"
	Gate to pasture is left open so cows can come back to the barn whenever they want.	Shut the gate.	What you need to know to set up a successful grazing system (69) Designing your grazing system (76)
COWS DROP IN MILK PRODUCTION AND BODY CONDITION SCORE DURING SUMMER GRAZING SEASON	Paddocks may be too small to provide enough DMI per cow.	Recalculate paddock sizes.	Calculating paddock sizes, DMI, and acreage needed (72)
	Pasture quality may be poor.	Sample forages and reformulate ration. Make a plan to improve pasture quality.	Ration balancing and pastures (88) Pasture pitfalls in poorly managed grazing systems (67) See also Chapter 6, "Managing Dairy Nutrition"
SOME OR MANY COWS WON'T BREED DURING SUMMER GRAZING SEASON	Pregrazing heightmay be too short to allow cows to get enough DMI.	Make sure that pastures are fully regrown so that they are tall enough when the cows go into each paddock.	Pasture pitfalls in poorly managed grazing systems (67) What you need to know to set up a successful grazing system (69)
	Supplemental feed in barn contains too much protein.	Reformulate ration to take out or reduce high protein forages such as haylage and replace high protein grain with energy grain. Test MUN.	Ration balancing and pastures (88) Pasture pitfalls in poorly managed grazing systems (67) Foraging behavior and diet selection (86) See also Chapter 6, "Managing Dairy Nutrition"

APPENDIX R

NUTRITION TROUBLESHOOTING GUIDE

SYMPTOM	PROBLEM	SOLUTION	CHAPTER 6 "MANAGING DAIRY NUTRITION" SECTIONS
MILK PRODUCTION DROPS OR IS VARIABLE	Forage quality has changed.	Sample forages and reformulate diet.	Forage Testing (142); Requirements and Rations (146)
	Amount of feed offered has changed.	Ensure feed mixing and/or delivery is consistent among people feeding.	Milk Production Trends (169)
	Pasture yield in paddocks is different.	Measure pasture yields and size of paddocks to provide equal pasture density; plan strategies to improve yields.	Chapter 4: Calculating Paddock Sizes, DMI, and Acreage Needed (72)
MILK COMPONENTS (FAT AND PROTEIN) DROP OR ARE VARIABLE	Not enough forage in diet and too much grain.	Increase forages fed and reduce grain.	Milk Components (170)
	Forage quality has changed.	Sample forages and reformulate diet; ensure adequate NDF.	Forage Testing (142); Fiber (139, 140); Requirements and Rations (146)
	Not enough protein in diet or amount fed is inconsistent.	Increase level of protein fed; ensure feed mixing is consistent.	Protein (140)
MILK UREA NITROGEN (MUN) LEVELS OVER 14	Too much protein in diet.	Sample forages and reformulate diet; reduce amount of protein fed.	MUN (171); Forage Testing (142); Requirements and Rations (146); Overview of Nutritional Concepts (138); pasture-based rations (144)
	Too little carbohydrate in diet.	Sample forages and reformulate diet; increase energy fed.	

SYMPTOM	PROBLEM	SOLUTION	CHAPTER 6 "MANAGING DAIRY NUTRITION" SECTIONS
RUMINATION ACTIVITY DECREASED	Not enough forage in diet and too much grain.	Increase forages fed and reduce grain.	Rumination (172)
	Forage quality has changed	Sample forages and reformulate diet; ensure adequate NDF.	Forage Testing (142); Fiber (139, 140); Requirements and Rations (146)
LOW BODY CONDITION	Lack of energy in diet or too much protein in diet	Reformulate diet to increase energy and reduce protein.	Body Condition Scoring (172); Requirements and Rations (146); Overview of Nutritional Concepts (138); Pasture-Based Rations (144)
	High milk production	Consider changing genetic base.	Breed characteristics (137)
LAME COWS	Rumen acidosis and laminitis	Increase forages fed and reduce grain.	Fiber (139); Appendix I, "Locomotion Scoring"
	Injury	Evaluate laneways, barnyards, barn for stones, wet areas.	Appendix I, "Locomotion Scoring"
MANURE TOO LOOSE OR TOO STIFF	Ration imbalance	Reformulate diet.	Manure scoring (172); Requirements and Rations (146)
	Serious illness	Call veterinarian!	Manure scoring (174)
GRAIN BILL TOO HIGH	Ration not feasible		Chapter 6, "Managing Dairy Nutrition"
	Purchasing majority of grains	Improve pasture and implement management intensive grazing (MIG); plan to grow more of your own organic feeds.	Chapter 4, "Pasture Management" and Chapter 5, "Organic Crop Management"

APPENDIX S

ORGANIC DAIRY HERD HEALTH TROUBLESHOOTING GUIDE

SYMPTOM	PROBLEM	SOLUTION	CHAPTER 7, "ORGANIC DAIRY HERD HEALTH" SECTIONS
CALVES NOT DOING WELL, NOT GAINING WEIGHT	Not enough milk	Increase amount of milk fed or increase number of feedings.	Youngstock (195)
	Internal parasites	Check fecal sample with vet. Increase milk or return to milk if shortly post-weaning. Try recommended alternative therapies.	Youngstock (195) Internal Parasites of Cattle (221)
	Scours	Take temperature and check hydration. Oral electrolyes. Improve hygiene. Call veterinarian.	Youngstock (195)
ANIMALS COUGHING OR HAVING TROUBLE BREATHING	Pneumonia Lungworms	Take temperature and listen to lungs. Improve ventilation in barn. Fecal sample Try recommended alternative therapies. Call veterinarian.	Youngstock (195) Adult Respiratory Disease (224)
COWS ARE LAME	Laminitis (acidosis)	Increase forages and decrease grain.	Chapter 6 "Managing Dairy Nutrition" Hoof Health and Lameness (217) Appendix I, "Locomotion Scoring Guide"
	Hairy heel wart	Foot baths	
	Digitial dermatitis Foot rot	Trimming, topical treatments, foot baths	
COWS ARE NOT CYCLING	Anestrus	Body Condition Score Increase energy in diet. Rectal palpation for persisted CLs	Reproduction (205) Appendix H, "Body Condition Scoring"
ABORTIONS	Infectious disease Toxins	Call veterinarian for diagnosis. Save aborted fetus for testing.	Reproduction (205) Biosecurity (228) Strategic Vaccination (226)
BULK MILK SOMATIC CELL COUNTS INCREASING	Mastitis	Culture bulk milk sample. Check cows individually with California Mastitis Test. Hold high-SCC cows out of tank. Culture Treat with recommended alternative therapies.	Udder Health and Milk Quality (209)
COW OFF-FEED	Ketosis	Check urine for ketones. Administer glycerin or sugar.	Reproduction (205) Care and Management of the Fresh Cow (207)
	Milk fever	Administer calcium, check potassium levels in dry-cow forage.	
	Displaced abomasum	Listen for ping on left and right sides. Call vet.	
	Infection (e.g., metritis, hardware, pneumonia)	Take temperature. Administer anti-inflammatories. Treat affected area alternatively. Call veterinarian.	

Boldface page references indicate photographs or illustrations. *Italic* references indicate boxed text, tables, figures or charts.

B

Bedding, 204, *205*
Body Condition Scoring, 172–73, *285–86,* **285, 286**

C

Calves
colostrum, 195, *195,* 197
diseases and disorders of, 199–202, *200, 201, 202,* 230–31, *301*
in grazing system, 79–80
housing, 197, *198*
nutrition and feeding, 163–64, *163, 165,* 195, *195,* 197–98, *197*
preventative healthcare, *195,* 198–99, *199*
troubleshooting, 301
Case studies
Arnold farm, *147,* 234–36, **234,** *235*
Bawden farm, 40–43, **40, 42–43,** *145*
Beidler farm, 95–98, **95, 96–97**
Blood farm, 24–25, **24, 25**
Brown farm, 237–39, **237, 238**
Englebert farm, 110, 131–33, **131, 132,** *133*
Fournier farm, *84,* 99–101, **99, 101,** *136*
Gardiner farm, *146, 156,* 183–85, **183, 184**
Griffin farm, 268–73, **268, 271, 272, 273**
Hardy farm, 64–65, **64, 65**
Huftalen farm, 265–67, **265, 266–67**
Knapp farm, 62–63, **62,** *63,* **63**
Miller Brothers' farm, 255–57, **255, 256–567**
Moore farm, *151, 163,* 179–82, **179, 180–81**
Sherman farm, *151,* 240–43, **240, 242–43**
Certification, 17–18, 274
Organic System Plan and, 18–23, *21, 22*
timeline for, 17, *18–19*
Compost management, 56–57, *57, 63,* **63**
Cover crops, 59–60, **59,** *112, 117–19*
Cows
assessing, *289–90,* **290**
pre-transition, 135–38, *136*
diseases and disorders of (*See* Diseases and disorders)
evaluating
body condition scoring (BCS) and, 172–73, *285–86,* **285, 286**
hock scoring, 204, *293,* **293**
locomotion scoring, 173–74, *287,* **287,** *288,* **288**
manure scoring, 174–76, **174, 175**
foraging behavior, *86–87*
housing, 203–4, *205,* 217
nutritional requirements (*See* Nutrition)
reproduction (*See* Reproduction)
Crop plan
components of, 103–8
assessing fields, 105, *106*
avoiding transition slump, 105–7, *107*
cash crops, 107–8
feed crops, 104–5
pasture, 103
evaluating, 129–30
transition points, 102
Crop rotation
cover crops and, 59–60, **59**
developing system for, 108–10, *108, 109, 111*
Crops
non-pasture
meeting feed needs through, 103–8, *104, 106, 107*
pros and cons of growing, 102–3
options for and management of, 110–11, *112–19*
pest management in
diseases, 127–29
insect pests, 127–29
principals of, 119–20
weeds in (*See* Weeds)
seeds for, 21, *22,* 277–78
soil management for tilled, 59–60
weeds and (*See* Weeds)

D

Diseases and disorders
Animal Health Toolbox, *297*
biosecurity and, 228–32, *228*
bovine leukemia virus, 232
bovine viral diarrhea, 231–32
digestive, 222–24
eyes, 220–21
of fresh cows, *210*
hoof health and lameness, 217, *218,* 219
Johne's disease, 174, **174,** 230–31
mastitis, 212–17, *213, 214–15, 216*
parasites
external, 219–20, *219, 220*
internal, 89, *90,* 221–22, *221*
reproductive, *208–9, 298, 301*
respiratory, 201–2, *202,* 224–25
scours (neonatal diarrhea), 199–201, *200, 201*
skin, 219–20, *219, 220*
teats and external udder, 212, *213*
troubleshooting guide, *296, 298–301*
vaccinations and, 226–28, *226, 227, 228*

F

Feed, 104–5. *See also* Nutrition
alternative, 138–39, *146*
assessing fields for growing, 105, *106*
finding organic, 143, 279–81
organic crop plan and, 21–22, 103–8, *104, 106, 107*
purchasing *vs.* growing, 104–5, *104*
Feeding programs and systems. *See also* Nutrition
evaluating
body condition scoring, *285–86,* **285, 286**
locomotion scoring, *287,* **287,** *288,* **288**
low-to-zero grain feeding system, 88–89
Feeding systems and methods,

134–36, *136*
considerations in, *144*
evaluating, 168–77, **174, 175**
body condition scoring, 172–73
locomotion scoring, 173–74
manure scoring, 174–76, **174, 175**
milk components and, 170–71
milk production trends, 169–70
MUN levels and, 171–72
overall nutrition assessment, 176–77
rumination (cud chewing) and, 172
freestall, *148–49, 151*
pasture-based, 144–45
tiestall, *145, 147, 148–49*
Fencing, 77, **77**
Forage
crops for, *112–19*
testing, 142, *142*
testing services, 282
Freestall facilities, 203–4
feeding systems for, *148–49, 151*
size of, *292*

G

Grazing systems. *See also* Pasture management
benefits of well-managed, 66–67
comparison of, *85*
crops for, *112–19*
designing, 76–80, **77, 78,** *79*
fencing, 77, **77**
guidelines summary, *79*
lanes, 78, **78**
paddock size, 72–76, **73,** *74, 75*
pitfalls of poorly-managed, 67
recordkeeping, *91, 92*
setting up, 69–72, *69, 70, 84–85, 84, 85*
shade, 79
types of, 67–68, *68*
continuous grazing, 69, *85,* 88
holistic planned, 68
large pasture, 68
Management Intensive Grazing (MIG), 68, *68,* 69–72, *85*
rotational grazing, 84–85, *85,* 88
water in, 77–78, **78**

H

Health. *See also* Diseases and disorders
Animal Health Toolbox, *297*
assessing, *289–90,* **290**
body condition scoring, 172–73, *285–86,* **285, 286**
hock scoring, 204, *293,* **293**
locomotion scoring, 173–74, *287,* **287,** *288,* **288**
basic concepts, 187
best management practices, 188–90, 195
immunology, 225
medicines (*See* Treatments)
preventative care of calves, *195,* 198–99, *199*
records, 189, *291*
risk assessment, 188
transition points, 186–87
troubleshooting guide, *296, 299–300, 301*
vaccinations, 226–28, *226, 227, 228*
veterinary-client-patient relation, 189–90
Heifers. *See also* Calves
in grazing system, 79–80
nutrition and rations for, 160–62, *161, 163*
Housing. *See also* Freestall; Tiestall
calves, 197, *198*
cows, 203–4, *205,* 217, *218*

I

Immunology, 225
Insect pests, 127–29

L

Lanes, in grazing systems, 78, **78**
Legumes
for pastures, 80–81
soil management and, 52–53, *54*

M

Management Intensive Grazing (MIG), 68, *68,* 69–72, *85*
Manure, *300*
scoring, 174–76, **174, 175**
soil management and, 54–56, *55, 56*
testing services, 282
Marketing. *See* Organic Milk Market
Mastitis, 212–17, *213, 214–15, 216*
Medicines. *See* Treatments
Milk
bacteria in, 209, 211–12, *296*
components, 170–71
marketing (*See* Organic milk market)
MUN levels, 171–72, *299*
production
drop in, *299*
goals, 143–44, *144*
per cow, *284*
trends, 169–70
quality of and udder health, 209–12, *210, 211, 212, 213, 235* (*See also* Mastitis)
Minerals
sources of, 279–81

N

Nutrition. *See also* Feeding programs and systems
basic, *140–41*
fiber, 139, *140–41,* 142
protein and energy, 88, *88, 140, 155*
vitamins and minerals, *141, 156*
concepts, 138–43, *140–41, 142*
goal identification, 143–44, *144*
guidelines, *159*
requirements and rations, 88, 146–47
for calves, 163–64, *163, 165,* 195, *195,* 197–98, *197*
for dry cows, 156–58, *159,* 160, *160*
for heifers, 160–62, *161, 163*
for late lactation period, 156, *159*
for milking cows, 147, 150, *150,* 152, *152, 153–54,* 154–56, *155, 159*
transition points, 134
seasonal diet shifts, *145, 146, 147, 156,* 164–68
transition points, 134
troubleshooting, *299–300*

O

Organic milk market
- alternatives to, 38–39
- choosing, 36–38
- growth of, 26–29, *28*
- how it works, 29–36, *34, 35*
  - contract negotiation, 33–34, *34, 35,* 36, 39
  - large buyers, 26–32, *35,* 283
  - pricing, 27, 29–31, 33–34, *34, 35,* 36
- transition points, 26

Organic System Plan (OSP), 18–23, *21, 22*

P

Parasites
- external, 219–20, *219, 220*
- internal, 89, *90,* 221–22, *221*

Pasture management. *See also* Grazing systems; Pastures
- benefits of good, 66–67
- grazing systems and methods in, 67–68, *68*
  - setting up a system, 69–72, *69, 70,* 84–85, *84, 85*
- Management Intensive Grazing (MIG), 68, *68,* 69–72, *85*
- overgrazing, 69–72, *69*
- paddock sizes, 72–76, **73,** *74, 75*
- pitfalls of poor, 67
- transition points, 66
- troubleshooting, *298*

Pastures, 103, *104,* 204. *See also* Pasture management
- certification standards for, 90, *91*
- converting cropland to, 84
- crops for, *112–19*
- improving, 82, 90, 92, *93*
- internal parasite management and, 89, *90*
- plants for, 80–83, *80,* **80,** *82*
- quality and condition score chart for, *93*
- ration balancing and, 88
- soil management in, 58–59, 83–84
- weeds and, 126–27

Personnel, 258–64, *261*

Pests. *See* Insect pests; Weeds

Planning
- assessing current operation, 15
- certification process, 17–18
- goal determination, 14–15
- identifying needed changes, 15–16, *16*
- implementing changes, 16
- transition timeline, 17, *18–19*

R

Recordkeeping systems
- choosing new, 246, *246*
- evaluating current, 245
- grazing, *91, 92*
- health, 189, *291*
- importance of, 23
- key points, 244–45
- supporting records, 246–54, *247, 248, 249, 250, 251, 252, 253*

Reproduction
- anatomy and estrous cycle, *294–95,* **294, 295**
- fresh cows
  - care and management of, 207, *209, 210*
- fundamentals of, 205–7, *208–9, 298*

Resources, 258–63, 275–76

S

Seeds, 21, *22,* 82–83, 125, 277–78

Soil
- amendments, 277–78
- cropland and, 59–60
- health of
  - assessing, 47–52, *48, 50, 51*
- identifying and mapping, 47–48, *48*
- organic management of, 44–46, 52–58, *54, 55, 56, 57*
  - amendments, 57–58
  - benefits of, *46*
  - compost and, 56–57, *57, 63,* **63**
  - evaluating, 60–61
  - key principles of, 44–47, **45,** *47*
  - legumes and, 52–53, *54*
  - manure and, 54–56, *55, 56*
  - supplying nutrients to, 52–58, *54, 55, 56, 57*
- organic matter and, 45–46, *46, 47*
- pastures and, 58–59, 83–84
- testing, 48–52, *50, 51*
- testing services, 282
- transition points, 44

T

Tiestall and freestall facilities
- size of, *292*

Tiestall facilities, 203–4
- feeding systems for, *145, 147, 148–49*

Transition points, key organic
- crop management, 102
- herd health, 186–87
- marketing, 26
- nutrition, 134–35
- pasture management, 66
- planning, 14
- recordkeeping, 244–45
- soil management, 44

Treatments
- alternative
  - botanicals, 190–91, *192–93*
  - evaluating, 194, *194*
  - homeopathy, 191, 193
  - immune stimulants, 193–94
- antibiotics, *188*
- homemade electrolyte solution, *200*
- permitted synthetics, *196*
- prohibited substances, 194–95

Troubleshooting, *208–9,* 210, 213, *296, 298, 299–301.* *See also* Diseases and disorders; Insect pests; Weeds

V

Vaccinations
- biosecurity, 228–32, *228*
- programs, 226–28, *226, 227, 228*

W

Water, 77–78, **78,** 203

Weeds
- field crops and, 120–26, **121,** *122,* **124,** *133*
- lifecycles of common, *122*
- pasture crops and, 126–27
- row crops and, 123, 124, **124,** *133*
- smother/cover crops and, 123
- sod/hay and, 123, *133*

Y

Youngstock. *See* Calves; Heifers